Library of Congress Cataloging-in-Publication Data

Craft, B. C. (Benjamin Cole)
    Applied petroleum reservoir engineering / B.C. Craft and M.F.
Hawkins. — 2nd ed./revised by Ronald E. Terry.
        p.   cm.
    Includes bibliographical references and index.
    ISBN 0-13-039884-5
    1. Oil reservoir engineering.   I. Hawkins, Murray F.
    II. Terry, Ronald E.   III. Title.
TN871.C67   1990
622'.338—dc20   90-47806
                                                            CIP

Editorial/production supervision and
    interior design: Fred Dahl and Rose Kernan
Pre-press buyer: Kelly Behr
Press Manufacturing Buyer: Susan Brunke

© 1991 by Prentice Hall PTR
Prentice-Hall, Inc.
A Simon & Schuster Company
Englewood Cliffs, New Jersey 07632

This book can be made available to businesses
and organizations at a special discount when
ordered in large quantities. For more information
contact:

Prentice-Hall, Inc.
Special Sales and Markets
College Division
Englewood Cliffs, N.J. 07632

Printed in the United States of America
10 9

ISBN 0-13-039884-5

Prentice-Hall International (UK) limited, *London*
Prentice-Hall of Australia Pty. Limited, *Sydney*
Prentice-Hall Canada Inc., *Toronto*
Prentice-Hall Hispanoamericana, S.A., *Mexico*
Prentice-Hall of India Private Limited, *New Delhi*
Prentice-Hall of Japan, Inc., *Tokyo*
Simon & Schuster Asia Pte. Ltd., *Singapore*
Editora Prentice-Hall do Brasil, Ltda., *Rio de Janeiro*

Applied Petroleum Reservoir Engineering
Second Edition

B. C. CRAFT
and
M. F. HAWKINS

Louisiana State University

Revised by
RONALD E. TERRY

Brigham Young University

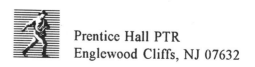

Prentice Hall PTR
Englewood Cliffs, NJ 07632

To Rebecca

Contents

Chapter 6
SATURATED OIL RESERVOIRS, 184

Chapter 7
SINGLE-PHASE FLUID FLOW IN RESERVOIRS, 210

Chapter 8
WATER INFLUX, 273

Preface

Shortly after undertaking the project of revising the text *Applied Petroleum Reservoir Engineering* by Ben Craft and Murray Hawkins, several colleagues expressed the wish that the revision retain the flavor and format of the original text. I am happy to say that I have attempted to do just that. The text contains many of the field examples that made the original text so popular and still more have been added. The revision includes a reorganization of the material as well as updated material in several chapters.

I have attempted to reorganize the chapters following a sequence that I have used for several years in teaching an undergraduate class in reservoir engineering. The first chapter contains a review of fluid and rock properties. Several new correlations are presented in this chapter that will assist those doing computer modeling. Chapter 2 contains a development of the general material balance equation. The next four chapters present information on the different reservoir types which the original text treated in the first four chapters. New material has been added in several places throughout these four chapters. Chapter 7 presents a discussion of one phase fluid flow. The radial diffusivity equation is derived and pressure transient analysis is introduced. Chapter 8 contains new material on water influx. Both edgewater and bottom-water drives are discussed. Chapter 9 is an update of the original Chapter 7 but also contains some new material on waterflooding and enhanced oil recovery techniques. Chapter 10 is a new chapter on history matching. This is a concept to which each reservoir engineer should have some exposure. The approach taken in the chapter is to provide an example of a history match by combining the Schilthuis material balance equation with a fluid flow equation.

There were some problems in the original text with units. I have attempted to eliminate these problems by using a consistent definition of terms. For example, formation volume factor is expressed in reservoir volume/surface condition volume. A consistent set of units is used throughout the text. The units used are ones standardized by the Society of Petroleum Engineers.

I would like to express my sincere appreciation to all those who have in some part contributed to the text. For their encouragement and helpful suggestions, I give special thanks to the following colleagues: John Lee at Texas A.M., James Smith formerly of Texas Tech, Don Green and Floyd Preston of the University of Kansas, and David Whitman and Jack Evers of the University of Wyoming.

Ronald E. Terry

Nomenclature

Normal Symbol	Definition	Units
A	areal extent of reservoir or well	acres or ft^2
A_c	cross sectional area perpendicular to fluid flow	ft^2
B'	water influx constant	bbl/psia
B_g	gas formation volume factor	ft^3/SCF or bbl/SCF
B_{gi}	gas formation volume factor at initial reservoir pressure	ft^3/SCF or bbl/SCF
B_{ga}	gas formation volume factor at abandonment pressure	ft^3/SCF or bbl/SCF
B_{Ig}	formation volume factor of injected gas	ft^3/SCF or bbl/SCF
B_o	oil formation volume factor	bbl/STB or ft^3/STB
B_{ofb}	oil formation volume factor at bubble point from separator test	bbl/STB or ft^3/STB
B_{oi}	oil formation volume factor at initial reservoir pressure	bbl/STB or ft^3/STB
B_{ob}	oil formation volume factor at bubble point pressure	bbl/STB or ft^3/STB
B_{odb}	oil formation volume factor at bubble point from differential test	bbl/STB or ft^3/STB

Normal Symbol	*Definition*	*Units*
B_t	two phase oil formation volume factor	bbl/STB or ft³/STB
B_w	water formation volume factor	bbl/STB or ft³/STB
c	isothermal compressibility	psi^{-1}
C_A	reservoir shape factor	unitless
c_f	formation isothermal compressibility	psi^{-1}
c_g	gas isothermal compressibility	psi^{-1}
c_o	oil isothermal compressibility	psi^{-1}
c_r	reduced isothermal compressibility	fraction, unitless
c_t	total compressibility	psi^{-1}
c_{ti}	total compressibility at initial reservoir pressure	psi^{-1}
c_w	water isothermal compressibility	psi^{-1}
E	overall recovery efficiency	fraction, unitless
E_d	microscopic displacement efficiency	fraction, unitless
E_i	vertical displacement efficiency	traction, unitless
E_o	expansion of oil (Havlena and Odeh method)	bbl/STB
$E_{f,w}$	expansion of formation and water (Havlena and Odeh method)	bbl/STB
E_g	expansion of gas (Havlena and Odeh method)	bbl/STB
E_s	areal displacement efficiency	fraction, unitless
E_v	macroscopic or volumetric displacement efficiency	fraction, unitless
f_g	gas cut of reservoir fluid flow	fraction, unitless
f_w	watercut of reservoir fluid flow	fraction, unitless
F	net production from reservoir (Havlena and Odeh method)	bbl
F_k	ratio of vertical to horizontal permeability	unitless
G	initial reservoir gas volume	SCF
G_a	remaining gas volume at abandonment pressure	SCF
G_f	volume of free gas in reservoir	SCF
G_I	volume of injected gas	SCF
G_{ps}	gas from primary separator	SCF
G_{ss}	gas from secondary separator	SCF
G_{st}	gas from stock tank	SCF
GE	gas equivalent of one STB of condensate liquid	SCF
GE_w	gas equivalent of one STB of produced water	SCF
GOR	gas-oil ratio	SCF/STB
h	formation thickness	ft
I	injectivity index	STB/day-psi
J	productivity index	STB/day-psi
J_s	specific productivity index	STB/day-psi-ft
J_{sw}	productivity index for a standard well	STB/day-psi
k	permeability	md
k'	water influx constant	bbl/day-psia
k_{avg}	average permeability	md
k_g	permeability to gas phase	md
k_o	permeability to oil phase	md

Normal Symbol	Definition	Units
k_w	permeability to water phase	md
k_{rg}	relative permeability to gas phase	fraction, unitless
k_{ro}	relative permeability to oil phase	fraction, unitless
k_{rw}	relative permeability to water phase	fraction, unitless
L	length of linear flow region	ft
m	ratio of initial reservoir free gas volume to initial reservoir oil volume	ratio, unitless
$m(p)$	real gas pseudo pressure	psia^2/cp
$m(pi)$	real gas pseudo pressure at initial reservoir pressure	psia^2/cp
$m(pwf)$	real gas pseudo pressure, flowing well	psia^2/cp
M	mobility ratio	ratio, unitless
M_w	molecular weight	lb/lb-mole
M_{wo}	molecular weight of oil	lb/lb-mole
n	moles	lb-mole
N	initial volume of oil in reservoir	STB
N_p	cumulative produced oil	STB
N_{vc}	capillary number	ratio, unitless
p	pressure	psia
p_b	pressure at bubble point	psia
p_c	pressure at critical point	psia
P_c	capillary pressure	psia
p_D	dimensionless pressure	ratio, unitless
p_e	pressure at outer boundary	psia
p_i	pressure at initial reservoir pressure	psia
p_{1hr}	pressure at one hour from transient time period on semilog plot	psia
p_{pc}	pseudocritical pressure	psia
p_{pr}	reduced pressure	ratio, unitless
p_R	pressure at a reference point	psia
p_{sc}	pressure at standard conditions	psia
p_w	pressure at wellbore radius	psia
p_{wf}	pressure at wellbore for flowing well	psia
$p_{wf(\Delta t=0)}$	pressure of flowing well just prior to shut in during a pressure build up test	psia
p_{ws}	shut in pressure at wellbore	psia
$\bar{p}$	volumetric average reservoir pressure	psia
$\Delta\bar{p}$	change in volumetric average reservoir pressure	psia
q	flow rate in standard condition units	STB/day (liquid)
q'_t	total flow rate in the reservoir in reservoir volume units	bbl/day
r	distance from center of well (radial dimension)	ft
r_D	dimensionless radius	ratio, unitless
r_e	distance from center of well to outer boundary	ft
r_R	distance from center of well to oil reservoir boundary	ft
r_w	distance from center of well to wellbore	ft

Normal Symbol	Definition	Units
R	instantaneous produced gas-oil ratio	SCF/STB
R'	universal gas constant	
R_p	cumulative produced gas-oil ratio	SCF/STB
R_{so}	solution gas-oil ratio	SCF/STB
R_{sob}	solution gas-oil ratio at bubble point pressure	SCF/STB
R_{sod}	solution gas-oil ratio, from differential liberation test	SCF/STB
R_{sodb}	solution gas-oil ratio at bubble point pressure, from differential liberation test	SCF/STB
R_{sofb}	solution gas-oil ratio, sum of separator gas and stock tank gas from separator test	SCF/STB
R_{soi}	solution gas-oil ratio at initial reservoir pressure	SCF/STB
R_{sw}	solution gas-water ratio for brine	SCF/STB
R_{swp}	solution gas-water ratio for deionized water	SCF/STB
R_1	solution gas-oil ratio for liquid stream out of separator	SCF/STB
R_3	solution gas-oil ratio for liquid stream out of stock tank	SCF/STB
RF	recovery factor	fraction, unitless
R.V.	relative volume from a flash liberation test	ratio, unitless
S	fluid saturation	fraction, unitless
S_g	gas saturation	fraction, unitless
S_{gr}	residual gas saturation	fraction, unitless
S_L	total liquid saturation	fraction, unitless
S_o	oil saturation	fraction, unitless
S_w	water saturation	fraction, unitless
S_{wi}	water saturation at initial reservoir conditions	fraction, unitless
t	time	hour
Δt	time of transient test	hour
t_o	dimensionless time	ratio, unitless
t_p	time of constant rate production prior to well shut in	hour
t_{pss}	time to reach pseudosteady state flow region	hour
T	temperature	°F or °R
T_c	temperature at critical point	°F or °R
T_{pc}	pseudocritical temperature	°F or °R
T_{pr}	reduced temperature	fraction, unitless
T_{ppr}	pseudo reduced temperature	fraction, unitless
T_{sc}	temperature at standard conditions	°F or °R
V	volume	ft^3
V_b	bulk volume of reservoir	ft^3 or acre-ft
V_p	pore volume of reservoir	ft^3
V_r	relative oil volume	ft^3
V_R	volume at some reference point	ft^3
W	width of fracture	ft
W_e	water influx	bbl

Normal Symbol	*Definition*	*Units*
W_{eD}	dimensionless water influx	ratio, unitless
W_{ei}	encroachable water in place at initial reservoir conditions	bbl
W_I	volume of injected water	STB
W_p	cumulative produced water	STB
z	gas deviation factor or gas compressibility factor	ratio, unitless
z_i	gas deviation factor at initial reservoir pressure	ratio, unitless

Greek Symbol	*Definition*	*Units*
α	90°-dip angle	degrees
ϕ	porosity	fraction, unitless
γ	specific gravity	ratio, unitless
γ_g	gas specific gravity	ratio, unitless
γ_o	oil specific gravity	ratio, unitless
γ_w	well fluid specific gravity	ratio, unitless
γ'	fluid specific gravity (always relative to water)	ratio, unitless
γ_1	specific gravity of gas coming from separator	ratio, unitless
γ_3	specific gravity of gas coming from stock tank	ratio, unitless
η	formation diffusivity	ratio, unitless
λ	mobility (ratio of permeability to viscosity)	md/cp
λ_g	mobility of gas phase	md/cp
λ_o	mobility of oil phase	md/cp
λ_w	mobility of water phase	md/cp
μ	viscosity	cp
μ_g	gas viscosity	cp
μ_i	viscosity at initial reservoir pressure	cp
μ_o	oil viscosity	cp
μ_{ob}	oil viscosity at bubble point	cp
μ_{od}	dead oil viscosity	cp
μ_w	water viscosity	cp
μ_{wI}	water viscosity at 14.7 psia and reservoir temperature	cp
μ_I	viscosity at 14.7 psia and reservoir temperature	cp
ν	apparent fluid velocity in reservoir	bbl/day-ft^2
ν_g	apparent gas velocity in reservoir	bbl/day-ft^2
ν_t	apparent total velocity in reservoir	bbl/day-ft^2
θ	contact angle	degrees
ρ	density	lb/ft^3
ρ_g	gas density	lb/ft^3
ρ_r	reduced density	ratio, unitless
$\rho_{o,API}$	oil density	°API
σ_{wo}	oil-brine interfacial tension	dynes/cm

Chapter 1

Introduction
to Reservoir Engineering

1. HISTORY OF RESERVOIR ENGINEERING

Crude oil, natural gas, and water are the substances that are of chief concern to petroleum engineers. Although these substances sometimes occur as solids or semisolids, usually at lower temperatures and pressures, as paraffin, gas-hydrates, ices, or high pour-point crudes, in the ground and in the wells they occur mainly as *fluids*, either in the *vapor* (gaseous) or in the *liquid* phase or, quite commonly, both. Even where solid materials are used, as in drilling, cementing, and fracturing, they are handled as fluids or slurries. The division of the well and reservoir fluids between the liquid and vapor phases depends mainly on the temperature and pressure. The state or phase of a fluid in the reservoir usually changes with pressure, the temperature remaining substantially constant. In many cases the state or phase in the reservoir is quite unrelated to the state of the fluid when it is produced at the surface. The precise knowledge of the behavior of crude oil, natural gas, and water, singly or in combination, under static conditions or in motion in the reservoir rock and in pipes and under changing temperature and pressure, is the main concern of petroleum engineers.

As early as 1928 petroleum engineers were giving serious consideration to gas-energy relationships and recognized the need for more precise information concerning physical conditions in wells and underground reservoirs. Early progress in oil recovery methods made it obvious that computations made from wellhead or surface data were generally misleading. Sclater and Stephenson described the first recording bottom-hole pressure gauge and thief for sampling fluids under pressure in wells.[*][1] It is interesting that this reference defines bottom-hole data as referring to positive measurements of pressure, temperature, gas-oil ratios, and the physical and chemical nature of the fluids. The need for accurate bottom-hole pressures was further emphasized when Millikan and Sidwell described the first precision pressure gauge and pointed out the fundamental importance of bottom-hole pressures to petroleum engineers in determining the most efficient methods of recovery and lifting procedures.[2] With this contribution the engineer was able to measure the most important basic information for reservoir performance calculations: *reservoir pressure.*

The study of the properties of the rocks and their relationship to the fluids they contain in both the static and flowing states is called *petro-physics.* Porosity, permeability, fluid saturations and distributions, electrical conductivity of both the rock and the fluids, pore structure, and radioactivity are some of the more important petrophysical properties of rocks. Fancher, Lewis, and Barnes made one of the earliest petrophysical studies of reservoir rocks in 1933, and 1934 Wycoff, Botset, Muskat, and Reed developed a method for measuring the permeability of reservoir rock samples based on the fluid flow equation discovered by Darcy in 1856.[3,4] Wycoff and Botset made a significant advance in their studies of the simultaneous flow of oil and water, and of gas and water in unconsolidated sands.[5] This work was later extended to consolidated sands and other rocks, and in 1940 Leverett and Lewis reported research on the three-phase flow of oil, gas, and water.[6]

It was early recognized by the pioneers in reservoir engineering that before the volumes of oil and gas in place could be calculated, the change in the physical properties of bottom-hole samples of the reservoir fluids with pressure would be required. Accordingly in 1935 Schilthuis described a bottom-hole sampler and a method of measuring the physical properties of the samples obtained.[7] These measurements included the pressure-volume-temperature relations, the saturation or bubble-point pressure, the total quantity of gas dissolved in the oil, the quantities of gas liberated under various conditions of temperature and pressure, and the shrinkage of the oil resulting from the release of its dissolved gas from solution. These data made the development of certain useful equations feasible, and they also provided an essential correction to the volumetric equation for calculating oil in place.

The next significant development was the recognition and measurement of connate water saturation, which was considered indigenous to the formation and remained to occupy a part of the pore space after oil or gas accumula-

---

[*] References throughout the text are given at the end of each chapter.

tion.[8,9] This development further explained the poor oil and gas recoveries in low permeability sands with high connate water saturation, and introduced the concept of water, oil, and gas saturations as percentages of the total pore space. The measurement of water saturation provided another important correction to the volumetric equation by correcting the pore volume to hydrocarbon pore space.

Although temperature and geothermal gradients had been of interest to geologists for many years, engineers could not make use of these important data until a precision subsurface recording thermometer was developed. Millikan pointed out the significance of temperature data in applications to reservoir and well studies.[10] From these basic data Schilthuis was able to derive a useful equation, commonly called the Schilthuis material balance equation.[11] It is a modification of an earlier equation presented by Coleman, Wilde, and Moore and is one of the most important tools of reservoir engineers.[12] Basically it is a statement of the conservation of matter and is a method of accounting for the volumes and quantities of fluids initially present in, produced from, injected into, and remaining in a reservoir at any state of depletion. Odeh and Havlena have shown how the material balance equation can be arranged into a form of a straight line and solved.[13] In reservoirs under water drive the volume of water which encroaches into the reservoir also enters into the material balance on the fluids. Although Schilthuis proposed a method of calculating water encroachment using the material-balance equation, it remained for Hurst and, later, van Everdingen and Hurst to develop methods for calculating water encroachment independent of the material balance equation, which applies to aquifers of either limited or infinite extent, in either steady-state or unsteady-state flow.[11,14,15] The calculations of van Everdingen and Hurst have been simplified by Fetkovich.[16] Following these developments for calculating the quantities of oil and gas initially in place or at any stage of depletion. Tarner and Buckley and Leverett laid the basis for calculating the oil recovery to be expected for particular rock and fluid characteristics.[17,18] Tarner and, later, Muskat[19] presented methods for calculating recovery by the internal or solution gas drive mechanism, and Buckley and Leverett presented methods for calculating the displacement of oil by external gas cap drive and water drive. These methods not only provided means for estimating recoveries for economic studies; they also explained the cause for disappointingly low recoveries in many fields. This discovery in turn pointed the way to improved recoveries by taking advantage of the natural forces and energies, and by supplying supplemental energy by gas and water injection, and by unitizing reservoirs to offset the losses that may be caused by competitive operations.

During the 1960s, the terms *reservoir simulation* and *reservoir mathematical modeling* became popular.[20,21,22] These terms are synonomous and refer to the ability to use mathematical formulas to predict the performance of an oil or gas reservoir. Reservoir simulation was aided by the development of large-scale, high-speed digital computers. Sophisticated numerical methods were also developed to allow the solution, of a large number of equations by finite-difference or finite-element techniques.

With the development of these techniques, concepts, and equations, reservoir engineering became a powerful and well-defined branch of petroleum engineering. *Reservoir engineering* may be defined as the application of scientific principles to the drainage problems arising during the development and production of oil and gas reservoirs. It has also been defined as "the art of developing and producing oil and gas fluids in such a manner as to obtain a high economic recovery."[23] The working tools of the reservoir engineer are subsurface geology, applied mathematics, and the basic laws of physics and chemistry governing the behavior of liquid and vapor phases of crude oil, natural gas, and water in reservoir rocks. Because reservoir engineering is the science of producing oil and gas, it includes a study of all the factors affecting their recovery. Clark and Wessely urge a joint application of geological and engineering data to arrive at sound field development programs.[24] Ultimately reservoir engineering concerns all petroleum engineers, from the drilling engineer who is planning the mud program to the corrosion engineer who must design the tubing string for the producing life of the well.

2. PETROLEUM RESERVOIRS AND PRODUCTION FROM PETROLEUM RESERVOIRS

Oil and gas accumulations occur in underground *traps* formed by structural and/or stratigraphic features.[25] Fortunately they usually occur in the more porous and permeable portions of beds, which are mainly sands, sandstones, limestones, and dolomites, in the intergranular openings, or in pore spaces caused by joints, fractures, and solution activity. A *reservoir* is that portion of a trap which contains oil and/or gas as a single hydraulically connected system. Many hydrocarbon reservoirs are hydraulically connected to various volumes of water-bearing rock called *aquifers*. Many reservoirs are located in large sedimentary basins and share a common aquifer. In this case the production of fluid from one reservoir will cause the pressure to decline in other reservoirs by fluid communication through the aquifer. In some cases the entire trap is filled with oil or gas, and in this case the trap and the reservoir are the same.

Under initial reservoir conditions, the hydrocarbon fluids are in either a single-phase or a two-phase state. The single phase may be a liquid phase in which all the gas present is dissolved in the oil. There are therefore dissolved natural gas reserves as well as crude oil reserves to be estimated. On the other hand, the single phase may be a gas phase. If there are hydrocarbons vaporized in this gas phase that are recoverable as natural gas liquids on the surface, the reservoir is called gas-condensate, or gas-distillate (the older name). In this case there are associated liquid (condensate or distillate) reserves as well as the gas reserves to be estimated. Where the accumulation is in a two-phase state, the vapor phase is called the *gas cap* and the underlying liquid phase, the *oil zone.* In this case there will be four types of reserves to be estimated: the free gas or associated gas, the dissolved gas, the oil in the oil zone, and the recoverable natural gas liquid from the gas cap.

Although the hydrocarbons in place are fixed quantities, which are referred to as the *resource,* the *reserves* depend on the method by which the reservoir is produced. In 1986 the Society of Petroleum Engineers (SPE) adopted the following definition for reserves:

> Reserves are estimated volumes of crude oil, condensate, natural gas, natural gas liquids, and associated marketable substances anticipated to be commercially recoverable and marketable from a given date forward, under existing economic conditions, by established operating practices, and under current government regulations.[26]

The amount of reserves is calculated from available engineering and geologic data. The estimate is updated over the producing life of the reservoir as more data become available. The SPE definition is further broken down into proved and unproved reserves. These definitions are fairly lengthy, and we encourage you to obtain a copy of the reference if you desire further information.

The initial production of hydrocarbons from the underground reservoir is accomplished by the use of natural reservoir energy and is referred to as *primary production.* The oil and gas are displaced to production wells under primary production by (a) fluid expansion, (b) fluid displacement, (c) gravitational drainage, and/or (d) capillary expulsion. When there is no aquifer and no fluid is injected into the reservoir, the hydrocarbon recovery is brought about mainly by fluid expansion; however, in the case of oil it may be materially aided by gravitational drainage. When there is water influx from the aquifer or when, in lieu of this, water is injected into selected wells, recovery is accomplished by the displacement mechanism, which again may be aided by gravitational drainage or capillary expulsion. Gas is injected as a displacing fluid to help in the recovery of oil and is also used in gas cycling to recover gas-condensate fluids.

The use of either a natural gas or a water injection scheme is called a *secondary recovery operation.* When a water injection sheme is used as a secondary recovery process, the process is referred to as *waterflooding.* The main purpose of either a natural gas or a water injection process is to maintain the reservoir pressure. Hence, the term *pressure maintenace program* is also used to describe a secondary recovery process.

Other displacement processes called *tertiary recovery processes* have been developed for application in situations in which secondary processes have become ineffective. However, the same processes have also been considered for reservoir applications when secondary recovery techniques were not used because of low recovery potential. In this latter case, the word *tertiary* is a misnomer. For some reservoirs, it is advantageous to begin a secondary or a tertiary process before primary production is completed. For these reservoirs, the term *improved recovery* was introduced and has become popular in referring to, in general, any recovery process that improves the recovery over what the natural reservoir energy would be expected to yield.

In many reservoirs several recovery mechanisms may be operating simultaneously, but generally one or two predominate. During the producing life of a reservoir, the predominance may shift from one mechanism to another, either naturally or because of operations planned by engineers. For example, the volumetric reservoir (no aquifer) may produce initially by fluid expansion. When its pressure is largely depleted, it may produce to the wells mainly by gravitational drainage, the fluid being lifted to the surface by pumps. Still later, water may be injected in some wells to drive additional oil to other wells. In this case, the cycle of the mechanisms is expansion-gravitational drainage-displacement. There are many alternatives in these cycles, and it is the object of reservoir engineering to plan these cycles for maximum recovery, usually in minimum time.

3. RESERVOIR TYPES DEFINED WITH REFERENCE TO PHASE DIAGRAMS

From a technical point of view, the various types of reservoirs can be defined by the location of the initial reservoir temperature and pressure with respect to the two-phase (gas and liquid) region as commonly shown on pressure-temperature (PT) phase diagrams. Figure 1.1 is the *PT* phase diagram of a particular reservoir fluid. The area enclosed by the bubble-point and dew-point lines to the lower left is the region of pressure-temperature combinations in which both gas and liquid phases will exist. The curves within the two-phase region show the percentage of the total hydrocarbon volume that is liquid for any temperature and pressure. Initially each hydrocarbon accumulation will have its own phase diagram, which depends only on the composition of the accumulation.

Consider a reservoir containing the fluid of Fig. 1.1 initially at 300°F and 3700 psia, point A. Since this point lies outside the two-phase region, it is originally in a one-phase state, commonly called gas as located at point A. Since the fluid remaining in the reservoir during production remains at 300°F, it is evident that it will remain in the single-phase or gaseous state as the pressure declines along path $\overline{AA_1}$. Furthermore, the composition of the produced well fluid will not change as the reservoir is depleted. This is true for any accumulation of this composition where the reservoir temperature exceeds the *cricondentherm,* or maximum two-phase temperature (250°F for the present example). Although the fluid left in the reservoir remains in one phase, the fluid produced through the wellbore and into surface separators, although the same composition, may enter the two-phase region owing to the temperature decline, as along line $\overline{AA_2}$. This accounts for the production of condensate liquid at the surface from a gas in the reservoir. Of course, if the

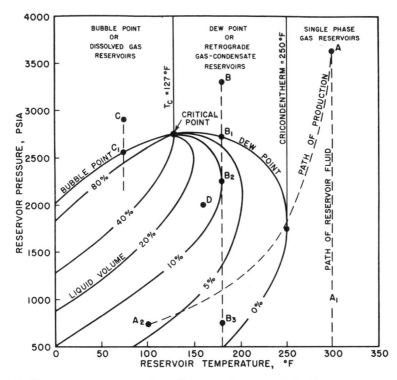

Fig. 1.1. Pressure-temperature phase diagram of a reservoir fluid

cricondentherm of a fluid is below say 50°F, then only gas will exist on the surface at usual ambient temperatures, and the production will be called *dry gas*. Nevertheless, it may contain liquid fractions that can be removed by low-temperature separation or by natural gasoline plants.

Next, consider a reservoir containing the same fluid of Fig. 1.1 but at a temperature of 180°F and an initial pressure of 3300 psia, point B. Here the fluid is also initially in the one-phase state, commonly called *gas*, where the reservoir temperature exceeds the critical temperature. As pressure declines because of production, the composition of the produced fluid will be the same as for reservoir A and will remain constant until the dew-point pressure is reached at 2700 psia, point B_1. Below this pressure a liquid condenses out of the reservoir fluid as a fog or dew, and this type of reservoir is commonly called a dew-point reservoir. This condensation leaves the gas phase with a lower liquid content. Because the condensed liquid adheres to the walls of the pore spaces of the rock, it is immobile. Thus the gas produced at the surface will have a lower liquid content, and the producing gas-oil ratio therefore rises. This process of *retrograde* condensation continues until a point of maximum liquid volume is reached, 10% at 2250 psia, point B_2. The term *retro-*

grade is used because generally vaporization, rather than condensation, occurs during isothermal expansion. Actually, after the dew point is reached, because the composition of the produced fluid changes, the composition of the remaining reservoir fluid also changes, and the phase envelope begins to shift. The phase diagram of Fig. 1.1 represents one and only one hydrocarbon mixture. Unfortunately for maximum liquid recovery, this shift is toward the right, and this further aggravates the retrograde liquid loss within the pores of the reservoir rock.

Neglecting for the moment this shift in the phase diagram, for qualitative purposes *vaporization* of the retrograde liquid occurs from B_2 to the abandonment pressure B_3. This revaporization aids liquid recovery and may be evidenced by decreasing gas-oil ratios on the surface. The overall retrograde loss will evidently be greater (a) for lower reservoir temperatures, (b) for higher abandonment pressures, and (c) for greater shift of the phase diagram to the right—this latter being a property of the hydrocarbon system. The retrograde liquid in the reservoir at any time is composed to a large extent of methane and ethane by volume, and so it is much larger than the volume of stable liquid that could be obtained from it at atmospheric temperature and pressure. The composition of this retrograde liquid is changing as pressure declines so that 4% retrograde liquid volume at, say, 750 psia might contain as much stable surface condensate as, say, 6% retrograde liquid volume at 2250 psia.

If the accumulation occurred at 2900 psia and 75°F, point C, the reservoir would be in a one-phase state, now called liquid, because the temperature is below the critical temperature. This type is called a bubble-point reservoir; as pressure declines, the bubble point will be reached, in this case at 2550 psia, point C_1. Below this point, bubbles, or a free-gas phase, will appear. Eventually the free gas evolved begins to flow to the well bore, and in ever increasing quantities. Conversely, the oil flows in ever decreasing quantities, and at depletion much unrecovered oil remains in the reservoir. Other names for this type of liquid (oil) reservoir are depletion, dissolved gas, solution gas drive, expansion, and internal gas drive.

Finally, if this same hydrocarbon mixture occurred at 2000 psia and 150°F, point D, it would be a two-phase reservoir, consisting of a liquid or oil zone overlain by a gas zone or cap. Because the composition of the gas and oil zones are entirely different from each other, they may be represented separately by individual phase diagrams that bear little relation to each other or to the composite. The liquid or oil zone will be at its bubble point and will be produced as a bubble-point reservoir modified by the presence of the gas cap. The gas cap will be at the dew point and may be either retrograde as shown in Fig. 1.2 (a) or nonretrograde, Fig. 1.2 (b).

From this technical point of view, hydrocarbon reservoirs are initially either in a single-phase state (A, B, and C) or a two-phase state (D), depending on their temperatures and pressures relative to their phase envelopes. On volumetric depletion (no water influx), these various one-phase reservoirs

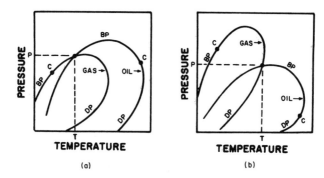

Fig. 1.2. Phase diagrams of a cap gas and oil zone fluid showing (a) retrograde cap gas and (b) nonretrograde cap gas

may behave as simple, single-phase gas reservoirs (A), in which reservoir temperature exceeds the cricondentherm; or retrograde condensate (dew-point) reservoirs (B), in which reservoir temperature lies between the critical temperature and the cricondentherm; or dissolved gas (bubble-point) reservoirs (C), in which reservoir temperature is below the critical temperature. When the pressure and temperature lie within the two-phase region, an oil zone with an overlying gas cap exists. The oil zone produces as a bubble-point oil reservoir and the gas cap either as a single-phase gas reservoir (A) or a retrograde gas reservoir (B).

4. REVIEW OF ROCK PROPERTIES

Properties discussed in this section include porosity, isothermal compressibility, and fluid saturation. Although permeability is a property of a rock matrix, because of its importance in fluid flow calculations, a discussion of permeability is postponed until Chapter 7, in which single phase fluid flow is considered.

4.1. Porosity

The porosity of a porous medium is given the symbol of ϕ and is defined as the ratio of void space, or pore volume, to the total bulk volume of the rock. This ratio is expressed either as a fraction or in pecent. When using a value of porosity in an equation it is nearly always expressed as a fraction. The term *hydrocarbon porosity* refers to that part of the porosity that contains hydrocarbon. It is the total porosity multiplied by the fraction of the pore volume that contains hydrocarbon.

The value of porosity is usually reported as either a total or an effective porosity, depending on the type of measurement used. The total porosity represents the total void space of the medium. The effective porosity is the amount of the void space that contributes to the flow of fluids. This is the type of porosity usually measured in the laboratory and used in calculations of fluid flow.

The laboratory methods of measuring porosity include Boyle's law, water-saturation, and organic-liquid saturation methods. Dotson, Slobod, McCreery, and Spurlock have described a porosity-check program made by five laboratories on 10 samples.[27] The average deviation of porosity from the average values was ±0.5 porosity %. The accuracy of the average porosity of a reservoir as found from core analysis depends on the quality and quantity of the data available and on the uniformity of the reservoir. The average porosity is seldom known more precisely than to 1 porosity % e.g., to 5% accuracy at 20% porosity. The porosity is also calculated from electric logs and neutron logs, often with the assistance of some core measurements. Logging techniques have the advantage of averaging larger volumes of rock than in core analysis. When calibrated with core data, they should provide average porosity figures in the same range of accuracy as core analysis. When there are variations in porosity across the reservoir, the average porosity should be found on a weighted volume basis.

4.2. Isothermal Compressibility

The isothermal compressibility for a substance is given by the following equation:

$$c = -\frac{1}{v}\frac{dv}{dp} \qquad (1.1)$$

where,

c = isothermal compressibility

v = volume

p = pressure

The equation describes the change in volume that a substance undergoes during a change in pressure while the temperature is held constant. The units are in reciprocal pressure units. When the internal fluid pressure within the pore spaces of a rock, which is subjected to a constant external (rock or overburden) pressure, is reduced, the bulk volume of the rock decreases while the volume of the solid rock material (e.g., the sand grains of a sandstone) increases. Both of these volume changes act to reduce the porosity of the rock slightly, of the order of 0.5% for a 1000 psi change in the internal fluid pressure (e.g., at 20% porosity to 19.9%).

Studies by van der Knaap indicate that this change in porosity for a given rock depends only on the *difference* between the internal and external pressures, and not on the absolute value of the pressures.[28] As with the volume of reservoir coils above the bubble point, however, the change in pore volume is nonlinear and the pore volume compressibility is not constant. The *pore volume compressibility* c_f at any value of external-internal pressure difference may be defined as the change in pore volume per unit of pore volume per unit change in pressure. The values for limestone and sandstone reservoir rocks lie in the range of 2×10^{-6} to 25×10^{-6} psi^{-1}. If the compressibility is given in terms of the change in pore volume per unit of *bulk* volume per unit change in pressure, dividing by the fractional porosity places it on a pore volume basis. For example, a compressibility of 1.0×10^{-6} pore volume per *bulk* volume per psi for a rock of 20% porosity is 50×10^{-6} pore volume per *pore* volume per psi.

Newman measured isothermal compressibility and porosity values in 79 samples of consolidated sandstones under hydrostatic pressure.[29] When he fit the data to hyperbolic equation, he obtained the following correlation:

$$c_f = \frac{97.3200(10)^{-6}}{(1 + 55.8721\phi)^{1.42859}} \tag{1.2}$$

This correlation was developed for consolidated sandstones having a range of porosity values from $0.02 < \phi < 0.23$. The average absolute error of the correlation over the entire range of porosity values was found to be 2.60%.

Newman also developed a similar correlation for limestone formations under hydrostatic pressure.[29] The range of porosity values included in the correlation was $0.02 < \phi < 0.33$, and the average absolute error was found to be 11.8%. The correlation for limestone formations is as follows:

$$c_f = \frac{0.853531}{(1 + 2.47664(10)^6\phi)^{0.92990}} \tag{1.3}$$

Even though the rock compressibilities are small figures, their effect may be important in some calculations on reservoirs or aquifers that contain fluids of compressibilites in the range of 3 to $25(10)^{-6}$ psi^{-1}. One application is given in Chapter 5 involving calculations above the bubble point. Geertsma points out that when the reservoir is not subjected to uniform external pressure, as are the samples in the laboratory tests of Newman, the effective value in the reservoir will be less than the measured value.[30]

4.3. Fluid Saturations

The ratio of the volume that a fluid occupies to the pore volume is called the *saturation* of that fluid. The symbol for oil saturation is S_o, where S refers to saturation and the subscript o refers to oil. Saturation is expressed as either a fraction or a percentage, but it is used as a fraction in equations. The saturations of all fluids present in a porous medium add to 1.

There are, in general, two ways of measuring original fluid saturations:

the direct approach and the indirect approach. The direct approach involves either the extraction of the reservoir fluids or the leaching of the fluids from a sample of the reservoir rock. The indirect approach relies on a measurement of some other property, such as capillary pressure, and the derivation of a mathematical relationship between the measured property and saturation.

Direct methods include retorting the fluids from the rock, distilling the fluids with a modified ASTM (American Society for Testing and Materials) procedure, and centrifuging the fluids. Each method relies on some procedure to remove the rock sample from the reservoir. Experience has found that it is difficult to remove the sample without altering the state of the fluids and/or rock. The indirect methods use logging or capillary pressure measurements. With either method, errors are built into the measurement of saturation. However, under favorable circumstances and with careful attention to detail, saturation values can be obtained within useful limits of accuracy.

5. REVIEW OF GAS PROPERTIES

5.1 Ideal Gas Law

Relationships that describe the pressure-volume-temperature (PVT) behavior of gases are called *equations of state*. The simplest equation of state is called the *ideal gas law* and is given by:

$$pV = nR'T \tag{1.4}$$

where,

p is absolute pressure

V is total volume

n is moles

T is absolute temperature

R' is the gas constant

When $R' = 10.73$, p must be in pounds per square inch absolute (psia), V in cubic feet, n in pound-moles (lb-moles), and T in degrees Rankine. The ideal gas law was developed from Boyle's and Charles's laws, which were formed from experimental observations.

The petroleum industry works with a set of standard conditions—usually 14.7 psia and 60°F. When a volume of gas is reported at these conditions, it is given the units of SCF (standard cubic feet). Sometimes the letter M will appear in the units (e.g., *MCF or M SCF*). This refers to 1000 standard cubic feet. The volume that 1 lb-mole occupies at standard conditions is 379.4 SCF. A quantity of a pure gas can be expressed as the number of cubic feet at a

McCain uses $380.7 \; SCF/lb \; mole$

specified temperature and pressure, the number of moles, the number of pounds, or the number of molecules. For practical measurement, the weighing of gases is difficult, so that gases are metered by volume at measured temperatures and pressures, from which the pounds or moles may be calculated. Example 1.1 illustrates the calculations of the contents of a tank of gas in each of four units.

Example 1.1. Calculating the contents of a tank of ethane in moles, pounds, molecules, and SCF.

Given: A 500 cu ft tank of ethane at 100 psia and 100°F.

SOLUTION: Assuming ideal gas behavior:

$$\text{Moles} = \frac{100 \times 500}{10.73 \times 560} = 8.32$$

$$\text{Pounds} = 8.32 \times 30.07 = 250.2 \quad \text{MW in } \text{lb/mole}$$

$$\text{Molecules} = 8.32 \times 2.733 \times 10^{26} = 22.75 \times 10^{26}$$

At 14.7 psia and 60°F:

$$\text{SCF/mole}$$

$$\text{SCF} = 8.32 \times 379.4 = 3157$$

Alternate solution using Eq. (1.4):

$$\text{SCF} = \frac{nR'T}{p} = \frac{8.32 \times 10.73 \times 520}{14.7} = 3158$$

5.2 Specific Gravity

Because the density of a substance is defined as mass per unit volume, the density of gas, ρ_g, at a given temperature and pressure can be derived as follows:

$$\text{density} = \frac{\text{mass}}{\text{volume}} = \frac{nM_w}{V}$$

$$\text{density} = \rho_g = \frac{\frac{pV}{R'T}M_w}{V} = \frac{\rho M_w}{R'T}$$

where,

M_w = molecular weight

Because it is more convenient to measure the specific gravity of gases than the gas density, *specific gravity* is more commonly used. *Specific gravity* is defined as the ratio of the density of a gas at a given temperature and pressure to the density of air at the same temperature and pressure, usually near 60°F and atmospheric pressure. Whereas the density of gases varies with temperature and pressure, the specific gravity is independent of temperature and pressure when the gas obeys the ideal gas law. By the previous equation, the density of air is

$$\rho_{air} = \frac{p \times 28.97}{R'T}$$

Then the specific gravity, γ_g, of as gas is

$$\gamma_g = \frac{\rho_g}{\rho_{air}} = \frac{\dfrac{pM_w}{R'T}}{\dfrac{p \times 28.97}{R'T}} = \frac{M_w}{28.97} \tag{1.5}$$

Equation (1.5) might also have been obtained from the previous statement that 379.4 cu ft of any ideal gas at 14.7 psia and 60°F is one mole, and therefore a weight equal to the molecular weight. Thus, by definition of specific gravity:

$$\gamma_g = \frac{\text{Weight of 379.4 cu ft of gas at 14.7 and 60°F}}{\text{Weight of 379.4 cu ft of air at 14.7 and 60°F}} = \frac{M_w}{28.97}$$

Thus if the specific gravity of a gas is 0.75, its molecular weight is 21.7 pounds per mole.

5.3 Real Gas Law

All that has just been said applies to a perfect or ideal gas. Actually there are no perfect gases; however, many gases near atmospheric temperature and pressure approach ideal behavior. All molecules of real gases have two tendencies: (1) to fly apart from each other because of their constant kinetic motion, and (2) to come together because of electrical attractive forces between the molecules. Because the molecules are quite far apart, the attractive forces are negligible, and the gas behaves close to ideal. Also at high temperatures the kinetic motion, being greater, makes the attractive forces comparatively negligible, and, again, the gas approaches ideal behavior.

Since the volume of a gas will be less than what the ideal gas volume would be, the gas is said to be supercompressible. The number, which is a measure of the amount the gas deviates from perfect behavior, is sometimes called the *supercompressibility factor,* usually shortened to the *compressibility factor.* More commonly it is called the *gas deviation factor,* symbol *z.* This dimensionless quantity varies usually between 0.70 and 1.20, a value of 1.00 representing ideal behavior.

At very high pressures, above about 5000 psia, natural gases pass from a supercompressible condition to one in which compression is more difficult than in the ideal gas. The explanation is that, in addition to the forces mentioned earlier, when the gas is highly compressed, the volume occupied by the molecules themselves becomes an appreciable portion of the total volume. Since it is really the space between the molecules which is compressed, and there is less compressible space, the gas appears to be more difficult to compress. In addition, as the molecules get closer together (i.e., at high pressure), repulsive forces begin to develop between the molecules. This is indicated by a gas deviation factor greater than unity. The gas deviation factor is by definition the ratio of the volume *actually* occupied by a gas at a given pressure and temperature to the volume it would occupy if it behaved ideally, or:

$$z = \frac{V_a}{V_i} = \frac{\text{Actual volume of } n \text{ moles of gas at } T \text{ and } p}{\text{Ideal volume of } n \text{ moles at same } T \text{ and } p} \qquad (1.6)$$

These theories qualitatively explain the behavior of nonideal or real gases. Equation (1.6) may be substituted in the ideal gas law, Eq. (1.4) to give an equation for use with nonideal gases

$$p\left(\frac{V_a}{z}\right) = nR'T \text{ or } pV_a = znR'T \qquad (1.7)$$

where V_a is the actual gas volume. The gas deviation factor must be determined for every gas and every combination of gases and at the desired temperature and pressure, for it is different for each gas or mixture of gases and for each temperature and pressure of that gas or mixture of gases. The omission of the gas deviation factor in gas reservoir calculations may introduce errors as large as 30%.[31] Figure 1.3 shows the gas deviation factors of two gases, one of 0.90 specific gravity and the other of 0.665 specific gravity. These curves show that the gas deviation factors drop from unity at low pressures to a minimum value near 2500 psia. They rise again to unity near 5000 psia and to values greater than unity at still higher pressures. In the range of 0 to 5000 psia, the deviation factors at the same temperature will be lower for the heavier gas, and for the same gas they will be lower at the lower temperature.

The deviation factor of natural gas is commonly measured in the laboratory on samples of surface gases. If there is condensate liquid at the point of sampling, the sample must be taken in such a way as to represent the single-phase reservoir gas. This may be accomplished with a special sampling nozzle or by recombining samples of separator gas, stock tank gas, and stock tank liquid in the proportions in which they are produced. The deviation factor of

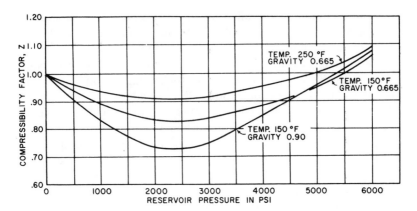

Fig. 1.3. Effect of pressure, temperature, and composition on the gas deviation factor.

solution gas is measured on samples evolved from solution in oil during the liberation process.

The gas deviation factor is commonly determined by measuring the volume of a sample at desired pressures and temperatures, and then measuring the volume of the same quantity of gas at atmospheric pressure and at a temperature sufficiently high so that all the material remains in the vapor phase. For example, a sample of the Bell Field gas has a measured volume of 364.6 cu cm at 213°F and 3250 psia. At 14.80 psia and 82°F it has a volume of 70,860 cu cm. Then by Eq. (1.7), assuming a gas deviation factor of unity at the lower pressure, the deviation factor at 3250 psia and 213°F is:

$$z = \frac{3250 \times 364.6}{460 + 213} \times \frac{1.00 \times (460 + 82)}{14.80 \times 70,860} = 0.910$$

If the gas deviation factor is not measured, it may be estimated from its specific gravity. Example 1.2 shows the method for estimating the gas deviation factor from its specific gravity. The method uses a correlation to estimate pseudocritical temperature and pressure values for a gas with a given specific gravity. The correlation was developed by Sutton on the basis of 264 different gas samples and is shown in Fig. 1.4.[32] Sutton also conducted a regression analysis on the raw data and obtained the following equations over the range of specific gas gravities with which he worked—$0.57 < \gamma_g < 1.68$:

$$p_{pc} = 756.8 - 131.0\gamma_g - 3.6\gamma_g^2 \tag{1.8}$$

$$T_{pc} = 169.2 - 349.5\gamma_g - 74.0\gamma_g^2 \tag{1.9}$$

Having obtained the pseudocritical values, the pseudoreduced pressure and temperature are calculated. The gas deviation factor is then found by using the correlation chart of Fig. 1.5.

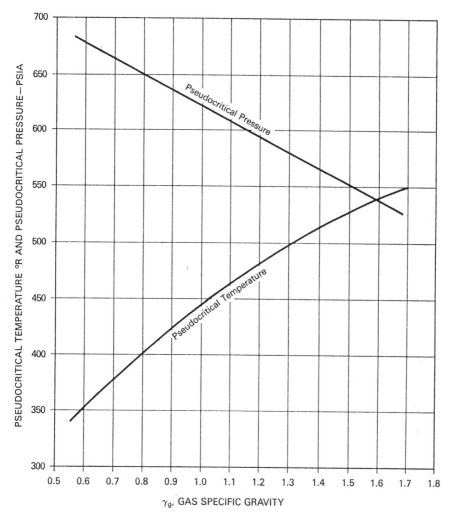

Fig. 1.4. Pseudocritical properties of natural gases. (*After* Sutton.[32])

Example 1.2. Calculating the gas deviation factor of the Bell Field gas from its specific gravity.

Given:

Specific gravity = 0.665
Reservoir temperature = 213°F
Reservoir pressure = 3250 psia

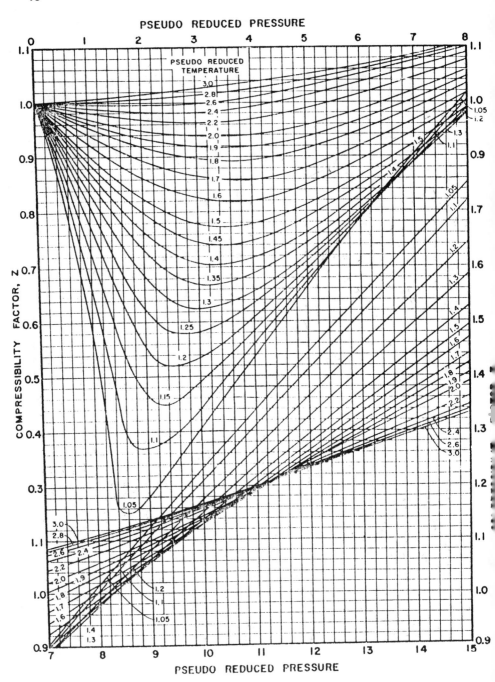

Fig. 1.5. Compressibility factors for natural gases. (*After* Standing and Katz, *Trans. AIM.*)

SOLUTION: From Fig. 1.4 the pseudocritical pressure and temperature
are

$$p_{pc} = 668 \text{ psia} \quad \text{and} \quad T_{pc} = 369°\text{R}$$

Using Eq. (1.8) and (1.9), the pseudocritical values are

$$p_{pc} = 756.8 - 131.0(.665) - 3.6(.665)^2 = 668 \text{ psia}$$
$$T_{pc} = 169.2 - 349.5(.665) - 74.0(.665)^2 = 369°R$$

For 3250 psia and 213°F, the pseudoreduced pressure and temperature are

$$p_{pr} = \frac{3250}{668} = 4.87, \quad T_{pr} = \frac{460 + 213}{369} = 1.82$$

Enter Fig. 1.5 with these values. Find $z = 0.918$.

In many reservoir engineering calculations, it is necessary to use the
assistance of a computer, and the chart of Standing and Katz then becomes
difficult to use. Dranchuk and Abou-Kassem fitted an equation of state to the
data of Standing and Katz in order to estimate the gas deviation factor in
computer routines.[34] Dranchuk and Abou-Kassem used 1500 data points and
found an average absolute error of 0.486% over ranges of pseudoreduced
pressure and temperature of:

$$0.2 < p_{pr} < 30$$
$$1.0 < T_{pr} < 3.0$$

and for

$$p_{pr} < 1.0 \text{ with } 0.7 < T_{pr} < 1.0$$

The Dranchuk and Abou-Kassem equation of state gives poor results for
$T_{pr} = 1.0$ and $p_{pr} > 1.0$. The form of the Dranchuk and Abou-Kassem equation
of state is as follows:

$$z = 1 + c_1(T_{pr})\rho_r + c_2(T_{pr})\rho_r^2 - c_3(T_{pr})\rho_r^5 + c_4(\rho_r, T_{pr}) \tag{1.10}$$

where,

$$\rho_r = 0.27\, p_{pr}/(z\, T_{pr}) \tag{1.10a}$$
$$c_1(T_{pr}) = A_1 + A_2/T_{pr} + A_3/T_{pr}^3 + A_4/T_{pr}^4 + A_5/T_{pr}^5 \tag{1.10b}$$
$$c_2(T_{pr}) = A_6 + A_7/T_{pr} + A_8/T_{pr}^2 \tag{1.10c}$$
$$c_3(T_{pr}) = A_9(A_7/T_{pr} + A_8/T_{pr}^2) \tag{1.10d}$$
$$c_4(\rho_r, T_{pr}) = A_{10}(1 + A_{11}\rho_r^2)(\rho_r^2/T_{pr}^3) \exp(-A_{11}\rho_r^2) \tag{1.10e}$$

where the constants $A_1 - A_{11}$ are as follows:

$A_1 = 0.3265$	$A_2 = -1.0700$	$A_3 = -0.5339$
$A_4 = 0.01569$	$A_5 = -0.05165$	$A_6 = 0.5475$
$A_7 = -0.7361$	$A_8 = 0.1844$	$A_9 = 0.1056$
$A_{10} = 0.6134$	$A_{11} = 0.7210$	

Because the z-factor is on both sides of the equation, a trial-and-error solution is necessary to solve the Dranchuk and Abou-Kassem equation of state. Any one of the number of iteration techniques can be used to assist in the trial-and-error procedure. One that is frequently used is the secant method[35] which has the following iteration formula:

$$x_{n+1} = x_n - f_n \left[\frac{x_n - x_{n-1}}{f_n - f_{n-1}} \right] \tag{1.11}$$

To apply the secant method to the foregoing procedure, Eq. (1.10) is rearranged to the form:

$$F(z) = z - (1 + c_1(T_{pr})\rho_r + c_2(T_{pr})\rho_r^2 - c_3(T_{pr})\rho_r^5 + c_4(\rho_r, T_{pr})) = 0 \tag{1.12}$$

The left-hand side of Eq. (1.12) becomes the function, f, and the z-factor becomes x. The iteration procedure is initiated by choosing two values of the z-factor and calculating the corresponding values of the function, f. The secant method provides the new guess for z, and the calculation is repeated until the function, f, is zero or within a specified tolerance (i.e., $\pm 10^{-4}$). This solution procedure is fairly easy to program on a computer.

Another popular iteration technique is the Newton-Raphson method, which has the following iteration formula:

$$x_{n+1} = x_n - \frac{f_n}{f'_n} \tag{1.13}$$

As can be seen in Eq. (1.13), the derivative, $f'(z)$, is required for the Newton-Raphson technique. The derivative of Eq. (1.12) with respect to z is as follows:

$$(\partial F(z)/\partial z)T_{pr} = 1 + c_1(T_{pr})\rho_r/z + 2c_2(T_{pr})\rho_r^2/z - 5c_3(T_{pr})\rho_r^5/z$$
$$+ \frac{2A_{10}\rho_r^2}{T_{pr}^3 z} (1 + A_{11}\rho_r^2 - (A_{11}\rho_r^2)^2) \exp(-A_{11}\rho_r^2) \tag{1.14}$$

A more accurate estimation of the deviation factor can be made when the analysis of the gas is available. This calculation assumes that each component contributes to the pseudocritical pressure and temperature in proportion to its volume percentage in the analysis and to the critical pressure and temperature, respectively, of that component. Table 1.1 gives the critical pressures and temperatures of the hydrocarbon compounds and others commonly found in natural gases.[36] It also gives some additional physical properties of these compounds. Example 1.3 shows the method of calculating the gas deviation factor from the composition of the gas.

TABLE 1.1.
Physical properties of the paraffin hydrocarbons and other compounds
(After *Eilerts*[36])　　　　　　　　　　　　　　　　　　GPM

Compound	Molecular Weight	Boiling Point at 14.7 psia °F	Critical Constants Pressure, p_e, psia	Temperature T_e, °R	Liquid Density 60°F, 14.7 psia G (Grams) per cc	Lb per Gal	Est. Part. Volume at 60°F, 14.7 psia, Gal per M SCF	Est. Part. Volume at 60°F, 14,4 psia, Gal per Lb-mole
Methane	16.04	−258.7	673.1	343.2	[a]0.348	2.90	14.6	5.53
Ethane	30.07	−127.5	708.3	549.9	[a]0.485	4.04	19.6	7.44
Propane	44.09	−43.7	617.4	666.0	[b]0.5077	4.233	27.46	10.417
Isobutane	58.12	10.9	529.1	734.6	[b]0.5631	4.695	32.64	12.380
n-Butane	58.12	31.1	550.1	765.7	[b]0.5844	4.872	31.44	11.929
Isopentane	72.15	82.1	483.5	829.6	0.6248	5.209	36.50	13.851
n-Pentane	72.15	96.9	489.8	846.2	0.6312	5.262	36.14	13.710
n-Hexane	86.17	155.7	440.1	914.2	0.6641	5.536	41.03	15.565
n-Heptane	100.2	209.2	395.9	972.4	0.6882	5.738	46.03	17.463
n-Octane	114.2	258.2	362.2	1024.9	0.7068	5.892	51.09	19.385
n-Nonane	128.3	303.4	334	1073	0.7217	6.017	56.19	21.314
n-Decane	142.3	345.4	312	1115	0.7341	6.121	61.27	23.245
Air	28.97	−317.7	547	239				
Carbon dioxide	44.01	−109.3	1070.2	547.5				
Helium	4.003	−452.1	33.2	9.5				
Hydrogen	2.106	−423.0	189.0	59.8				
Hydrogen sulfide	34.08	−76.6	1306.5	672.4				
Nitrogen	28.02	−320.4	492.2	227.0				
Oxygen	32.00	−297.4	736.9	278.6				
Water	18.02	212.0	3209.5	1165.2	0.9990	8.337		

[a] Basis partial volume in solution.
[b] At bubble-point pressure and 60°F.

Example 1.3. Calculating the gas deviation factor of the Bell Field gas from its composition.

Given: The composition Col. (2), and the physical data Cols. (3) to (5) taken from Table 1.1.

(1) Component	(2) Comp., Mole Fract.	(3) Mol. Wt.	(4) p_c	(5) T_c	(6) (2) × (3)	(7) (2) × (4)	(8) (2) × (5)
Methane	0.8612	16.04	673	343	13.81	579.59	295.39
Ethane	0.0591	30.07	708	550	1.78	41.84	32.51
Propane	0.0358	44.09	617	666	1.58	22.09	23.84
Butane	0.0172	58.12	550	766	1.00	9.46	13.18
Pentanes	0.0050	72.15	490	846	0.36	2.45	4.23
CO_2	0.0010	44.01	1070	548	0.04	1.07	0.55
N_2	0.0207	28.02	492	227	0.58	10.18	4.70
	1.0000				19.15	666.68	374.40

SOLUTION: The specific gravity may be obtained from the sum of Col. (6), which is the average molecular weight of the gas,

$$\gamma_g = \frac{19.15}{28.97} = 0.661$$

The sums of Cols. (7) and (8) are the pseudocritical pressure and temperature, respectively. Then at 3250 psia and 213°F, the pseudoreduced pressure and temperature are

$$p_{pr} = \frac{3250}{666.68} = 4.87 \qquad T_{pr} = \frac{673}{374.4} = 1.80$$

The gas deviation factor using Fig. 1.5 is $z = 0.91$.

Wichert and Aziz have developed a correlation to account for inaccuracies in the Standing and Katz chart when the gas contains significant fractions of carbon dioxide (CO_2) and hydrogen sulfide (H_2S).[37] The Wichert and Aziz correlation modifies the values of the pseudocritical constants of the natural gas. Once the modified constants are obtained, they are used to calculate pseudoreduced properties as described in Ex. 1.2 and the z-factor is determined from Fig. 1.5 or Eq. (1.10). The Wichert and Aziz correlation equation is as follows:

$$\varepsilon = 120(A^{0.9} - A^{1.6}) + 15(B^{0.5} - B^4) \qquad (1.15)$$

where,

A = sum of the mole fractions of CO_2 and H_2S in the gas mixture

B = mole fraction of H_2S in the gas mixture

The modified pseudocritical properties are given by:

$$T'_{pc} = T_{pc} - \varepsilon \tag{1.15a}$$

$$p'_{pc} = \frac{p_{pc}T'_{pc}}{(T_{pc} + B(1-B)\varepsilon)} \tag{1.15b}$$

Wichert and Aziz found their correlation to have an average absolute error of 0.97% over the following ranges of data: $154 < p$ (psia) < 7026 and $40 < T$ (°F) < 300. The correlation is good for concentrations of $CO_2 < 54.4\%$ (mole %) and $H_2S < 73.8\%$ (mole %).

5.4. Formation Volume Factor and Density

Gas volume factors, symbol B_g, relate the volume of gas in the reservoir to the volume on the surface (i.e., at standard conditions, p_{sc} and T_{sc}). They are generally expressed in either cubic feet or barrels of reservoir volume per standard cubic foot of gas. Assuming a gas deviation factor of unity for the standard conditions, the reservoir volume of one standard cubic foot ($V_{sc} = 1.00$) at reservoir pressure p and temperature T by Eq. (1.7) is

$$B_g = \frac{p_{sc}zT}{T_{sc}p} \tag{1.16}$$

Where p_{sc} is 14.7 psia and T_{sc} is 60°F,

$$B_g = 0.02829\frac{zT}{p} \text{ cu ft/SCF}$$

$$= 0.00504\frac{zT}{p} \text{ bbl/SCF} \tag{1.17}$$

The constants in Eqs. (1.17) are for 14.7 psia and 60°F *only*, and different constants must be calculated for other standards. Thus for the Bell Field gas at a reservoir pressure of 3250 psia, a temperature of 213°F, and a gas deviation factor of 0.910, the gas volume factor is

$$B_g = \frac{0.02829 \times 0.910 \times 673}{3250} = 0.00533 \text{ cu ft/SCF}$$

These gas volume factors mean that 1 std cu ft (at 14.7 psia and 60°F) will occupy 0.00533 cu ft of space in the reservoir at 3250 psia and 213°F. Because oil is usually expressed in barrels and gas in cubic feet, when calculations are made on combination reservoirs containing both gas and oil, either the oil volume must be expressed in cubic feet or the gas volume in barrels. The foregoing gas volume factor expressed in barrels is 0.000949 bbl/SCF. Then

1000 cu ft of reservoir pore volume in the Bell Field gas reservoir at 3250 psia contains

$$G = 1000 \text{ cu ft} \div 0.00533 \text{ cu ft/SCF} = 188 \text{ M SCF}$$

Equation (1.7) may also be used to calculate the density of a reservoir gas. The moles of gas in 1 cu ft of reservoir gas pore space is p/zRT. By Eq. (1.5) the molecular weight of a gas is $28.97 \times \gamma_g$ lb per mole. Therefore the pounds contained in 1 cu ft, that is the *reservoir* gas density, symbol ρ_g, is

$$\rho_g = \frac{28.97 \times \gamma_g \times p}{zR'T} \tag{1.18}$$

For example, the density of the Bell Field reservoir gas with a gas gravity of 0.665 is

$$\rho_g = \frac{28.97 \times 0.665 \times 3250}{0.910 \times 10.73 \times 673} = 9.530 \text{ lb/cu ft}$$

5.5 Isothermal Compressibility

The change in volume with pressure for gases under *isothermal* conditions, which is closely realized in reservoir gas flow, is expressed by the real gas law:

$$V = \frac{znR'T}{p} \text{ or } V = \text{constant} \times \frac{z}{p} \tag{1.18}$$

Sometimes it is useful to introduce the concept of *gas compressibility*. This must not be confused with the gas deviation factor, which is also referred to as the gas *compressibility factor*. The above equation may be differentiated with respect to pressure at constant temperature to give

$$\frac{dV}{dp} = \frac{nR'T}{p}\frac{dz}{dp} - \frac{znR'T}{p^2}$$

$$= \left(\frac{znR'T}{p}\right)\frac{1}{z}\frac{dz}{dp} - \left(\frac{znR'T}{p}\right) \times \frac{1}{p}$$

$$\frac{1}{V} \times \frac{dV}{dp} = \frac{1}{z}\frac{dz}{dp} - \frac{1}{p}$$

Finally, because

$$c = -\frac{1}{V}\frac{dV}{dp}$$

$$c_g = \frac{1}{p} - \frac{1}{z}\frac{dz}{dp} \tag{1.19}$$

For an ideal gas $z = 1.00$ and $dz/dp = 0$ and the compressibility is simply the reciprocal of the pressure. An ideal gas at 1000 psia, then, has a compressibility of 1/1000 or 1000×10^{-6} psi^{-1}. Example 1.4 shows the calculation of the compressibility of a gas from the gas deviation factor curve of Fig. 1.6 using Eq. (1.19).

Example 1.4. To find the compressibility of a gas from the gas deviation factor curve.

Given: The gas deviation factor curve for a gas at 150°F, Fig. 1.6.

SOLUTION: At 1000 psia, the slope dz/dp is shown graphically in Fig. 1.6 as -127×10^{-6}. Note that this is a negative slope. Then, because $z = 0.83$:

$$c_g = \frac{1}{1000} - \frac{1}{0.83}\,(-127 \times 10^{-6})$$

$$= 1000 \times 10^{-6} + 153 \times 10^{-6} = 1153 \times 10^{-6}\,\text{psi}^{-1}$$

At 2500 psia the slope dz/dp is zero, so the compressibility is simply:

$$c_g = \frac{1}{2500} = 400 \times 10^{-6}\,\text{psi}^{-1}$$

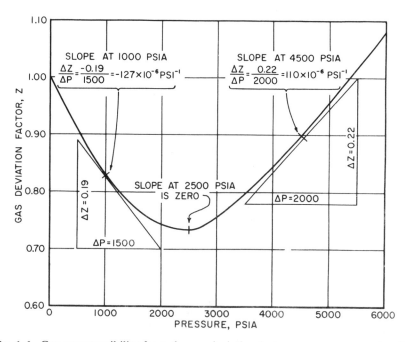

Fig. 1.6. Gas compressibility from the gas deviation factor versus pressure plot. (See Example 1.4)

At 4500 psia the slope dz/dp is positive and as shown in Fig. 1.6 is equal to $110 \times 10^{-6}\text{psi}^{-1}$. Since $z = 0.90$ at 4500 psia:

$$c_g = \frac{1}{4500} - \frac{1}{0.90} \, (110 \times 10^{-6})$$

$$= 222 \times 10^{-6} - 122 \times 10^{-6} = 100 \times 10^{-6}\text{psi}^{-1}$$

Trube has replaced the pressure in Eq. (1.19) by the product of the pseudocritical and the pseudoreduced pressures, or $p = p_{pc}(p_{pr})$ and $dp = p_{pc}dp_{pr}$.[38] This obtains

$$c_g = \frac{1}{p_{pc}p_{pr}} - \frac{1}{zp_{pc}}\frac{dz}{dp_{pr}} \tag{1.20}$$

Multiplying through by the pseudocritical pressure, the product $c_g(p_{pc})$ is obtained, which Trube defined as the pseudoreduced compressibility, c_r, or

$$c_r = c_g p_{pc} = \frac{1}{p_{pr}} - \frac{1}{z}\frac{dz}{dp_{pr}} \tag{1.21}$$

Mattar, Brar, and Aziz developed an analytical expression for calculating the pseudoreduced compressibility.[39] The expression is

$$c_r = \frac{1}{p_{pr}} - \frac{0.27}{z^2 T_{pr}}\left[\frac{(\partial z/\partial \rho_r)_{T_{pr}}}{1 + (\rho_r/z)(\partial z/\partial \rho_r)_{T_{pr}}}\right] \tag{1.22}$$

Taking the derivative of Eq. (1.10), the equation of state developed by Dranchuk and Abou-Kassem,[34] the following are obtained:

$$\left(\frac{\partial z}{\partial \rho_r}\right)_{T_{pr}} = c_1(T_{pr}) + 2c_2(T_{pr})\rho_r - 5c_3(T_{pr})\rho_r^4 + \frac{\partial}{\partial \rho_r}[c_4(T_{pr},\rho_r)] \tag{1.23}$$

and

$$\frac{\partial}{\partial \rho_r}[c_4(T_{pr},\rho_r)] = \frac{2A_{10}\rho_r}{T_{pr}^3}[1 + A_{11}\rho_r^2 - (A_{11}\rho_r^2)^2]\exp(-A_{11}\rho_r^2) \tag{1.24}$$

Using Eqs. (1.22) to (1.24) and the definition of the pseudoreduced compressibility, the gas compressibility can be calculated for any gas as long as the gas pressure and temperature are within the ranges specified for the Dranchuk and Abou-Kassem correlation. Using these equations, Blasingame, Johnston, and Poe generated Figs. 1.7 and 1.8.[40] In these figures, the product of $c_r T_{pr}$ is plotted as a function of the pseudoreduced properties, p_{pr} and T_{pr}. Example 1.5 illustrates how to use these figures. Because they are logarithmic in nature, better accuracy can be obtained by using the equations directly.

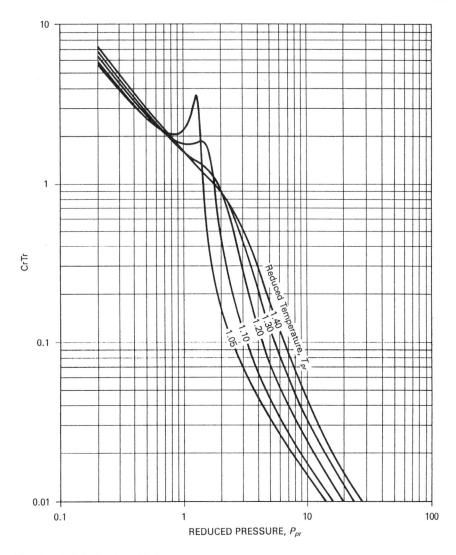

Fig. 1.7. Variation in $c_r T_{pr}$ for natural gases for $1.05 \le T_{pr} \le 1.4$. (*After* Blasingame.)

Example 1.5. To find the compressibility of a 0.90 specific gravity gas at 150°F and 4500 psia using the Mattar, Brar, and Aziz method.

SOLUTION: From Fig. 1.4, $p_{pc} = 636$ psia $T_{pc} = 424°$ R

Then

$$p_{pr} = \frac{4500}{636} = 7.08 \text{ and } T_{pr} = \frac{150 + 460}{424} = 1.44$$

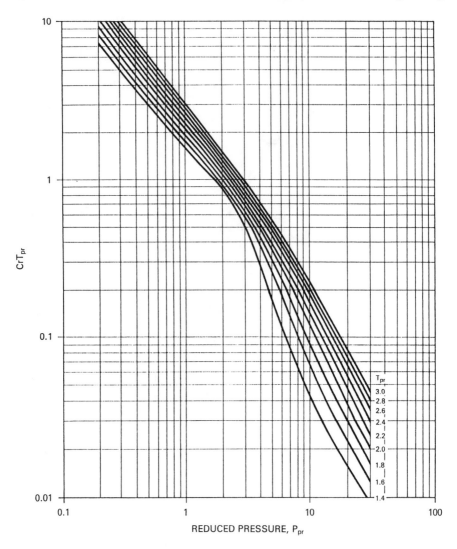

Fig. 1.8. Variation in $c_r T_{pr}$ for natural gases for $1.4 \leq T_{pr} \leq 3.0$. (*After* Blasingame.)

From Fig. 1.8, $c_r T_{pr} = 0.088$:

Then

$$c_r = \frac{c_r T_{pr}}{T_{pr}} = \frac{0.088}{1.44} = 0.061$$

$$c_g = \frac{c_r}{p_{pc}} = \frac{0.061}{636} = 95.9(10)^{-6} \text{ psi}^{-1}$$

5.6 Viscosity

The viscosity of natural gas depends on the temperature, pressure, and composition of the gas. It has units of centipoise (cp). It is not commonly measured in the laboratory because it can be estimated with good precision. Carr, Kobayashi, and Burrows have developed correlation charts, Fig. 1.9 and 1.10, for estimating the viscosity of natural gas from the pseudoreduced critical temperature and pressure.[41] The pseudoreduced temperature and pressure may be estimated from the gas gravity or calculated from the composition of the gas. The viscosity at one atmosphere and reservoir temperature (Fig. 1.9) is multiplied by the viscosity ratio (Fig. 1.10) to obtain the viscosity at reservoir temperature and pressure. The inserts of Fig. 1.9 are corrections to be added to the atmospheric viscosity when the gas contains nitrogen, carbon dioxide, and/or hydrogen sulfide. Example 1.6 illustrates the use of the estimation charts.

Example 1.6. Use of correlation charts to estimate reservoir gas viscosity.

Given:

Reservoir pressure = 2680 psia

Reservoir temperature = 212°F

Well fluid specific gravity = 0.90 (Air = 1.00)

Pseudocritical temperature = 420°R

Pseudocritical pressure = 670 psia

Carbon dioxide content = 5 mole %

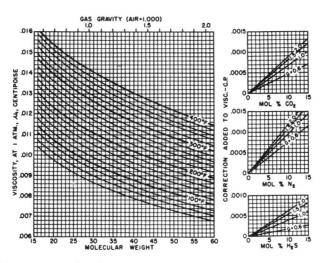

Fig. 1.9. The viscosity of hydrocarbon gases at one atmosphere and reservoir temperatures, with corrections for nitrogen, carbon dioxide, and hydrogen sulfide. (*After* Carr, Kobayashi, and Burrows,[41] *Trans.* AIME.)

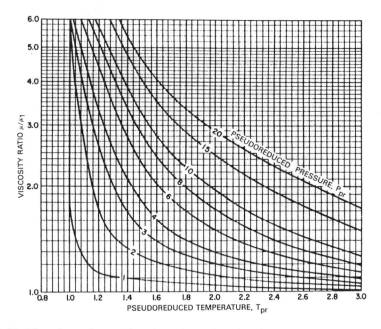

Fig. 1.10. Viscosity ratio as a function of pseudoreduced temperature and pressure. (*After* Carr, Kobayashi, and Burrows,[41] *Trans.* AIME.)

SOLUTION:

$$\mu_1 = 0.0117 \text{ cp at one atmosphere (Fig. 1.9)}$$

Correction for $CO_2 = 0.0003$ cp (Fig. 1.9, insert)

$$\mu_1 = 0.0117 + 0.0003 = 0.0120 \text{ cp (corrected for } CO_2)$$

$$T_{pr} = \frac{672}{420} = 1.60 \qquad P_{pr} = \frac{2680}{670} = 4.00$$

$$\mu/\mu_1 = 1.60 \text{ (Fig. 1.10)}$$

$$\mu = 1.60 \times 0.0120 = 0.0192 \text{ cp at } 212°F \text{ and } 2608 \text{ psia}$$

Lee, Gonzalez, and Eakin developed a semiempirical method that gives an accurate estimate of gas viscosity for most natural gases if the z-factor has been calculated to include the effect of contaminants.[42] For the data from which the correlation was developed, the standard deviation in the calculated gas viscosity was 2.69%. The ranges of variables used in the correlation were $100 < p \text{ (psia)} < 8000$, $100 < T \text{ (°F)} < 340$, and $0.90 < CO_2 \text{ (mole \%)} < 3.20$. In addition to the gas temperature and pressure, the method requires the z-

factor and molecular weight of the gas. The following equations are used in the calculation for the gas viscosity:

$$\mu_g = (10^{-4})K \exp(X\rho^Y) \tag{1.25}$$

where

$$\rho = 1.4935(10)^{-3}\frac{pM_w}{zT} \tag{1.25a}$$

$$K = \frac{(9.4 + 0.02M_w)T^{1.5}}{(209 + 19M_w + T)} \tag{1.25b}$$

$$X = 3.5 + \frac{986}{T} + 0.01M_w \tag{1.25c}$$

$$Y = 2.4 - 0.2X \tag{1.25d}$$

where,

ρ = gas density, g/cc

p = pressure, psia

T = temperature, $°R$

M_w = gas molecular weight

6. REVIEW OF CRUDE OIL PROPERTIES

The next few sections contain information on crude oil properties, including several correlations that can be used to estimate values for the properties. However, these crude oil property correlations are in general, not as reliable as the correlations that have been presented earlier for gases. There are two main reasons for the oil correlations being less reliable. The first is that oils usually consist of many more components than do gases. Whereas gases are mostly made up of alkanes, oils can be made up of several different classes of compounds (e.g., aromatics and paraffins). The second reason is that mixtures of liquid components exhibit more nonidealities than do mixtures of gas components. These nonidealities can lead to errors in extrapolating correlations that have been developed for a certain database of samples to particular applications outside the database. Before using any of the correlations, the engineer should make sure that the application of interest fits within the range of parameters for which a correlation was developed. As long as this is done, then, the correlations used for estimating liquid properties will be adequate and can be expected to yield accurate results.

6.1. Solution Gas-Oil Ratio, R_{so}

The solubility of natural gas in crude oil depends on the pressure, the temperature, and the composition of the gas and the crude oil. For a particular gas and crude oil at constant temperature, the quantity of solution gas increases with pressure; and at constant pressure the quantity decreases with increasing temperature. For any temperature and pressure, the quantity of solution gas increases as the compositions of the gas and crude oil approach each other; that is, it will be greater for higher specific gravity gases and higher API gravity crudes. Unlike the solubility of, say, sodium chloride in water, gas is infinitely soluble in crude oil, the quantity being limited only by the pressure or by the quantity of gas available.

A crude oil is said to be *saturated* with gas at any pressure and temperature if on a slight reduction in pressure some gas is released from solution. Conversely, if no gas is released from solution, the crude oil is said to be *undersaturated* at that pressure. The undersaturated state implies that there is a deficiency of gas present and that had there been an abundance of gas present, the oil would be saturated at that pressure. The undersaturated state further implies that there is no free gas in contact with the crude oil (i.e., there is no gas cap).

Gas solubility under isothermal conditions is generally expressed in terms of the increase in solution gas per unit of oil per unit increase in pressure (e.g., SCF/STB/psi or dR_{so}/dp). Although for many reservoirs this solubility figure is approximately constant over a considerable range of pressures, for precise reservoir calculations the solubility is expressed in terms of the total gas in solution at any pressure (e.g., SCF/STB or R_{so}). It will be shown that the volume of crude oil increases appreciably because of the solution gas, and for this reason the quantity of solution gas is usually referred to as a unit of stock tank oil, and the *solution gas-oil ratio, R_{so},* is expressed in standard cubic feet per stock tank barrel. Figure 1.11 shows the variation of solution gas with pressure for the Big Sandy reservoir fluid at reservoir temperature of 160°F. At the initial reservoir pressure of 3500 psia there is 567 SCF/STB of solution gas. The graph indicates that no gas is evolved from solution when the pressure drops from the initial pressure at 2500 psia. Thus the oil is undersaturated in this region, and there is no free gas phase (cap) in the reservoir. The pressure 2500 psia is called the *bubble-point pressure,* for at this pressure bubbles of free gas first appear. At 1200 psia the solution gas is 337 SCF/STB, and the *average* solubility between 2500 and 1200 psia is:

$$\text{Average solubility} = \frac{567 - 337}{2500 - 1200} = 0.177 \text{ SCF/STB/psi}$$

The data of Fig. 1.11 were obtained from a laboratory PVT study of a bottom-hole sample of the Big Sandy reservoir fluid using the flash liberation process.

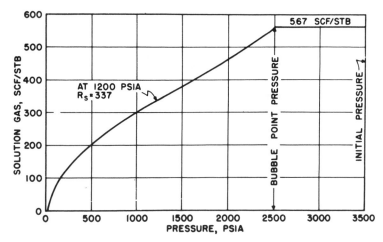

Fig. 1.11. Solution gas-oil ratio of the Big Sandy Field reservoir oil, by flash liberation at reservoir temperature of 160°F.

In Chapter 6 it will be shown that the solution gas-oil ratio and other fluid properties depend on the manner by which the gas is liberated from the oil. The nature of the phenomenon is discussed together with the complications it introduces into certain reservoir calculations. For the present, however, for the sake of simplicity this phenomenon is ignored, and a stock tank barrel of oil is identified with a barrel of residual oil following a flash, liberation process, and the solution gas-oil ratios by flash liberation is used.

When laboratory analyses of the reservoir fluids are not available, it is often possible to estimate with reasonable accuracy the solution gas-oil ratio. Standing gives a correlation method from which the solution gas-oil ratio may be estimated from the reservoir pressure, the reservoir temperature, the API gravity of the tank oil, and the specific gravity of the produced gas.[43] Beggs presents Standing's correlation in equation form for pressures less than or equal to the bubble point pressure.[44] That equation is:

$$R_{so} = \gamma_g \left(\frac{p}{18(10)^{Y_g}} \right)^{1.204} \tag{1.26}$$

where,

$Y_g = 0.00091T - 0.0125 \gamma_{o, API}$

T = temperature, °F

p = pressure, psia

For 105 samples tested, the absolute error found in this correlation was 4.8%. The database contained the following ranges of parameters:

$$130 < p_b \text{ (psia)} < 7000$$
$$100 < T \text{ (°F)} < 258$$
$$20 < \text{gas-oil ratio (SCF/STB)} < 1425$$
$$16.5 < \rho_{o, \text{API}} \text{ (°API)} < 63.8$$
$$0.59 < \gamma_g < 0.95$$
$$1.024 < \text{oil formation volume factor (bbl/STB)} < 2.05$$

6.2. Formation Volume Factor, B_o

It was observed earlier that the solution gas causes a considerable increase in the volume of the crude oil. Figure 1.12 shows the variation in the volume of the reservoir liquid of the Big Sandy reservoir as a function of pressure at reservoir temperature of 160°F. Because no gas is released from solution when the pressure drops from the initial pressure of 3500 psia to the bubble-point pressure at 2500 psia, the reservoir fluid remains in a single (liquid) state; however, because liquids are slightly compressible, the volume increases from 1.310 bbl/STB at 3500 psia to 1.333 bbl/STB at 2500 psia. Below 2500 psia this liquid expansion continues but is masked by a much larger effect: the decrease in the liquid volume due to the release of gas from solution. At 1200 psia the volume has decreased to 1.210 bbl/STB, and at atmospheric pressure and 160°F to 1.040 bbl/STB. The coefficient of temperature expansion for the 30°

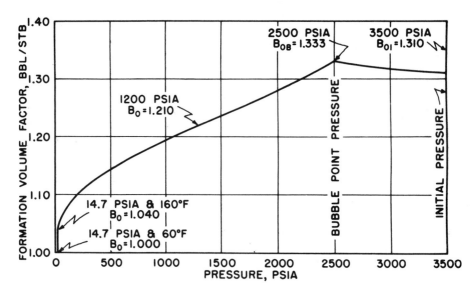

Fig. 1.12. Formation volume factor of the Big Sandy Field reservoir oil, by flash liberation at reservoir temperature of 160°F.

API stock tank oil of the Big Sandy reservoir is close to 0.00040 per °F; therefore, one stock tank barrel at 60°F will expand to about 1.04 bbl at 160°F as calculated from:

$$V_T = V_{60}[1 + \beta(T - 60)]$$

$$V_{160} = 1.00[1 + 0.00040 \ (160 - 60)] = 1.04 \ \text{bbl} \tag{1.27}$$

where β is the temperature coefficient of expansion of the oil.

The *formation volume factor,* symbol B_o and the abbreviation FVF, at any pressure may be defined as the volume in barrels that one stock tank barrel occupies in the *formation* (reservoir) at reservoir temperature and with the solution gas which can be held in the oil at that pressure. Because both the temperature and the solution gas increase the volume of the stock tank oil, the factor will always be greater than 1. When all the gas present is in solution in the oil (i.e., at the bubble-point pressure), a further increase in pressure decreases the volume at a rate that depends on the compressibility of the liquid. One obvious implication of the formation volume factor is that for every 1.310 bbl of reservoir liquid in the Big Sandy reservoir, only 1.000 bbl, or 76.3%, can reach the stock tank. This figure, 76.3%, or 0.763, is the reciprocal of the formation volume factor and is called the *shrinkage factor.* Just as the formation volume factor is multiplied by the stock tank volume to find the reservoir volume, the shrinkage factor is multiplied by the reservoir volume to find the stock tank volume. Although both terms are in use, petroleum engineers have almost universally adopted *formation volume factor.* As mentioned previously, the formation volume factors depend on the type of gas liberation process, a phenomenon that we ignore until Chapter 6.

In some equations it has been found convenient to use a term called the *two-phase formation volume factor,* symbol B_t, which may be defined as the volume in barrels one stock tank barrel *and its initial complement of dissolved gas* occupies at any pressure and reservoir temperature. In other words, it includes the liquid volume, B_o, plus the volume of the *difference* between the initial solution gas-oil ratio, R_{soi}, and the solution gas-oil ratio at the specified pressure, R_{so}. If B_g is the gas volume factor in barrels per standard cubic foot of the solution gas, then the two-phase formation volume factor may be expressed as

$$B_t = B_o + B_g(R_{soi} - R_{so}) \tag{1.28}$$

Above the bubble-point, pressure $R_{soi} = R_{so}$ and the single-phase and two-phase factors are equal. Below the bubble point, however, while the single-phase factor decreases as pressure decreases, the two-phase factor increases owing to the release of gas from solution and the continued expansion of the gas released from solution.

The single-phase and two-phase volume factors for the Big Sandy reservoir fluid may be visualized by referring to Fig. 1.13, which is based on data from Figs. 1.11 and 1.12. Fig. 1.13 (A) shows a cylinder fitted with a piston that initially contains 1.310 bbl of the initial reservoir fluid (liquid) at the initial pressure of 3500 psia and 160°F. As the piston is withdrawn, the volume increases and the pressure consequently must decrease. At 2500 psia, which is the bubble-point pressure, the liquid volume has expanded to 1.333 bbl. Below 2500 psia, a gas phase appears and continues to grow as the pressure declines, owing to the release of gas from solution and the expansion of gas already released; conversely, the liquid phase shrinks because of loss of solution gas, to 1.210 bbl at 1200 psia. At 1200 psia and 160°F the liberated gas has a deviation factor of 0.890, and therefore the gas volume factor with reference to standard conditions of 14.7 psia and 60°F is

$$\frac{379.4 \text{ SCF}}{1 \text{ lb-mole}}$$

$$B_g = \frac{znR'T}{p} = \frac{0.890 \times 10.73 \times 620}{379.4 \times 1200}$$

$$= 0.01300 \text{ cu ft/SCF} \qquad \left(\text{See } 1.17\right)$$

$$= 0.002316 \text{ bbl/SCF}$$

Figure 1.11 shows an initial solution gas of 567 SCF/STB and at 1200 psia 337 SCF/STB, the difference 230 SCF being the gas liberated down to 1200 psia. The volume of these 230 SCF is

$$V_g = 230 \times 0.01300 = 2.990 \text{ cu ft}$$

This free gas volume, 2.990 cu ft or 0.533 bbl, plus the liquid volume, 1.210 bbl, is the total volume, or 1.743 bbl/STB—the two-phase volume factor at 1200 psia. It may also be obtained by Eq. (1.28) as:

$$B_t = 1.210 + 0.002316 \, (567 - 337)$$

$$= 1.210 + 0.533 = 1.743 \text{ bbl/STB}$$

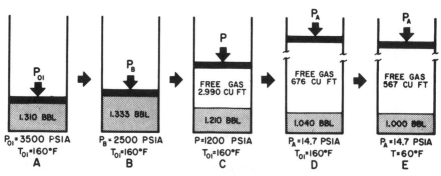

Fig. 1.13. Visual conception of the change in single-phase and in two-phase formation volume factors for the Big Sandy reservoir fluid.

Figure 1.13 (C) shows these separate and total volumes at 1200 psia. At 14.7 psia and 160°F (D), the gas volume has increased to 676 cu ft and the oil volume has decreased to 1.040 bbl. The total liberated gas volume, 676 cu ft, at 160°F and 14.7 psia, is converted to standard cubic feet at 60°F and 14.7 psia using the ideal gas law to give 567 SCF/STB as shown in (E). Correspondingly, 1.040 bbl at 160°F is converted to stock tank conditions of 60°F as shown by Eq. (1.27) to give 1.000 STB, also shown in (E).

The single-phase formation volume factor may be estimated from the solution gas, the gravity of the solution gas, the API gravity of the tank oil, and reservoir temperature using a correlation prepared by Standing.[43] Beggs[44] presents Standing's correlation for the oil formation volume factor in equation form as:

For $p \leq p_b$:

$$B_o = 0.972 + 0.000147F^{1.175} \tag{1.29}$$

where,

$$F = R_{so}\left(\frac{\gamma_g}{\gamma_o}\right)^{.5} + 1.25T$$

$$\gamma_o = \text{oil specific gravity} = \frac{141.5}{131.5 + \rho_{o,\,API}}$$

$T = \text{temperature, }°F$

The average error determined from this correlation with the same database used by Standing and Beggs in the solution gas-ratio correlation was 1.17%.

For $p > p_b$;

$$B_o = B_{ob} \exp\left[c_o(p_b - p)\right] \tag{1.30}$$

where,

B_{ob} = oil formation volume factor at the bubble-point pressure

c_o = oil compressibility, psi^{-1}

Col (2) of Table 1.2 shows the variation in the volume of a reservoir fluid relative to the volume at the bubble point of 2695 psig, as measured in the laboratory. These relative volume factors (RVF) may be converted to formation volume factors if the formation volume factor at the bubble point is known. For example, if B_{ob} = 1.391 bbl/STB, then the formation volume factor at 4100 psig is

$$B_o \text{ at } 4100 \text{ psig} = 1.391(0.9829) = 1.367 \text{ bbl/STB}$$

TABLE 1.2.

Relative volume data

(1) Pressure, psig	(2) RVF[a] V_r
5000	0.9739
4700	0.9768
4400	0.9799
4100	0.9829
3800	0.9862
3600	0.9886
3400	0.9909
3200	0.9934
3000	0.9960
2900	0.9972
2800	0.9985
2695	1.0000

[a] V_r = volume relative to the volume at the bubble-point pressure V_b, laboratory data.

6.3. Isothermal Compressibility

Sometimes it is desirable to work with values of the liquid compressibility rather than the formation or relative volume factors. The compressibility, or the bulk modulus of elasticity of a liquid, is defined by:

$$c_o = -\frac{1}{V}\frac{dV}{dp} \qquad (1.1)$$

Because dV/dp is a negative slope, the negative sign converts the compressibility c_o into a positive number. Because the values of the volume V and the slope of dV/dp are different at each pressure, the compressibility is different at each pressure, being higher at the lower pressure. Average compressibilities may be used by writing Eq. (1.1) in the difference form as:

$$c_o = -\frac{1}{V}\times\frac{(V_1 - V_2)}{(p_1 - p_2)} \qquad (1.31)$$

The reference volume V in Eq. (1.31) may be V_1, V_2, or an average of V_1 and V_2. It is commonly reported for reference to the smaller volume—that is, the volume at the higher pressure. Then the average compressibility of the fluid of Table 1.2 between 5000 psig and 4100 psig is

$$c_o = \frac{0.9829 - 0.9739}{0.9739\,(5000 - 4100)} = 10.27\times 10^{-6}\ \text{psi}^{-1}$$

Between 4100 psig and 3400 psig,

$$c_o = \frac{0.9909 - 0.9829}{0.9829 \, (4100 - 3400)} = 11.63 \times 10^{-6} \text{ psi}^{-1}$$

And between 3400 psig and 2695 psig,

$$c_o = \frac{1.0000 - 0.9909}{0.9909 \, (3400 - 2695)} = 13.03 \times 10^{-6} \text{ psi}^{-1}$$

A compressibility of 13.03×10^{-6} psi^{-1} means that the volume of 1 million barrels of *reservoir* fluid will increase by 13.03 bbls for a reduction of 1 psi in pressure. The compressibility of undersaturated oils ranges from 5 to 100×10^{-6} psi^{-1}, being higher for the higher API gravities, for the greater quantity of solution gas, and for higher temperatures.

Villena-Lanzi developed a correlation to estimate c_o for black oils.[45] The correlation is good for pressures below the bubble-point pressure and is given by:

$$1n(c_o) = -0.664 - 1.430 \, 1n(p) - 0.395 \, 1n(p_b) + 0.390 \, 1n(T)$$
$$+ 0.455 \, 1n(R_{sob}) + 0.262 \, 1n(\rho_{o, \text{ API}}) \qquad (1.32)$$

where,

$$T = {}^\circ F$$

The correlation was developed from a database containing the following ranges:

$$31.0(10)^{-6} < c_o \text{ (psia)} < 6600(10)^{-6}$$
$$500 < p \text{ (psig)} < 5300$$
$$763 < p_b \text{ (psig)} < 5300$$
$$78 < T \text{ (}^\circ F) < 330$$
$$1.5 < \text{GOR, gas-oil ratio (SCF/STB)} < 1947$$
$$6.0 < \rho_{o, \text{ API}} \text{ (}^\circ \text{API)} < 52.0$$
$$0.58 < \gamma_g < 1.20$$

Vasquez and Beggs presented a correlation for estimating the compressibility for pressures above the bubble point pressure.[44, 46] This correlation is

$$c_o = (5 \, R_{sob} + 17.2T - 1180\gamma_g + 12.61\rho_{o, \text{ API}} - 1433)/(p \times 10^5) \qquad (1.33)$$

The correlation gives good results for the following ranges of data:

$$126 < p \text{ (psig)} < 9500$$
$$1.006 < B_o \text{ (bbl/STB)} < 2.226$$
$$9.3 < \text{GOR, gas-oil ratio (SCF/STB)} < 2199$$
$$15.3 < \rho_{o, API} \text{ (°API)} < 59.5$$
$$0.511 < \gamma_g < 1.351$$

6.4. Viscosity

The viscosity of oil under reservoir conditions is commonly measured in the laboratory. Fig. 1.14 shows the viscosities of four oils at reservoir tempera-ture, above and below bubble-point pressure. Below the bubble point, the viscosity decreases with increasing pressure owing to the thinning effect of gas entering solution, but above the bubble point, the viscosity increases with increasing pressure.

When it is necessary to estimate the viscosity of reservoir oils, correla-tions have been developed for both above and below the bubble-point pres-sure. Egbogah presented a correlation which is accurate to an average error of 6.6% for 394 different oils.[47] The correlation is for what is referred to as "dead" oil, which simply means it does not contain solution gas. A second correlation is used in conjuction with the Egbogah correlation to include the

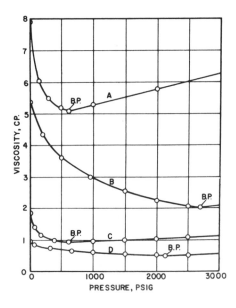

Fig. 1.14. The viscosity of four crude oil samples under reservoir conditions.

effect of solution gas. Egbogah's correlation for dead oil at pressures less than or equal to the bubble-point pressure is

$$\log_{10}[\log_{10}(\mu_{od} + 1)] = 1.8653 - 0.025086\rho_{o,\,API} - 0.5644\log(T) \tag{1.34}$$

where,

μ_{od} = dead oil viscosity, cp
T = temperature, °F

The correlation was developed from a database containing the following ranges:

$$59 < T \text{ (°F)} < 176$$
$$-58 < T_{pour} \text{ (°F)} < 59$$
$$5.0 < \rho_{o,\,API} \text{ (°API)} < 58.0$$

Beggs and Robinson[44, 48] developed the live oil viscosity correlation that is used in conjunction with the dead oil correlation given in Eq. (1.34)

$$\mu_o = A\mu_{od}^B \tag{1.35}$$

where,

$A = 10.715 \ (R_{so} + 100)^{-0.515}$
$B = 5.44 \ (R_{so} + 150)^{-0.338}$

The average absolute error found by Beggs and Robinson while working with 2073 oil samples was 1.83%. The oil samples contained the following ranges:

$$0 < p \text{ (psig)} < 5250$$
$$70 < T \text{ (°F)} < 295$$
$$20 < \text{GOR, gas-oil ratio (SCF/STB)} < 2070$$
$$16 < \rho_{o,\,API} \text{ (°API)} < 58$$

For pressures above the bubble point, the oil viscosity can be estimated by the following correlation developed by Vasquez and Beggs:

$$\mu_o = \mu_{ob} \ (p/p_b)^m \tag{1.36}$$

where,

$m = 2.6 \ p^{1.187} \exp[-11.513 - 8.98(10)^{-5} p]$
μ_{ob} = oil viscosity at the bubble point-pressure, cp

Vasquez and Beggs found an average absolute error for this correlation of 7.54% for 3143 oil samples that involved the following ranges:

$$126 < p \text{ (psig)} < 9500$$
$$0.117 < \mu_o \text{ (cp)} < 148.0$$
$$9.3 < \text{GOR, gas-oil ratio (SCF/STB)} < 2199$$
$$15.3 < \rho_{o, \text{API}} \text{ (°API)} < 59.5$$
$$0.511 < \gamma_g < 1.351$$

The following example problem illustrates the use of the correlations that have been presented for the various oil properties.

Example 1.7. Use of correlations to estimate values for liquid properties at pressures of 2000 and 4000 psia.

Given:
$$T = 180°F$$
$$p_b = 2500 \text{ psia}$$
$$\gamma_g = 0.80$$
$$\rho_{o, \text{API}} = 40°\text{API}$$
$$\gamma_o = 0.85$$

SOLUTION: Solution gas-oil ratio, R_{so}:

$$p = 4000 \text{ psia } (p > p_b)$$

For pressures greater than the bubble-point pressure, $R_{so} = R_{sob}$, therefore from Eq. (1.26):

$$R_{sob} = R_{so} = \gamma_g \left[\frac{p}{18(10)^{Y_g}} \right]$$

$$Y_g = 0.00091T - 0.0125\rho_{o, \text{API}} = 0.00091(180) - 0.0125(40) = -0.336$$

$$R_{sob} = R_{so} = 0.80 \left[\frac{2500}{18(10)^{-0.336}} \right]^{1.204} = 772 \text{ SCF/STB}$$

$p = 2000 \text{ psia } (p < p_b)$

$$R_{so} = \gamma_g \left[\frac{p}{18(10)^{Y_g}} \right] = 0.80 \left[\frac{2000}{18(10)^{-0.336}} \right]^{1.204} = 590 \text{ SCF/STB}$$

Isothermal compressibility, c_o:

$p = 4000$ psia $(p > p_b)$

From Eq. (1.33):

$$c_o = (5R_{sob} + 17.2T - 1180\gamma_g + 12.61 \rho_{o, API} - 1433)/(p(10)^5)$$
$$c_o = (5(772) + 17.2(180) - 1180(0.80) + 12.61(40) - 1433)/(4000(10)^5)$$
$$= 12.7(10)^{-6} \text{ psi}^{-1}$$

$p = 2000$ psia $(p < p_b)$

From Eq. (1.32):

$$1n\ c_o = -0.664 - 1.430\ 1n(p) - 0.395 1n(p_b) + 0.390\ 1n(T)$$
$$+ 0.455\ 1n(R_{sob}) + 0.262\ 1n(\rho_{o, API})$$
$$1n\ c_o = -0.664 - 1.430\ 1n(2000) - 0.395\ 1n(2500) + 0.390\ 1n(180)$$
$$+ 0.455\ 1n(772) + 0.262\ 1n(40)$$
$$c_o = 183\ (10)^{-6} \text{ psi}^{-1}$$

Formation volume factor, B_o:

$$p = 4000 \text{ psia } (p > p_b)$$

From Eq. (1.30):

$$B_o = B_{ob} \exp[c_o(p_b - p)]$$

B_{ob} is calculated from Eq. (1.29)

$$B_{ob} = 0.972 + 0.000147F^{1.175}$$

where,

$$F = R_{sob}\left(\frac{\gamma_g}{\gamma_o}\right)^{0.5} + 1.25T$$
$$F = 772\left(\frac{0.80}{0.825}\right)^{0.5} + 1.25(180) = 985$$
$$B_{ob} = 0.972 + 0.000147(985)^{1.175} = 1.456 \text{ bbl/STB}$$
$$B_o = 1.456 \exp[12.7(10)^{-6}(2500 - 4000)] = 1.429 \text{ bbl/STB}$$
$$p = 2000 \text{ psia } (p < p_b)$$

From Eq. (1.29):

$$B_o = 0.972 + 0.000147F^{1.175}$$

$$F = R_{so}\left(\frac{\gamma_g}{\gamma_o}\right)1.25T = 590\left(\frac{0.80}{0.825}\right)^{0.5} + 1.25(180) = 806$$

$$B_o = 0.972 + 0.000147(806)^{1.175} = 1.354 \text{ bbl/STB}$$

Viscosity, μ_o:

$$p = 4000 \text{ psia } (p > p_b)$$

From Eq. (1.36):

$$\mu_o = \mu_{ob} (p/p_b)^m$$

$$m = 2.6p^{1.187} \exp[-11.513 - 8.98(10)^{-5}p]$$

$$m = 2.6(4000)^{1.187} \exp[-11.513 - 8.98(10)^{-5}(4000)] = 0.342$$

From Eq. (1.34):

$$\log_{10}[\log_{10}(\mu_{obd} + 1)] = 1.8653 - 0.025086\rho_{o,API} - 0.5644\log(T)$$

$$\log_{10}[\log_{10}(\mu_{obd} + 1)] = 1.8653 - 0.025086(40) - 0.5644\log(180)$$

$$\mu_{obd} = 1.444 \text{ cp}$$

From Eq. (1.35):

$$\mu_{ob} = A \mu_{obd}^B$$

$$A = 10.715 (R_{sob} + 100)^{-0.515} = 10.715 (772 + 100)^{-0.515} = 0.328$$

$$B = 5.44 (R_{sob} + 150)^{-0.338} = 5.44 (772 + 150)^{-0.338} = 0.541$$

$$\mu_{ob} = 0.328(1.444)^{0.541} = 0.400 \text{ cp}$$

$$\mu_o = 0.400(4000/2500)^{0.342} = 0.470 \text{ cp}$$

$p = 2000 \text{ psia } (p < p_b)$

From Eq. (1.34), μ_{od} will be the same as μ_{obd}:

$$\mu_{od} = 1.444 \text{ cp}$$

$$\mu_o = A \mu_{od}^B$$

$$A = 10.715 (R_{so} + 100)^{-0.515} = 10.715(590 + 100)^{-0.515} = 0.370$$

$$B = 5.44 (R_{so} + 150)^{-0.338} = 5.44(590 + 150)^{-0.338} = 0.583$$

$$\mu_o = 0.370(1.444)^{0.583} = 0.458 \text{ cp}$$

7. REVIEW OF RESERVOIR WATER PROPERTIES

The properties of formation waters, like crude oils but to a much smaller degree, are affected by temperature, pressure, and the quantity of solution gas and dissolved solids. The compressibility of the formation, or connate, water contributes materially in some cases to the production of volumetric reservoirs above the bubble point and accounts for much of the water influx in water-drive reservoirs. When the accuracy of other data warrants it, the properties of the connate water should be entered into the material-balance calculations on reservoirs. The following sections contain a number of correlations adequate for use in engineering applications.

7.1. Formation Volume Factor

McCain[49, 50] developed the following correlation for the water formation volume factor, B_w (bbl/STB):

$$B_w = (1 + \Delta V_{wt})(1 + \Delta V_{wp}) \tag{1.37}$$

where,

$$\Delta V_{wt} = -1.00010 \times 10^{-2} + 1.33391 \times 10^{-4}\ T + 5.50654 \times 10^{-7}\ T^2$$
$$\Delta V_{wp} = -1.95301 \times 10^{-9}\ pT - 1.72834 \times 10^{-13}\ p^2T - 3.58922 \times 10^{-7}\ p$$
$$\quad\quad - 2.25341 \times 10^{-10}\ p^2$$
$$T = \text{temperature, } °F$$
$$p = \text{pressure, psia}$$

For the data used in the development of the correlation, the correlation was found to be accurate to within 2%. The correlation does not account for the salinity of normal reservoir brines explicitly, but McCain observed that variations in salinity caused offsetting errors in the terms ΔV_{wt} and ΔV_{wp}. The offsetting errors cause the correlation to be within engineering accuracy for the estimation of the B_w of reservoir brines.

7.2. Solution Gas-Water Ratio

McCain has also developed a correlation for the solution gas-water ratio, R_{sw} (SCF/STB).[49, 50] The correlation is:

$$\frac{R_{sw}}{R_{swp}} = 10^{(-0.0840655\ S\ T^{-0.285854})} \tag{1.38}$$

where

   S = salinity, % by weight solids

   T = temperature, °F

   R_{swp} = solution gas to pure water ratio, SCF/STB

R_{swp} is given by another correlation developed by McCain as:

$$R_{swp} = A + Bp + Cp^2 \tag{1.39}$$

where,

   $A = 8.15839 - 6.12265 \times 10^{-2}\ T + 1.91663 \times 10^{-4}\ T^2 - 2.1654 \times 10^{-7}\ T^3$

   $B = 1.01021 \times 10^{-2} - 7.44241 \times 10^{-5}\ T + 3.05553 \times 10^{-7}\ T^2$

   $\quad - 2.94883 \times 10^{-10}\ T^3$

   $C = -10^{-7}\ (9.02505 - 0.130237\ T + 8.53425 \times 10^{-4}\ T^2$

   $\quad - 2.34122 \times 10^{-6}\ T^3 + 2.37049 \times 10^{-9}\ T^4)$

   T = temperature, °F

The correlation of Eq. (1.39) was developed for the following range of data and found to be within 5% of the published data:

$$1000 < p\ (\text{psia}) < 10,000$$
$$100 < T\ (°F) < 340$$

Eq. (1.38) was developed for the following range of data and found to be accurate to within 3% of published data:

$$0 < S\ (\%) < 30$$
$$70 < T\ (°F) < 250$$

7.3. Isothermal Compressibility

Osif developed a correlation for the water isothermal compressibility, c_w, for pressures greater than the bubble point pressure.[51] The equation is

$$c_w = -\frac{1}{B_w}\left(\frac{\partial B_w}{\partial p}\right)_T = \frac{1}{[7.033p + 541.5C_{NaCl} - 537.0T + 403,300]} \tag{1.40}$$

where,

   C_{NaCl} = salinity, g NaCl/liter

   T = temperature, °F

The correlation was developed for the following range of data:

$$1000 < p \ (\text{psig}) < 20{,}000$$

$$0 < C_{\text{NaCl}} \ (g \ \text{NaCl/liter}) < 200$$

$$200 < T \ (\text{°F}) < 270$$

The water isothermal compressibility has been found to be strongly affected by the presence of free gas. Therefore, McCain proposed using the following expression for estimating c_w for pressures below or equal to the bubble-point pressure:

$$c_w = -\frac{1}{B_w}\left(\frac{\partial B_w}{\partial p}\right)_T + \frac{B_g}{B_w}\left(\frac{\partial R_{swp}}{\partial p}\right)_T \qquad (1.41)$$

The first term on the right-hand side of Eq. (1.41) is simply the expression for c_w in Eq. (1.40). The second term on the right-hand side is found by differentiating Eq. (1.39) with respect to pressure, or:

$$\left(\frac{\partial R_{swp}}{\partial p}\right)_T = B + 2Cp$$

where

B and C are defined in Eq. (1.39).

In proposing Eq. (1.41), McCain suggested that B_g should be estimated using a gas with a gas gravity of 0.63, which represents a gas composed mostly of methane and a small amount of ethane. McCain could not verify this expression by comparing calculated values of c_w with published data, so there is no guarantee of accuracy. This suggests that Eq. (1.41) should be used only for gross estimations of c_w.

7.4. Viscosity

The viscosity of water increases with decreasing temperature and in general with increasing pressure and salinity. Pressure below about 70°F causes a reduction in viscosity, and some salts (e.g., KCl) reduce the viscosity at some concentrations and within some temperature ranges. The effect of dissolved gases is believed to cause a minor reduction in viscosity. McCain developed the following correlation for water viscosity at atmospheric pressure and reservoir temperature:

$$\mu_{w1} = AT^B \qquad (1.42)$$

where

$$A = 109.574 - 8.40564\ S + 0.313314\ S^2 + 8.72213 \times 10^{-3}\ S^3$$

$$B = -1.12166 + 2.63951 \times 10^{-2}\ S - 6.79461 \times 10^{-4}\ S^2 - 5.47119 \times 10^{-5}\ S^3$$
$$+ 1.55586 \times 10^{-6}\ S^4$$

T = temperature, °F

S = salinity, % by weight solids

Eq. (1.42) was found to be accurate to within 5% over the following range of data:

$$100 < T\ (°F) < 400$$
$$0 < S\ (\%) < 26$$

The water viscosity can be adjusted to reservoir pressure by the following correlation, again developed by McCain:

$$\frac{\mu_w}{\mu_{w1}} = 0.9994 + 4.0295(10)^{-5}p + 3.1062(10)^{-9}p^2 \qquad (1.43)$$

This correlation was found to be accurate to within 4% for pressures below 10,000 psia and within 7% for pressures between 10,000 and 15,000 psia. The temperature range for which the correlation was developed was between 86 and 167 °F.

8. SUMMARY

The correlations presented in this chapter are valid for estimating properties providing the parameters fall within the specified ranges for the particular property in question. The correlations were presented in the form of equations to facilitate their implementation into computer programs.

PROBLEMS

   1.1   Calculate the volume 1 lb-mole of ideal gas will occupy at:
   (a) 14.7 psia and 60°F
   (b) 14.7 psia and 32°F
   (c) 14.7 plus 10 oz and 80°F
   (d) 15.025 psia and 60°F

1.2 A 500 cu ft tank contains 10 lb of methane and 20 lb of ethane at 90°F.

(a) How may moles are in the tank?

(b) What is the pressure of the tank is psia?

(c) What is the molecular weight of the mixture?

(d) What is the specific gravity of the mixture?

1.3 What are the molecular weight and specific gravity of a gas that is one third each of methane, ethane, and propane by volume?

1.4 A 10 lb block of dry ice is placed in a 50 cu ft tank that contains air at atmospheric pressure 14.7 psia and 75°F. What will be the final pressure of the sealed tank when all the dry ice has evaporated and cooled the gas to 45°F?

1.5 A welding apparatus for a drilling rig uses acetylene (C_2H_2), which is purchased in steel cylinders containing 20 lb of gas, costs $4.50 exclusive of the cylinder. If a welder is using 200 cu ft per day measured at 16 oz gauge and 85°F, what is the daily cost of acetylene? What is the cost per MCF at 14.7 psia and 60°F?

1.6 (a) A 55,000 bbl (nominal) pipeline tank has a diameter of 110 ft and a height of 35 ft. It contains 25 ft of oil at the time suction is taken on the oil with pumps that handle 20,000 bbl per day. The breather and safety valves have become clogged so that a vacuum is drawn on the tank. If the roof is rated to withstand 3/4 oz per sq in. pressure, how long will it be before the roof collapses? Barometric pressure is 29.1 in. Neglect the fact that the roof is peaked and that there may be some leaks.

(b) Calculate the total force on the roof at the time of collapse.

(c) If the tank had contained more oil, would the collapse time have been greater or less?

1.7 (a) What percentage of methane by weight does a gas of 0.65 specific gravity contain if it is composed only of methane and ethane? What percentage by volume?

(b) Explain why the percentage by volume is greater than the percentage by weight.

1.8 A 50 cu ft tank contains gas at 50 psia and 50°F. It is connected to another tank that contains gas at 25 psia and 50°F. When the valve between the two is opened, the pressure equalizes at 35 psia at 50°F. What is the volume of the second tank?

1.9 Gas was contracted at $1.10 per MCF at contract conditions of 14.4 psia and 80°F. What is the equivalent price at a legal temperature of 60°F and pressure of 15.025 psia?

1.10 A cylinder is fitted with a leakproof piston and calibrated so that the volume within the cylinder can be read from a scale for any position of the piston. The cylinder is immersed in a constant temperature bath maintained at 160°F, which is the reservoir temperature of the Sabine Gas Field. Forty-five thousand cc of the gas, measured at 14.7 psia and 60°F, is charged into the cylinder. The volume is decreased in the steps indicated below, and the corresponding pressures are read with a dead weight tester after temperature equilibrium is reached.

V, cc	2529	964	453	265	180	156.5	142.2
p, psia	300	750	1500	2500	4000	5000	6000

(a) Calculate and place in tabular form the gas deviation factors and the ideal volumes that the initial 45,000 cc occupies at 160°F and at each pressure.

(b) Calculate the gas volume factors at each pressure, in units of ft³/SCF.

(c) Plot the deviation factor and the gas volume factors calculated in part (b) versus pressure on the same graph.

1.11 (a) If the Sabine Field gas gravity is 0.65, calculate the deviation factors from zero to 6000 psia at 160°F, in 1000 lb increments, using the gas gravity correlation of Fig. 1.4.

(b) Using the critical pressures and temperatures in Table 1.1, calculate and plot the deviation factors for the Sabine gas at several pressures and 160°F. The gas analysis is as follows:

Component	C_1	C_2	C_3	iC_4	nC_4	iC_5
Mole Fraction	0.875	0.083	0.021	0.006	0.008	0.003

Component	nC_5	C_6	C_{7+}
Mole Fraction	0.002	0.001	0.001

Use the molecular weight and critical temperature and pressure of octane for the heptanes-plus. Plot the data of Prob. 1.10(a) and Prob. 1.11(a) on the same graph for comparison.

(c) Below what pressure at 160°F may the ideal gas law be used for the gas of the Sabine Field if errors are to be kept within 2%?

(d) Will a reservoir contain more SCF of a real or of an ideal gas at similar conditions? Explain.

1.12 A high-pressure cell has a volume of 0.330 cu ft and contains gas at 2500 psia and 130°F, at which conditions its deviation factor is 0.75. When 43.6 SCF measured at 14.7 psia and 60°F were bled from the cell through a wet test meter, the pressure dropped to 1000 psia, the temperature remaining at 130°F. What is the gas deviation factor at 1000 psia and 130°F?

1.13 A dry gas reservoir is initially at an average pressure of 6000 psia and temperature of 160°F. The gas has a specific gravity of 0.65. What will the average reservoir pressure be when one-half of the original gas (in SCF) has been produced? Assume the volume occupied by the gas in the reservoir remains constant. If the reservoir originally contained 1 MM ft³ of reservoir gas, how much gas has been produced at a final reservoir pressure of 500 psia?

1.14 A reservoir gas has the following gas deviation factors at 150°F?

p, psia	0	500	1000	2000	3000	4000	5000
z	1.00	0.92	0.86	0.80	0.82	0.89	1.00

Plot z versus p and graphically determine the slopes at 1000 psia, 2200 psia, and 4000 psia. Then, using Eq. (1.19), find the gas compressibility at these pressures.

1.15 Using Figs. 1.4 and 1.5, find the compressibility of a 70% specific gravity gas at 5000 psia and 203°F.

1.16 Using Eq. (1.22) and the generalized chart for gas deviation factors, Fig. 1.5, find the pseudoreduced compressibility of a gas at a pseudoreduced temperature of 1.30 and a pseudoreduced pressure of 4.00. Check this value on Fig. 1.7.

1.17 Estimate the viscosity of a gas condensate fluid at 7000 psia and 220°F. It has a specific gravity of 0.90 and contains 2% nitrogen, 4% carbon dioxide, and 6% hydrogen sulfide.

1.18 Experiments were made on a bottom-hole sample of the reservoir liquid taken from the LaSalle Oil Field to determine the solution gas and the formation volume factor as functions of pressure. The initial bottom-hole pressure of the reservoir was 3600 psia, and bottom-hole temperature was 160°F; so all measurements in the laboratory were made at 160°F. The following data, converted to practical units, were obtained from the measurements:

Pressure psia	Solution gas SCF/STB at 14.7 psia and 60°F	Formation Volume Factor, bbl/STB
3600	567	1.310
3200	567	1.317
2800	567	1.325
2500	567	1.333
2400	554	1.310
1800	436	1.263
1200	337	1.210
600	223	1.140
200	143	1.070

(a) What factors affect the solubility of gas in crude oil?

(b) Plot the gas in solution versus pressure.

(c) Was the reservoir initially saturated or undersaturated? Explain.

(d) Does the reservoir have an initial gas cap?

(e) In the region of 200 to 2500 psia, determine the solubililty of the gas from your graph in SCF/STB/psi.

(f) Suppose 1000 SCF of gas had accumulated with each stock tank barrel of oil in this reservoir instead of 567 SCF. Estimate how much gas would have been in solution at 3600 psia. Would the reservoir oil then be called saturated or undersaturated?

1.19 From the bottom-hole sample given in Prob. 1.18:

(a) Plot the formation volume factor versus pressure.

(b) Explain the break in the curve.

(c) Why is the slope above the bubble-point pressure negative and smaller than the positive slope below the bubble-point pressure?

(d) If the reservoir contains 250 MM reservoir barrels of oil initially, what is the initial number of STB in place?

(e) What is the initial volume of dissolved gas in the reservoir?

(f) What will be the formation volume factor of the oil when the bottom-hole pressure is essentially atmospheric if the coefficient of expansion of the stock tank oil is 0.0006 per °F?

1.20 If the gravity of the stock tank oil of the Big Sandy reservoir is 30°API and the gravity of the solution gas is 0.80, estimate the solution gas-oil ratio and the single-phase formation volume factor at 2500 psia and 165°F. The bubble-point pressure is 2800 psia.

1.21 A 1000 cu ft tank contains 85 STB of crude and 20,000 SCF of gas, all at 120°F. When equilibrium is established, (i.e., when as much gas has dissolved in the oil as will), the pressure in the tank is 500 psia. If the solubility of the gas in the crude is 0.25 SCF/STB/psi and the deviation factor for the gas at 500 psia and 120°F is 0.90, find the liquid formation volume factor at 500 psia and 120°F.

1.22 A crude oil has a compressibility of 20×10^{-6} psi^{-1} and a bubble point of 3200 psia. Calculate the relative volume factor at 4400 psia (i.e., the volume relative to its volume at the bubble point), assuming constant compressibility.

1.23 (a) Estimate the viscosity of an oil at 3000 psia and 130°F. It has a stock tank gravity of 35°API at 60°F and contains an estimated 750 SCF/STB of solution gas at the initial bubble point of 3000 psia.

(b) Estimate the viscosity at the initial reservoir pressure of 4500 psia.

(c) Estimate the viscosity at 1000 psia if there is an estimated 300 SCF/STB of solution gas at that pressure.

1.24 Given the following laboratory data:

Cell Pressure (psia)	Oil Volume in cell (cc)	Gas Volume in cell (cc)	Cell Temperature (°F)
2000	650	0	195
1500 = P_{bp}	669	0	195
1000	650	150	195
500	615	700	195
14.7	500	44,500	60

Evaluate R_{so}, B_o, and B_t at the stated pressures. The gas deviation factors at 1000 psia and 500 psia have been evaluated as 0.91 and 0.95, respectively.

1.25 (a) Find the compressibility for a connate water that contains 20,000 parts per million of total solids at a reservoir pressure of 4000 psia and temperature of 150°F.

(b) Find the formation volume factor of the formation water of part (a).

1.26 (a) What is the approximate viscosity of pure water at room temperature and atmospheric pressure?

(b) What is the approximate viscosity of pure water at 200°F?

1.27 A container has a volume of 500 cc and is full of pure water at 180°F and 6000 psia.

(a) How much water would be expelled if the pressure were reduced to 1000 psia?

(b) What would be the volume of water expelled if the salinity were 20,000 ppm and there were no gas in solution?

(c) Rework part (b) assuming that the water is initially saturated with gas and that all the gas is evolved during the pressure change.

(d) Estimate the viscosity of the water.

REFERENCES

[1] K. C. Sclater and B. R. Stephenson, "Measurements of Original Pressure, Temperature and Gas-Oil Ratio in Oil Sands," *Trans.* AIME (1928–29), **82,** 119.

[2] C. V. Millikan and Carrol V. Sidwell, "Bottom-hole Pressures in Oil Wells," *Trans.* AIME (1931), **92,** 194.

[3] G. H. Fancher, J. A. Lewis, and K. B. Barnes, "Some Physical Characteristics of Oil Sands," *The Pennsylvania State College,* Bull. 12 (1933), p. 65.

[4] R. D. Wyckoff, H. G. Botset, M. Muskat, and D. W. Reed, "Measurement of Permeability of Porous Media," Bull. AAPG (1934), **18,** No. 2, p. 161.

[5] R. D. Wyckoff and H. G. Botset, "The Flow of Gas-Liquid Mixtures Through Unconsolidated Sands," *Physics* (1936), **7,** 325.

[6] M. C. Leverett and W. B. Lewis, "Steady Flow of Oil-Gas-Water Mixtures Through Unconsolidated Sands," *Trans.,* AIME (1941), **142,** 107.

[7] Ralph J. Schilthuis, "Technique of Securing and Examining Sub-Surface Samples of Oil and Gas," *Drilling and Production Practice,* API (1935), pp. 120–126.

[8] Howard C. Pyle and P. H. Jones, "Quantitative Determination of the Connate Water Content of Oil Sands," *Drilling and Production Practice,* API (1936), pp. 171–180.

[9] Ralph J. Schilthuis, "Connate Water in Oil and Gas Sands," *Trans.* AIME (1938), **127,** 199–214.

[10] C. V. Millikan, "Temperature Surveys in Oil Wells," *Trans.* AIME (1941), **142,** 15.

[11] Ralph J. Schilthuis, "Active Oil and Reservoir Energy," *Trans.* AIME (1936), **118,** 33.

[12] Stewart Coleman, H. D. Wilde, Jr., and Thomas W. Moore, "Quantitative Effects of Gas-Oil Ratios on Decline of Average Rock Pressure," *Trans.* AIME (1930), **86,** 174.

[13] A. S. Odeh and D. Havlena, "The Material Balance As an Equation of a Straight Line," *Journal of Petroleum Technology* (July 1963), 896–900.

[14] W. Hurst, "Water Influx into a Reservoir and Its Application to the Equation of Volumetric Balance," *Trans.* AIME (1943), **151,** 57.

[15] A. F. van Everdingen and W. Hurst, "Application of the LaPlace Transformation to Flow Problems in Reservoirs," *Trans.* AIME (1949), **186,** 305.

[16] M. J. Fetkovich, "A Simplified Approach to Water Influx Calculations—Finite Aquifer Systems," *Jour. of Petroleum Technology* (July 1971), 814–828.

[17] J. Tarner, "How Different Size Gas Caps and Pressure Maintenance Programs Affect Amount of Recoverable Oil," *Oil Weekly,* June 12, 1944, **144,** No. 2, 32–44.

[18] S. E. Buckley, and M. C. Leverett, "Mechanism of Fluid Displacement in Sands," *Trans.* AIME (1942), **146**, 107–117.

[19] M. Muskat, "The Petroleum Histories of Oil Producing Gas-Drive Reservoirs," *Jour. of Applied Physics* (1945), **16**, 147.

[20] A. Odeh, "Reservoir Simulation—What Is It?" *Jour. of Petroleum Technology* (Nov. 1969), 1383–1388.

[21] K. H. Coats, "Use and Misuse of Reservoir Simulation Models," *Jour. of Petroleum Technology* (Nov. 1969), 1391–1398.

[22] K. H. Coats, "Reservoir Simulation: State of the Art," *Jour. of Petroleum Technology* (Aug. 1982), 1633–1642.

[23] T. V. Moore, "Reservoir Engineering Begins Second 25 Years," *Oil and Gas Jour.* (1955), **54**, No. 29, 148.

[24] Norman J. Clark and Arthur J. Wessely, "Coordination of Geology and Reservoir Engineering—A Growing Need for Management Decisions." Presented before API, *Division of Production,* March, 1957.

[25] *Principles of Petroleum Conservation,* Engineering Committee, Interstate Oil Compact Commission (1955), p. 2.

[26] "Proved Reserves Definitions," *Jour. of Petroleum Technology* (Nov. 1981), 2113–2114.

[27] B. J. Dotson, R. L. Slobod, P. N. McCreery, and James W. Spurlock, "Porosity-Measurement Comparisons by Five Laboratories," *Trans.* AIME (1951), **192**, 344.

[28] W. van der Knaap, "Non-linear Elastic Behavior of Porous Media," Presented before Society of Petroleum Engineers of AIME, October 1958, Houston, TX.

[29] G. H. Newman, "Pore-Volume Compressibility of Consolidated, Friable, and Unconsolidated Reservoir Rocks Under Hydrostatic Loading," *Jour. of Petroleum Technology* (Feb. 1973), 129–134.

[30] J. Geertsma, "The Effect of Fluid Pressure Decline on Volumetric Changes of Porous Rocks," *Jour. of Petroleum Technology* (1957), **11**, No. 12, 332.

[31] Henry J. Gruy and Jack A. Crichton, "A Critical Review of Methods Used in the Estimation of Natural Gas Reserves," *Trans.* AIME (1949), **179**, 249–263.

[32] R. P. Sutton, "Compressibility Factors for High-Molecular-Weight Reservoir Gases." Paper SPE 14265 presented at the 1985 SPE Annual Technical Meeting and Exhibition, Las Vegas, NV, Sept. 22–25.

[33] Marshall B. Standing and Donald L. Katz, "Density of Natural Gases," *Trans.* AIME (1942), **146**, 144.

[34] P. M. Dranchuk and J. H. Abou-Kassem, "Calculation of Z Factors for Natural Gases Using Equations of State," *Jour. of Petroleum Technology* (July–Sept. 1975), 34–36.

[35] R. W. Hornbeck, *Numerical Methods.* New York: Quantum Publishers, 1975.

[36] C. Kenneth Eilerts and Others, *Phase Relations of Gas-Condensate Fluids,* U.S. Bureau of Mines Monograph 10, Vol. I (New York: American Gas Association, 1957), 427–434.

[37] E. Wichert and K. Aziz, "Calculate Z's for Sour Gases," *Hyd. Proc.* (May 1972), 119–122.

[38] A. S. Trube, "Compressibility of Natural Gases," *Trans.* AIME (1957), **210,** 61.

[39] L. Mattar, G. S. Brar, and K. Aziz, "Compressibility of Natural Gases," *JCPT* (Oct.–Dec. 1975), 77–80.

[40] T. A. Blasingame, J. L. Johnston, and R. D. Poe, Jr.: "Properties of Reservoir Fluids," Texas A&M University (1989).

[41] N. L. Carr, R. Kobayashi and D. B. Burrows: "Viscosity of Hydrocarbon Gases Under Pressure," *Trans.,* AIME (1954), **201,** 264–272.

[42] A. L. Lee, M. H. Gonzalez and B. E. Eakin, "The Viscosity of Natural Gases," *Jour. of Petroleum Technology* (Aug. 1966), 997–1000; *Trans.,* AIME (1966), 237.

[43] M. B. Standing, "A Pressure-Volume-Temperature Correlation for Mixtures of California Oils and Gases," *Drill. and Prod. Prac.,* API (1947), 275–287.

[44] H. D. Beggs, "Oil System Correlations," *Petroleum Engineering Handbook,* H. C. Bradley (ed.), SPE, Richardson, TX (1987) 1, Chap. 22.

[45] A. J. Villena-Lanzi, "A Correlation for the Coefficient of Isothermal Compressibility of Black Oil at Pressures Below the Bubble Point," M.S. Thesis, Texas A&M University, College Station, TX (1985).

[46] M. Vasquez and H. D. Beggs, "Correlations for Fluid Physical Property Prediction," *Jour. of Petroleum Engineering* (June 1980), 968–970.

[47] E. O. Egbogah, "An Improved Temperature-Viscosity Correlation for Crude Oil Systems," Paper 83-34-32 presented at the 1983 Annual Technical Meeting of the Petroleum Society of CIM, Banff, Alberta, May 10–13, 1983.

[48] H. D. Beggs and J. R. Robinson, "Estimating the Viscosity of Crude Oil Systems," *Jour. of Petroleum Technology* (Sept. 1975) 1140–41.

[49] W. D. McCain, Jr. *The Properties of Petroleum Fluids,* 2nd ed. Tulsa, OK: PennWell Publishing Co., 1988.

[50] W. D. McCain, Jr. "Reservoir Water Property Correlations—State of The Art," Paper SPE 18573 submitted for publication (1988).

[51] T. L. Osif, "The Effects of Salt, Gas, Temperature, and Pressure on the Compressibility of Water," Paper SPE 13174 presented at the 1984 SPE Annual Technical Conference and Exhibition, Houston, TX, Sept. 16–19.

Chapter **2**

The General Material Balance Equation

1. INTRODUCTION

A general material balance equation that can be applied to all reservoir types is developed in this chapter. From this general equation, each of the individual equations for the reservoir types, defined in Chapter 1 and discussed in subsequent chapters, can easily be derived.

The general material balance equation was first developed by Schilthuis in 1936.[*,1] Since that time, the use of the computer and sophisticated multidimensional mathematical models has replaced the zero dimensional Schilthuis equation in many applications.[2] However, the Schilthuis equation, if fully understood, can provide great insight for the practicing reservoir engineer. Following the derivation of the general material balance equation, a method of using the equation, discussed in the literature by Havlena and Odeh, is presented.[3,4]

2. DERIVATION OF MATERIAL BALANCE EQUATION

When an oil and gas reservoir is tapped with wells, oil and gas, and frequently some water, are produced, thereby reducing the reservoir pressure and causing the remaining oil and gas to expand to fill the space vacated by the fluids

* References throughout the text are given at the end of each chapter.

56

removed. When the oil- and gas-bearing strata are hydraulically connected with water-bearing strata, or aquifers, water encroaches into the reservoir as the pressure drops owing to production, as illustrated in Fig. 2.1. This water encroachment decreases the extent to which the remaining oil and gas expand and accordingly retards the decline in reservoir pressure. Inasmuch as the temperature in oil and gas reservoirs remains substantially constant during the course of production, the degree to which the remaining oil and gas expand depends only on the pressure. By taking bottom-hole samples of the reservoir fluids under pressure and measuring their relative volumes in the laboratory at reservoir temperature and under various pressures, it is possible to predict how these fluids behave in the reservoir as reservoir pressure declines.

In a subsequent chapter it is shown that, although the connate water and formation compressibilities are quite small, they are, relative to the compressibility of reservoir fluids above their bubble points, significant, and they account for an appreciable fraction of the production above the bubble point. Table 2.1 gives a range of values for formation and fluid compressibilities from which it may be concluded that water and formation compressibilities are less significant in gas and gas cap reservoirs and in undersaturated reservoirs below the bubble point where there is appreciable gas saturation. Because of this and the complications they would introduce in already fairly complex equations, water and formation compressibilities are generally neglected, except in un-

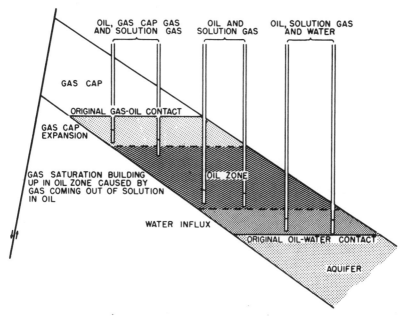

Fig. 2.1. Cross section of a combination drive reservoir. (*After* Woody and Moscrip,[5] *Trans.* AIME.)

TABLE 2.1
Range of compressibilities

Formation rock	$3 - 10 \times 10^{-6}$ psi^{-1}
Water	$2 - 4 \times 10^{-6}$ psi^{-1}
Undersaturated oil	$5 - 100 \times 10^{-6}$ psi^{-1}
Gas at 1000 psi	$900 - 1300 \times 10^{-6}$ psi^{-1}
Gas at 5000 psi	$50 - 200 \times 10^{-6}$ psi^{-1}

dersaturated reservoirs producing above the bubble point. A term accounting for the change in water and formation volumes owing to their compressibilities is included in the material balance derivation; the engineer can choose to eliminate them for particular applications. The gas in solution in the formation water is neglected, and in many instances the volume of the produced water is not known with sufficient accuracy to justify the use of a formation volume factor with the produced water.

The general material balance equation is simply a volumetric balance, which states that since the volume of a reservoir (as defined by its initial limits) is a constant, the algebraic sum of the volume changes of the oil, free gas, water, and rock volumes in the reservoir must be zero. For example, if both the oil and gas reservoir volumes decrease, the sum of these two decreases must be balanced by changes of equal magnitude in the water and rock volumes. If the assumption is made that complete equilibrium is attained at all times in the reservoir between the oil and its solution gas, it is possible to write a generalized material balance expression relating the quantities of oil, gas, and water produced, the average reservoir pressure, the quantity of water that may have encroached from the aquifer, and finally the initial oil and gas content of the reservoir. In making these calculations the following production, reservoir, and laboratory data are involved:

1. The initial reservoir pressure and the average reservoir pressure at successive intervals after the start of production.

2. The stock tank barrels of oil produced, measured at 1 atm and 60°F, at any time or during any production interval.

3. The total standard cubic feet of gas produced. When gas is injected into the reservoir, this will be the difference between the total gas produced and that returned to the reservoir.

4. The ratio of the *initial* gas cap volume to the initial oil volume, symbol m.

$$m = \frac{\text{Initial reservoir free gas volume}}{\text{Initial reservoir oil volume}}$$

If this value can be determined with reasonable precision, there is only one unknown (N) in the material balance on volumetric gas cap reservoirs, and two (N and W_e) in water-drive reservoirs. The value of m is determined from log and core data and from well completion data, which frequently helps to locate the gas-oil and water-oil contacts. The ratio m is known in many instances much more accurately than the absolute values of the gas cap and oil zone volumes. For example, when the rock in the gas cap and that in the oil zone appear to be essentially the same, it may be taken as the ratio of the net or even the gross volumes, without knowing the average connate water or average porosity, or when gross volumes are used, the factors for reducing gross to net productive volumes.

5. The gas and oil volume factors and the solution gas-oil ratios. These are obtained as functions of pressure by laboratory measurements on bottom-hole samples by the differential and flash liberation methods.

6. The quantity of water that has been produced.

7. The quantity of water that has been encroached into the reservoir from the aquifer.

For simplicity, the derivation is divided into the changes in the oil, gas, water, and rock volumes that occur between the start of production and any time t. The change in the rock volume is expressed as a change in the void space volume, which is simply the negative of the change in the rock volume. In the development of the general material balance equation, the following terms are used:

N	Initial reservoir oil, STB
B_{oi}	Initial oil formation volume factor, bbl/STB
N_p	Cumulative produced oil, STB
B_o	oil formation volume factor, bbl/STB
G	Initial reservoir gas, SCF
B_{gi}	Initial gas formation volume factor, bbl/SCF
G_f	Amount of free gas in the reservoir, SCF
R_{soi}	Initial solution gas-oil ratio, SCF/STB
R_p	Cumulative produced gas-oil ratio, SCF/STB
R_{so}	Solution gas-oil ratio, SCF/STB
B_g	Gas formation volume factor, bbl/SCF
W	Initial reservoir water, bbl
W_p	Cumulative produced water, STB
B_w	Water formation volume factor, bbl/STB
W_e	Water influx into reservoir, bbl
c_w	Water isothermal compressibility, psi^{-1}
$\Delta \overline{p}$	Change in average reservoir pressure, psia
S_{wi}	Initial water saturation
V_f	Initial void space, bbl
c_f	Formation isothermal compressibility, psi^{-1}

Change in the Oil Volume:

$$\text{Initial reservoir oil volume} = NB_{oi} \quad \text{bbl}$$

$$\text{Oil volume at time } t \text{ and pressure } p = (N - N_p)B_o \quad \text{bbl}$$

$$\text{Change in oil volume} = NB_{oi} - (N - N_p)B_o \quad \text{bbl} \tag{2.1}$$

Change in Free Gas Volume:

$$\left[\begin{array}{c}\text{Ratio of initial free gas} \\ \text{to initial oil volume}\end{array}\right] = m = \frac{GB_{gi}}{NB_{oi}}$$

Initial free gas volume $= GB_{gi} = NmB_{oi}$

$$\left[\begin{array}{c}\text{SCF free} \\ \text{gas at } t\end{array}\right] = \left[\begin{array}{c}\text{SCF initial gas,} \\ \text{free and dissolved}\end{array}\right] - \left[\begin{array}{c}\text{SCF gas} \\ \text{produced}\end{array}\right] - \left[\begin{array}{c}\text{SCF remaining} \\ \text{in solution}\end{array}\right]$$

$$G_f = \left[\frac{NmB_{oi}}{B_{gi}} + NR_{soi}\right] - [N_pR_p] - [(N - N_p)R_{so}] \quad \text{SCF}$$

$$\left[\begin{array}{c}\text{Reservoir free gas} \\ \text{volume at time } t\end{array}\right] = \left[\frac{NmB_{oi}}{B_{gi}} + NR_{soi} - N_pR_p - (N - N_p)R_{so}\right]B_g \quad \text{bbl}$$

$$\left[\begin{array}{c}\text{Change in free} \\ \text{gas volume}\end{array}\right] = NmB_{oi} - \left[\frac{NmB_{oi}}{B_{gi}} + NR_{soi} - N_pR_p - (N - N_p)R_{so}\right]B_g$$
$$\text{bbl} \tag{2.2}$$

Change in Water Volume:

Initial reservoir water volume $= W$
Cumulative water produced at $t = W_p$
Reservoir volume of cumulative produced water $= B_wW_p$
Volume of water encroached at $t = W_e$

$$\left[\begin{array}{c}\text{Change in} \\ \text{water volume}\end{array}\right] = W - (W + W_e - B_wW_p + Wc_w\Delta\bar{p}) = -W_e + B_wW_p - Wc_w\Delta\bar{p}) \tag{2.3}$$

Change in the Void Space Volume:

$$\text{Initial void space volume} = V_t$$

$$\left[\begin{array}{c}\text{Change in void} \\ \text{space volume}\end{array}\right] = V_f - [V_f - V_fc_f\Delta\bar{p}] = V_fc_f\Delta\bar{p}$$

Or, because the change in void space volume is the negative of the change in rock volume:

$$\left[\begin{array}{c} \text{Change in} \\ \text{rock volume} \end{array}\right] = -V_f c_f \Delta\bar{p} \qquad (2.4)$$

Combining the changes in water and rock volumes into a single term, yields the following:

$$= -W_e + B_w W_p - W c_w \Delta\bar{p} - V_f c_f \Delta\bar{p}$$

Recognizing that $W = V_f S_{wi}$ and that $V_f = \dfrac{NB_{oi} + NmB_{oi}}{1 - S_{wi}}$ and substituting, the following is obtained:

$$= -W_e + B_w W_p - \left[\frac{NB_{oi} + NmB_{oi}}{1 - S_{wi}}\right](c_w S_{wi} + c_f)\, \Delta\bar{p}$$

or

$$= -W_e + B_w W_p - (1 + m)\, NB_{oi}\left[\frac{c_w S_{wi} + c_f}{1 - S_{wi}}\right]\Delta\bar{p} \qquad (2.5)$$

Equating the changes in the oil and free gas volumes to the negative of the changes in the water and rock volumes and expanding all terms

$$N B_{oi} - N B_o + N_p B_o + NmB_{oi} - \left[\frac{NmB_{oi}B_g}{B_{gi}}\right] - N R_{soi}B_g + N_p R_p B_g$$

$$+ NB_g R_{so} - N_p B_g R_{so} = W_e - B_w W_p + (1 + m)N\, B_{oi}\left[\frac{c_w S_{wi} + c_f}{1 - S_{wi}}\right]\Delta\bar{p}$$

Now adding and subtracting the term $N_p B_g R_{soi}$

$$N B_{oi} - N B_o + N_p B_o + NmB_{oi} - \left[\frac{NmB_{oi}B_g}{B_{gi}}\right] - N R_{soi}B_g + N_p R_p B_g + NB_g R_{so}$$

$$- N_p B_g R_{so} + N_p B_g R_{soi} - N_p B_g R_{soi} = W_e - B_w W_p + (1 + m)\, N\, B_{oi}\left[\frac{c_w S_{wi} + c_f}{1 - S_{wi}}\right]\Delta\bar{p}$$

Then grouping terms:

$$NB_{oi} + NmB_{oi} - N[B_o + (R_{soi} - R_{so})B_g] + N_p[B_o + (R_{soi} - R_{so})B_g]$$

$$+ (R_p - R_{soi})B_g N_p - \left[\frac{NmB_{oi}B_g}{B_{gi}}\right] = W_e - B_w W_p + (1 + m)\, N\, B_{oi}\left[\frac{c_w S_{wi} + c_f}{1 - S_{wi}}\right]\Delta\bar{p}$$

Now writing $B_{oi} = B_{ti}$ and $[B_o + (R_{soi} - R_{so})B_g] = B_t$, where B_t is the two phase formation volume factor, as defined by Eq. 1.28:

$$N(B_{ti} - B_t) + N_p[B_t + (R_p - R_{oi})B_g] + NmB_{ti}\left(1 - \frac{B_g}{B_{gi}}\right)$$

$$= W_e - B_w W_p + (1 + m)\, N\, B_{ti}\left[\frac{c_w S_{wi} + c_f}{1 - S_{wi}}\right]\Delta\bar{p} \qquad (2.6)$$

This is the general volumetric material balance equation. It can be rearranged into the following form that is useful for discussion purposes.

$$N(B_t - B_{ti}) + \frac{NmB_{ti}}{B_{gi}}(B_g - B_{gi}) + (1 + m)\, N\, B_{ti}\left[\frac{c_w S_{wi} + c_f}{1 - S_{wi}}\right]\Delta\bar{p} + W_e$$

$$= N_p[B_t + (R_p - R_{soi})B_g] + B_w W_p \qquad (2.7)$$

Each term on the left-hand side of Eq. (2.7) accounts for a method of fluid production, and each term on the right-hand side represents an amount of hydrocarbon or water production. The first two terms on the left-hand side account for the expansion of any oil and/or gas zones that might be present. The third term accounts for the change in void space volume, which is the expansion of the formation and connate water. The fourth term is the amount of water influx that has occurred into the reservoir. On the right-hand side, the first term represents the production of oil and gas and the second term represents the water production.

Equation (2.7) can be arranged to apply to any of the different reservoir types discussed in Chapter 1. Without eliminating any terms, Eq. (2.7) is used for the case of a saturated oil reservoir with an associated gas cap. These reservoirs are discussed in Chapter 6. When there is no original free gas, such as in an undersaturated oil reservoir (discussed in Chapter 5), $m = 0$ and Eq. (2.7) reduces to:

$$N(B_t - B_{ti}) + N\, B_{ti}\left[\frac{c_w S_{wi} + c_f}{1 - S_{wi}}\right]\Delta\bar{p} + W_e$$

$$= N_p[B_t + (R_p - R_{soi})B_g] + B_w W_p \qquad (2.8)$$

For gas reservoirs, Eq. (2.7) can be modified by recognizing that $N_p R_p = G_p$ and that $NmB_{ti} = GB_{gi}$ and substituting these terms into Eq. (2.7):

$$N(B_t - B_{ti}) + G\,(B_g - B_{gi}) + (N\, B_{ti} + GB_{gi})\left[\frac{c_w S_{wi} + c_f}{1 - S_{wi}}\right]\Delta\bar{p} + W_e$$

$$= N_p B_t + (G_p - NR_{soi})\, B_g + B_w W_p \qquad (2.9)$$

When working with gas reservoirs, there is no initial oil amount; therefore, N and N_p are equal to zero. The general material balance equation for a gas reservoir can then be obtained

$$G(B_g - B_{gi}) + GB_{gi}\left[\frac{c_w S_{wi} + c_f}{1 - S_{wi}}\right]\Delta\bar{p} + W_e = G_p B_g + B_w W_p \qquad (2.10)$$

This equation is discussed in conjunction with gas and gas-condensate reservoirs in Chapters 3 and 4.

In the study of reservoirs that are produced simultaneously by the three major mechanisms of depletion drive, gas cap drive, and water drive, it is of practical interest to determine the relative magnitude of each of these mechanisms that contribute to the production. Pirson rearranged the material balance Equation (2.7) as follows to obtain three fractions, whose sum is one, which he called the depletion drive index (DDI), the segregation (gas cap) index (SDI), and the water-drive index (WDI).[6]

When all three drive mechanisms are contributing to the production of oil and gas from the reservoir, the compressibility term in Eq. (2.7) is negligible and can be ignored. Moving the water production term to the left-hand side of the equation, the following is obtained:

$$N(B_t - B_{ti}) + \frac{NmB_{ti}}{B_{gi}}(B_g - B_{gi}) + (W_e - B_w W_p) = N_p[B_t + (R_p - R_{soi})B_g]$$

Dividing through by the term on the right hand side of the equation:

$$\frac{N(B_t - B_{ti})}{N_p[B_t + (R_p - R_{soi})B_g]} + \frac{\dfrac{NmB_{ti}}{B_{gi}}(B_g - B_{gi})}{N_p[B_t + (R_p - R_{soi})B_g]}$$

$$+ \frac{(W_e - B_w W_p)}{N_p[B_t + (R_p - R_{soi})B_g]} = 1 \qquad (2.11)$$

The numerators of these three fractions that result on the left-hand side of Eq. (2.11) are the expansion of the initial oil zone, the expansion of the initial gas zone, and the net water influx, respectively. The common denominator is the reservoir volume of the cumulative gas and oil production expressed at the lower pressure, which evidently equals the sum of the gas and oil zone expansions plus the net water influx. Then, using Pirson's abbreviations:

$$DDI + SDI + WDI = 1$$

Calculations are performed in Chapter 6 to illustrate how these drive indexes are used.

3. USES AND LIMITATIONS OF THE MATERIAL BALANCE METHOD

The material balance equation derived in the previous section has been in general use for many years mainly for the following:

1. Determining the initial hydrocarbon in place.
2. Calculating water influx.
3. Predicting reservoir pressures.

Although in some cases it is possible to solve simultaneously to find the initial hydrocarbon and the water influx, generally one or the other must be known from data or methods that do not depend on the material balance calculations. One of the most important uses of the equations is predicting the effect of production rate and/or injection rates (gas or water) on reservoir pressure; therefore, it is very desirable to know in advance the initial oil and the ratio m from good core and log data. The presence of a water drive is usually indicated by geologic evidence; however, the material balance may be used to detect the existence of water drives by calculating the value of the initial hydrocarbon at successive production periods, assuming zero water influx. Unless other complicating factors are present, the constancy in the calculated value of N and/or G indicates a volumetric reservoir, and continually changing values of N and G indicate a water drive.

The precision of the calculated values depends on the accuracy of the data available to substitute in the equation and on the several assumptions that underlie the equations. One such assumption is the attainment of thermodynamic equilibrium in the reservoir, mainly between the oil and its solution gas. Wieland and Kennedy have found a tendency for the liquid phase to remain supersaturated with gas as the pressure declines.[7] Values in the range of 19 psi for East Texas Field fluid and cores and 25 psi for the Slaughter Field fluid and cores were measured. The effect of supersaturation causes the reservoir pressures to be lower than they would be had equilibrium been attained.

It is also implicitly assumed that the PVT data used in the material balances are obtained using gas liberation processes that closely duplicate the gas liberation processes in the reservoir, in the well, and in separators on the surface. This matter is discussed in detail in Chapter 6, and it is only stated here that in some cases the PVT data are based on gas liberation processes that vary widely from those obtaining for the reservoir fluid, and therefore produce results in the material balances that are in considerable error.

Another source of error is introduced in the determination of average reservoir pressure at the end of any production interval. Aside from instrument errors and those introduced by difficulties in obtaining true static or final build-up pressures (see Chapter 7), there is often the problem of correctly weighting or averaging the individual well pressures. For thicker formations of higher permeabilities and oils of lower viscosities, when final build-up pressures are readily and accurately obtained, and when there are only small pressure differences across the reservoir, reliable values of average reservoir pressure are easily obtained. On the other hand, for thinner formations of lower permeability and oils of higher viscosity, difficulties are met in obtaining accurate final build-up pressures, and there are often large pressure variations throughout the reservoir. These are commonly averaged by preparing isobaric maps superimposed on isopachous maps. This method usually provides reliable results unless the measured well pressures are erratic and therefore cannot be accurately contoured. These differences may be due to variations in formation thickness and permeability and in well production and producing rates. Also, difficulties are encountered when producing formations are comprised of two or more zones or strata of different permeabilities. In this case, the pressures are generally higher in the strata of low permeability, and, because the measured pressures are nearer to those in the zones of high permeability, the measured static pressures tend to be lower, and the reservoir behaves as if it contained less oil. Schilthuis explained this phenomenon by referring to the oil in the more permeable zones as active oil and by observing that the calculated active oil usually increases with time because the oil and gas in the zones of lower permeability slowly expand to help offset the pressure decline. This is also true of fields that are not fully developed, because the average pressure can be that of the developed portion only whereas the pressure is higher in the undeveloped portions.

The effect of pressure errors on calculated values of initial oil or water influx depends on the size of the errors in relation to the reservoir pressure decline. This is true because pressure enters the material balance equation mainly as differences $(B_o - B_{oi})$, $(R_{si} - R_s)$, and $(B_g - B_{gi})$. Because water influx and gas cap expansion tend to offset pressure decline, the pressure errors are more serious than for the undersaturated depletion reservoirs. In the case of very active water drives and gas caps that are large compared with the associated oil zone, the material balance is useless to determine the initial oil in place because of the very small pressure decline. Hutchinson emphasized the importance of obtaining accurate values of static well pressures, in his quantitative study of the effect of data errors on the values of initial gas or of initial oil in volumetric gas or undersaturated oil reservoirs, respectively.[8]

Uncertainties in the ratio of the initial free gas volume to the initial reservoir oil volume also affect the calculations. The error introduced in the calculated values of initial oil, water influx, or pressure increases with the size of this ratio because, as explained in the previous paragraph, larger gas caps reduce the effect of pressure decline. For quite large caps relative to the oil zone, the material balance approaches a gas balance modified slightly by production from the oil zone. The value of m is obtained from core and log data used to determine the net productive bulk gas and oil volumes and their average porosities and interstitial water. Because there is frequently oil saturation in the gas cap, the oil zone must include this oil, which correspondingly diminishes the initial free gas volume. Well tests are often useful in locating gas-oil and water-oil contacts in the determination of m. In some cases, these contacts are not horizontal planes but are tilted owing to water movement in the aquifer, or dish-shaped owing to the effect of capillarity in the less permeable boundary rocks of volumetric reservoirs.

Whereas the cumulative oil production is generally known quite precisely, the corresponding gas and water production is usually much less accurate, and therefore introduces additional sources of errors. This is particularly true when the gas and water production is not directly measured but must be inferred from periodic tests to determine the gas-oil ratios and water cuts of the individual wells. When two or more wells completed in different reservoirs are producing to common storage, unless there are individual meters on the wells, only the aggregate production is known and not the individual oil production from each reservoir. Under the circumstances that exist in many fields, it is doubtful that the cumulative gas and water production is known to within 10% and in some instances the errors may be larger. With the growing importance of natural gas and because more of the gas associated with the oil is being sold, better values of gas production are becoming available.

4. THE HAVLENA AND ODEH METHOD OF APPLYING THE MATERIAL BALANCE EQUATION

As early as 1953, van Everdingen, Timmerman, and McMahon recognized a method of applying the material balance equation as a straight line.[9] But it wasn't until Havlena and Odeh published their work that the method became fully exploited.[3,4] Normally, when using the material balance equation, an engineer considers each pressure and the corresponding production data as being separate points from other pressure values. From each separate point, a calculation for a dependent variable is made. The results of the calculations are sometimes averaged. The Havlena-Odeh method uses all the data points, with the further requirement that these points must yield solutions to the material balance equation that behave linearly to obtain values of the independent variable.

The straight-line method begins with the material balance equation written as:

$$N_p[B_t + (R_p - R_{soi})B_g] + B_w W_p - W_I - G_I B_{Ig}$$
$$= N\left[(B_t - B_{ti}) + B_{ti}(1 + m)\left(\frac{c_w S_{wi} + c_f}{1 - S_{wi}}\right)\Delta\bar{p} + \frac{mB_{ti}}{B_{gi}}(B_g - B_{gi})\right] + W_e \quad (2.12)$$

The terms W_I, cumulative water injection, G_I, cumulative gas injection, and B_{Ig}, formation volume factor of the injected gas have been added to Eq. (2.7). In Havlena and Odeh's original development, they chose to neglect the effect of the compressibilities of the formation and connate water in the gas cap portion of the reservoir. That is, in their development the compressibility term is multiplied by N and not by $N(1 + m)$. In Eq. (2.12), the compressibility term is multiplied by $N(1 + m)$ for completeness. You may choose to ignore the $(1 + m)$ multiplier in particular applications. Havlena and Odeh defined the following terms and rewrote Eq. (2.12) as:

$$F = N_p[B_t + (R_p - R_{soi})B_g] + B_w W_p - W_I - G_I B_{Ig}$$

$$E_o = B_t - B_{ti}$$

$$E_{f,w} = \left[\frac{c_w S_{wi} + c_f}{1 - S_{wi}}\right] \Delta \bar{P}$$

$$E_g = B_g - B_{gi}$$

$$F = NE_o + N(1 + m)B_{ti} E_{f,w} + \left[\frac{NmB_{ti}}{B_{gi}}\right] E_g + W_e \qquad (2.13)$$

In Eq. (2.13) F represents the net production from the reservoir. E_o, $E_{f,w}$, and E_g represent the expansion of oil, formation and water, and gas, respectively. Havlena and Odeh examined several cases of varying reservoir types with this equation and found that the equation can be rearranged into the form of a straight line. For instance, consider the case of no original gas cap, no water influx, and negligible formation and water compressibilities. With these assumptions, Eq. (2.13) reduces to:

$$F = NE_o \qquad (2.14)$$

This would suggest that a plot of F as the y coordinate and E_o as the x coordinate would yield a straight line with slope N and intercept equal to zero. Additional cases can be derived as will be shown in Chapter 6.

Once a linear relationship has been obtained, the plot can be used as a predictive tool for estimating future production. Examples are shown in subsequent chapters to illustrate the application of the Havelena-Odeh method.

REFERENCES

[1] Ralph J. Schilthuis, "Active Oil and Reservoir Energy," *Trans.* AIME (1936), **118**, 33.

[2] L. P. Dake, *Fundamentals of Reservoir Engineering,* 1st ed. Amsterdam: Elsevier, 1978, pp. 73–102.

[3] D. Havlena and A. S. Odeh, "The Material Balance as an Equation of a Straight Line," Part I. *Jour. of Petroleum Technology* (Aug. 1963), 896–900.

[4] D. Havlena and A. S. Odeh, "The Material Balance as an Equation of a Straight Line. Part II—Field Cases," *Jour. of Petroleum Technology* (July 1964), 815–822.

[5] L. D. Woody, Jr. and Robert Moscrip III, "Performance Calculations for Combination Drive Reservoirs," *Trans*. AIME (1956), **207,** 129.

[6] Sylvain J. Pirson, *Elements of Oil Reservoir Engineering,* 2nd ed. New York: McGraw-Hill, 1958, pp. 635–693.

[7] Denton R. Wieland and Harvey T. Kennedy, "Measurements of Bubble Frequency in Cores," *Trans*. AIME (1957), **210,** 125.

[8] Charles A. Hutchinson, "Effect of Data Errors on Typical Engineering Calculations." Presented at the Oklahoma City Meeting, *Petroleum Branch,* AIME, 1951.

[9] A. F. van Everdingen, E. H. Timmerman, and J. J. McMahon, "Application of the Material Balance Equation to a Partial Water-Drive Reservoir," *Trans*. AIME (1953), **198, 51.**

Chapter **3**

Single-Phase Gas Reservoirs

1. INTRODUCTION

This chapter discusses single-phase gas reservoirs (refer to Fig. 1.1). The reservoir fluid, usually called a natural gas, in a single-phase gas reservoir remains as nonassociated gas during the entire producing life of the reservoir. This type of reservoir is frequently referred to as a dry gas reservoir because no condensate is formed in the reservoir during the life of production. However, this does not preclude the production of condensate, because the temperature and pressure in the producing well and at the surface can be significantly different from the reservoir temperature and pressure. This change in conditions can cause some components in the producing gas phase to condense and be produced as liquid. The amount of condensation is a function of not only the pressure and temperature but also the composition of the natural gas, which typically consists primarily of methane and ethane. The tendency for condensate to form on the surface increases as the concentration of heavier components increases in the reservoir fluid.

We begin with a consideration of volumetric calculations for initial gas in place.

2. CALCULATING GAS IN PLACE BY THE VOLUMETRIC METHOD

The standard cubic feet of gas in a reservoir with a gas pore volume of V_g cu ft is simply V_g/B_g, where B_g is expressed in units of cubic feet per standard cubic foot. As the gas volume factor B_g changes with pressure (see Eq. 1.17, the gas in place also changes as the pressure declines. The gas pore volume V_g may also be changing, owing to water influx into the reservoir. The gas pore volume is related to the bulk, or total, reservoir volume by the average porosity ϕ and the average connate water S_w. The bulk reservoir volume V_b is commonly expressed in acre-feet, and the standard cubic feet of gas in place, G, is given by:

$$G = \frac{43{,}560 \ V_b \ \phi (1 - S_w)}{B_g} \qquad (3.1)$$

The areal extent of the Bell Field gas reservoir was 1500 acres. The average thickness was 40 ft, so that the initial bulk volume was 60,000 ac-ft. Average porosity was 22%, and average connate water was 23%. B_g at the initial reservoir pressure of 3250 psia was calculated to be 0.00533 cu ft/SCF. Therefore the initial gas in place was

$$G = 43{,}560 \times 60{,}000 \times 0.22 \times (1 - 0.23) \div 0.00533$$

$$= 83.1 \ MMM \ SCF$$

Because the gas volume factor was calculated using 14.7 psia and 60°F as standard conditions, the initial gas in place is also expressed at these conditions.

The volumetric method uses subsurface and isopachous maps based on the data from electric logs, cores, and drill-stem and production tests.[*,1,2] A *subsurface contour* map shows lines connecting points of equal elevations on the top of a marker bed and therefore shows geologic structure. A net *isopachous* map shows lines connecting points of equal net formation thickness; and the individual lines connecting points of equal thickness are called *isopach* lines. The reservoir engineer uses these maps to determine the bulk productive volume of the reservoir. The contour map is used in preparing the isopachous maps when there is an oil-water, gas-water, or gas-oil contact. The contact line is the zero isopach line. The volume is obtained by planimetering the areas between the isopach lines of the entire reservoir or of the individual units under consideration. The principal problems in preparing a map of this type are the proper interpretation of net sand thickness from the well logs and the outlining of the productive area of the field as defined by the fluid contacts, faults, or permeability barriers on the subsurface contour map.

---

* References throughout the text are given at the end of each chapter.

Two equations are commonly used to determine the approximate volume of the productive zone from the planimeter readings. The volume of the frustum of a pyramid is given by:

$$\Delta V_b = \frac{h}{3}(A_n + A_{n+1} + \sqrt{A_n A_{n+1}}) \qquad (3.2)$$

where ΔV_b is the bulk volume in acre-feet, A_n is the area enclosed by the lower isopach line in acres, A_{n+1} is the area enclosed by the upper isopach line in acres, and h is the interval between the isopach lines in feet. This equation is used to determine the volume between successive isopach lines, and the total volume is the sum of these separate volumes. The volume of a trapezoid is:

$$\Delta V_b = \frac{h}{2}(A_n + A_{n+1})$$

or for a series of successive trapezoids:

$$V_b = \frac{h}{2}(A_0 + 2A_1 + 2A_2 \ldots 2A_{n-1} + A_n) + t_{avg}A_n \qquad (3.3)$$

A_0 is the area enclosed by the zero isopach line in acres; $A_1, A_2 \ldots A_n$ are the areas enclosed by successive isopach lines in acres; t_{avg} is the average thickness above the top or maximum thickness isopach line in feet; and h is the isopach interval.

For best accuracy the pyramidal formula should be used. Because of its simpler form, however, the trapezoidal formula is commonly used, but it introduces an error of 2% when the ratio of successive areas is 0.50. Therefore, a commonly adopted rule in unitization programs is that wherever the ratio of the areas of any two successive isopach lines is smaller than 0.5, the pyramidal formula is applied. Whenever the ratio of the areas of any two successive isopach lines is found to be larger than 0.5, the trapezoidal formula is applied. Example 3.1 shows the method of calculating the volume of a gas reservoir from an isopachous map, Fig. 3.1. The volume between areas A_4 and A_5 by the trapezoidal equation is 570 ac-ft, compared with the more accurate figure of 558 ac-ft by the pyramidal equation. When the formation is rather uniformly developed and there is good well control, the error in the net bulk reservoir volume should not exceed a few percentage points.

Example 3.1. Calculating the net volume of an idealized reservoir from the isopachous map.

Given: The planimetered areas in Fig. 3.1 within each isopach line, A_0, A_1, A_2, etc. and the planimeter constant.

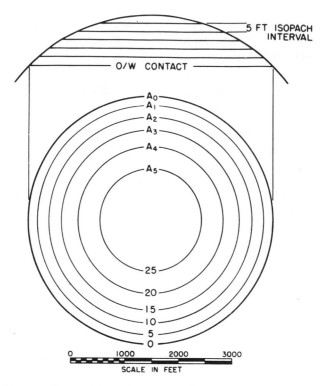

Fig. 3.1. Cross section and isopachous map of an idealized reservoir.

SOLUTION:

Productive Area	Planimeter Area* sq in.	Area Acres*	Ratio of Areas	Interval h, feet	Equation	ΔV ac-ft
A_0	19.64	450				
A_1	16.34	375	0.83	5	Trap.	2063
A_2	13.19	303	0.80	5	Trap.	1695
A_3	10.05	231	0.76	5	Trap.	1335
A_4	6.69	154	0.67	5	Trap.	963[a]
A_5	3.22	74	0.48	5	Pyr.	558[b]
A_6	0.00	0	0.00	4	Pyr.	99[c]
						6713 ac-ft

*For a map scale of 1 in. = 1000 ft; 1 sq in. = 22.96 ac.

[a] $\Delta V = \dfrac{5}{2}(231 + 154) = 963$ ac-ft

[b] $\Delta V = \dfrac{5}{3}(154 + 74 + \sqrt{154 \times 74}) = 558$ ac-ft

[c] $\Delta V = \dfrac{4}{3}(74) = 99$ ac-ft

The water in the oil- and gas-bearing parts of a petroleum reservoir above the transition zone is called *connate*, or *interstitial*, water. The two terms are used more or less interchangeably. Connate water is important primarily because it reduces the amount of pore space available to oil and gas and it also affects their recovery. It is generally not uniformly distributed throughout the reservoir but varies with the permeability and lithology as shown in Fig. 3.2 and with the height above the free water table as shown in Fig. 3.3.

Another problem in any volumetric or material balance calculation is that of obtaining the average reservoir pressure at any time after initial production. Figure 3.4 is a static reservoir pressure survey of the Jones sand in the Schuler Field.[4] Because of the large reservoir pressure gradient from east to west, some averaging technique must be used to obtain an average reservoir pressure. This can be calculated either as an average well pressure, average areal pressure, or average volumetric pressure as follows:

$$\text{Well average pressure} = \frac{\sum\limits_{0}^{n} p_i}{n} \tag{3.4}$$

$$\text{Areal average pressure} = \frac{\sum\limits_{0}^{n} p_i A_i}{\sum\limits_{0}^{n} A_i} \tag{3.5}$$

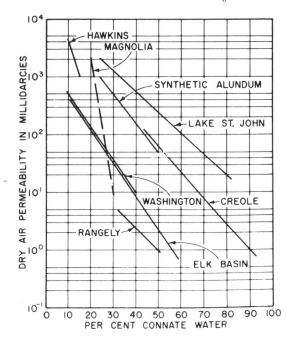

Fig. 3.2. Connate water versus permeability. (*After* Bruce and Welge,[3] Bruce *Trans.* AIME.)

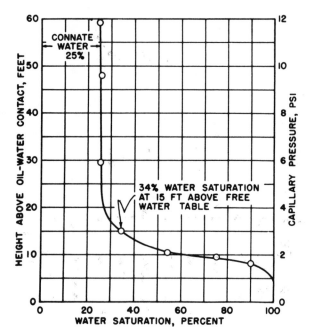

Fig. 3.3. Typical capillary pressure curve.

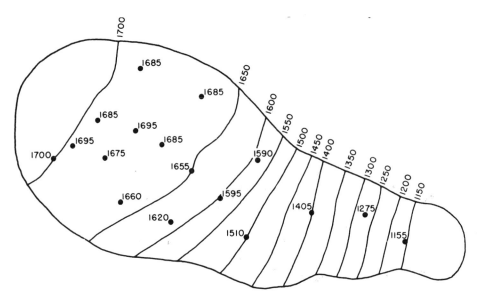

Fig. 3.4. Reservoir pressure survey showing isobaric lines drawn from the measured bottom-hole pressures. (*After* Kaveler,[4] *Trans.* AIME.)

$$\text{Volumetric average pressure} = \frac{\sum_{0}^{n} p_i A_i h_i}{\sum_{0}^{n} A_i h_i} \qquad (3.6)$$

where n is the number of wells in Eq. (3.4) and the number of reservoir units in Eqs. (3.5) and (3.6). Because we are interested in obtaining the average pressure of the hydrocarbon contents, the volumetric average, Eq. (3.6), should be used in the volumetric and material balance calculations. Where the pressure gradients in the reservoir are small, the average pressures obtained with Eqs. (3.4) and (3.5) will be very close to the volumetric average. Where the gradients are large, there may be considerable differences. For example, the average volumetric pressure of the Jones sand survey in Fig. 3.4 is 1658 psia as compared with 1598 psia for the well average pressure.

The calculations in Table (3.1) show how the average pressures are obtained. The figures in Col. (3) are the estimated drainage areas of the wells, which in some cases vary from the well spacing because of the reservoir limits. Owing to the much smaller gradients, the three averages are much closer together than in the case of the Jones sand.

Most engineers prefer to prepare an isobaric map as illustrated in Fig. (3.5) and to planimeter the areas between the isobaric lines and the isopach lines as shown in Fig. 3.5. Table 3.2, using data taken from Fig. 3.5, illustrates the method of obtaining the average volumetric pressure from this type of map.

TABLE 3.1.
Calculation of average reservoir pressure

Well No.	Pressure psia	Drainage Area Acres	p × A	Est. Sd. Thick.	p × A × h	A × h
1	2750	160	440,000	20	8,800,000	3200
2	2680	125	335,000	25	8,375,000	3125
3	2840	190	539,600	26	14,029,600	4940
4	2700	145	391,500	31	12,136,500	4495
	10,970	620	1,706,100		43,341,100	15,760

$$\text{Well average pressure} = \frac{10,970}{4} = 2743 \text{ psia}$$

$$\text{Areal average pressure} = \frac{1,706,100}{620} = 2752 \text{ psia}$$

$$\text{Volumetric average pressure} = \frac{43,341,100}{15,760} = 2750 \text{ psia}$$

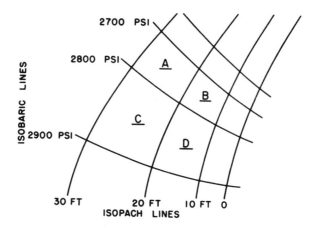

Fig. 3.5. Section of an isobaric and isopachous map.

TABLE 3.2.
Volumetric calculations of reservoir pressure

(1)	(2)	(3)	(4)	(5)	(6)
		Pressure			
Area	Acres[a]	psia	h ft	$A \times h$	$p \times A \times h$
A	25.5	2750	25	637.5	175,313,000
D	15.1	2750	15	226.5	62,288,000
C	50.5	2850	25	1262.5	359,813,000
D	30.2	2850	15	453.0	129,105,000
				2579.5	726,519,000

[a] Planimetered areas of Fig. 3.5

$$\text{Average pressure on a volume basis} = \frac{726,519,000}{2579.5}$$

$$= 2817 \text{ psia}$$

3. CALCULATION OF UNIT RECOVERY FROM VOLUMETRIC GAS RESERVOIRS

In many gas reservoirs, particularly during the development period, the bulk volume is not known. In this case it is better to place the reservoir calculations on a unit basis, usually 1 ac-ft of bulk reservoir rock. Then one unit, or 1 ac-ft, of bulk reservoir rock contains:

$$\text{Connate water:} \quad 43,560 \times \phi \times S_w \text{ cu ft}$$

$$\text{Reservoir gas volume:} \quad 43,560 \times \phi \ (1 - S_w) \text{ cu ft}$$

$$\text{Reservoir pore volume:} \quad 43,560 \times \phi \text{ cu ft}$$

The initial standard cubic feet of gas in place in the unit is:

$$G = \frac{43,560\ (\phi)(1 - S_{wi})}{B_{gi}}\ SCF/ac\text{-}ft \tag{3.7}$$

G is in standard cubic feet when the gas volume factor B_{gi} is in cubic feet per standard cubic foot Eq. (1.17). The *standard* conditions are those used in the calculation of the gas volume factor, and they may be changed to any other standard by means of the ideal gas law. The porosity ϕ is expressed as a fraction of the bulk volume, and the initial connate water S_{wi} as a fraction of the pore volume. For a reservoir under volumetric control, there is no change in the interstitial water, so the *reservoir* gas volume remains the same. If B_{ga} is the gas volume factor at the abandonment pressure, then the standard cubic feet of gas remaining at abandonment is:

$$G_a = \frac{43,560\ (\phi)(1 - S_{wi})}{B_{ga}}\ SCF/ac\text{-}ft \tag{3.8}$$

Unit *recovery* is the difference between the initial gas in place and that remaining at abandonment pressure (i.e., that produced at abandonment pressure), or:

$$\text{Unit recovery} = 43,560(\phi)(1 - S_{wi})\left[\frac{1}{B_{gi}} - \frac{1}{B_{ga}}\right]\ SCF/ac\text{-}ft \tag{3.9}$$

The unit recovery is also called the *initial unit reserve*, which is generally lower than the initial unit in-place gas. The remaining reserve at any stage of depletion is the difference between this initial reserve and the unit production at that stage of depletion. The fractional recovery or recovery factor expressed in a percentage of the initial in-place gas is

$$\text{Recovery factor} = \frac{100(G - G_a)}{G} = \frac{100\left[\dfrac{1}{B_{gi}} - \dfrac{1}{B_{ga}}\right]}{\dfrac{1}{B_{gi}}}\ \% \tag{3.10}$$

or

$$\text{Recovery factor} = 100\left[1 - \frac{B_{gi}}{B_{ga}}\right]$$

Experience with volumetric gas reservoirs indicates that the recoveries will range from 80 to 90%. Some gas pipeline companies use an abandonment pressure of 100 psi per 1000 ft of depth.

The gas volume factor in the Bell Gas Field at initial reservoir pressure is 0.00533 cu ft/SCF and at 500 psia it is 0.03623. The initial unit reserve or unit recovery based on volumetric performance at an abandonment pressure of 500 psia is

$$\text{Unit recovery} = 43{,}560 \times 0.22 \times (1 - 0.23) \times \left[\frac{1}{0.00533} - \frac{1}{0.03623}\right]$$

$$= 1180\text{M SCF/ac-ft}$$

$$\text{Recovery factor} = 100 \left[1 - \frac{0.00533}{0.03623}\right]$$

$$= 85\%$$

These recovery calculations are valid provided the unit neither drains nor is drained by adjacent units.

4. CALCULATION OF UNIT RECOVERY FROM GAS RESERVOIRS UNDER WATER DRIVE

Under initial conditions one unit (1 ac-ft) of bulk reservoir rock contains

$$\text{Connate water:} \quad 43{,}560 \times \phi \times S_{wi} \text{ cu ft}$$

$$\text{Reservoir gas volume:} \quad 43{,}560 \times \phi \times (1 - S_{wi}) \text{ cu ft}$$

$$\text{Surface units of gas:} \quad 43{,}560 \times \phi \times (1 - S_{wi}) \div B_{gi} \text{ SCF}$$

In many reservoirs under water drive, the pressure suffers an initial decline, after which water enters the reservoir at a rate to equal the production, and the pressure stabilizes. In this case the stabilized pressure is the abandonment pressure. If B_{ga} is the gas volume factor at the abandonment pressure and S_{gr} is the *residual* gas saturation, expressed as a fraction of the pore volume, after water invades the unit, then under abandonment conditions a unit (1 ac-ft) of the reservoir rock contains

$$\text{Water volume:} \quad 43{,}560 \times \phi \times (1 - S_{gr}) \text{ cu ft}$$

$$\text{Reservoir gas volume:} \quad 43{,}560 \times \phi \times S_{gr} \text{ cu ft}$$

$$\text{Surface units of gas:} \quad 43{,}560 \times \phi \times S_{gr} \div B_{ga} \text{ SCF}$$

Unit recovery is the difference between the initial and the residual surface units of gas, or:

$$\text{Unit recovery in } SCF/ac\text{-}ft = 43{,}560 \, (\phi) \left[\frac{1 - S_{wi}}{B_{gi}} - \frac{S_{gr}}{B_{ga}}\right] \quad (3.11)$$

The recovery factor expressed in a percentage of the initial gas in place is

$$\text{Recovery factor} = \frac{100\left[\dfrac{1 - S_{wi}}{B_{gi}} - \dfrac{S_{gr}}{B_{ga}}\right]}{\left[\dfrac{1 - S_{wi}}{B_{gl}}\right]} \tag{3.12}$$

Suppose the Bell Gas Field is produced under a water drive such that the pressure stabilizes at 1500 psia. If the residual gas saturation is 24% and the gas volume factor at 1500 psia is 0.01122 cu ft/SCF, then the initial unit *reserve* or unit recovery is

$$\text{Unit recovery} = 43,560 \times 0.22 \times \left[\frac{(1 - 0.23)}{0.00533} - \frac{0.24}{0.0112}\right]$$

$$= 1180 \text{ M SCF/ac-ft}$$

The recovery factor under these conditions is

$$\text{Recovery factor} = \frac{100\left[\dfrac{1 - 0.23}{0.00533} - \dfrac{0.24}{0.0112}\right]}{\left[\dfrac{1 - 0.23}{0.00533}\right]} = 85\%$$

Under these particular conditions, the recovery by water drive is the same as the recovery by volumetric depletion, illustrated in Sect. 3. If the water drive is very active so that there is essentially no decline in reservoir pressure, unit recovery and the recovery factor become

$$\text{Unit recovery} = 43,560 \times \phi \times (1 - S_{wi} - S_{gr}) \div B_{gi} \text{ SCF/ac-ft} \tag{3.13}$$

$$\text{Recovery factor} = \frac{100(1 - S_{wi} - S_{gr})}{(1 - S_{wi})} \% \tag{3.14}$$

For the Bell Gas Field, assuming a residual gas saturation of 24%:

$$\text{Unit recovery} = 43,560 \times 0.22 \times (1 - 0.23 - 0.24) \div 0.00533$$

$$= 953 \text{ M SCF/ac-ft}$$

$$\text{Recovery factor} = \frac{100(1 - 0.23 - 0.24)}{(1 - 0.23)}$$

$$= 69\%$$

Because the residual gas saturation is independent of the pressure, the recovery will be greater for the lower stabilization pressure.

The residual gas saturation can be measured in the laboratory on representative core samples. Table 3.3 gives the residual gas saturations that were measured on core samples from a number of producing horizons and on some synthetic laboratory samples. The values, which range from 16 to 50% and average near 30%, help to explain the disappointing recoveries obtained in some water-drive reservoirs. For example, a gas reservoir with an initial water saturation of 30% and a residual gas saturation of 35% has a recovery factor of only 50% if produced under an active water drive (i.e., where the reservoir pressure stabilizes near the initial pressure). When the reservoir permeability is uniform, this recovery factor should be representative, except for a correction to allow for the efficiency of the drainage pattern and water coning or cusping. When there are well-defined continuous beds of higher and lower permeability, the water will advance more rapidly through the more permeable beds so that when a gas well is abandoned owing to excessive water production, considerable unrecovered gas remains in the less permeable beds. Because of these factors, it may be concluded that generally gas recoveries by water drive are lower than by volumetric depletion; however, the same conclusion does not apply to oil recovery, which is discussed separately. Water-drive gas reservoirs do have the advantage of maintaining higher flowing wellhead pressures and higher well rates compared with depletion gas reservoirs. This is due, of course, to the maintenance of higher reservoir pressure as a result of the water influx.

TABLE 3.3.

Residual gas saturation after water flood as measured on core plugs
(*After* Geffen, Parish, Haynes, and Morse[5].)

Porous Material	Formation	Residual Gas Saturation, Percentage of Pore Space	Remarks
Unconsolidated sand		16	(13-ft Column)
Slightly consolidated sand (synthetic)		21	(1 Core)
Synthetic consolidated materials	Selas Porcelain	17	(1 Core)
	Norton Alundum	24	(1 Core)
Consolidated sandstones	Wilcox	25	(3 Cores)
	Frio	30	(1 Core)
	Nellie Bly	30–36	(12 Cores)
	Frontier	31–34	(3 Cores)
	Springer	33	(3 Cores)
	Frio	30–38	(14 Cores)
			(Average 34.6)
	Torpedo	34–37	(6 Cores)
	Tensleep	40–50	(4 Cores)
Limestone	Canyon Reef	50	(2 Cores)

In calculating the gas reserve of a particular lease or unit, the gas that can be recovered by the well(s) on the lease is important rather than the total recoverable gas *initially* underlying the lease, some of which may be recovered by adjacent wells. In volumetric reservoirs where the recoverable gas beneath each lease (well) is the same, the recoveries will be the same only if all wells are produced at the same rate. On the other hand, if wells are produced at equal rates when the gas beneath the leases (wells) varies, as from variable formation thickness, the initial gas reserve of the lease where the formation is thicker will be less than the initial recoverable gas underlying the lease.

In water-drive gas reservoirs, when the pressure stabilizes near the initial reservoir pressure, the lowest well on structure will divide its initial recoverable gas with all updip wells in line with it. For example, if three wells in line along the dip are drilled at the updip edge of their units, which are presumed equal, and if they all produce at the same rate, the lowest well on structure will recover approximately one-third of the gas initially underlying it. If the well is drilled further downstructure near the center of the unit, it will recover still less. If the pressure stabilizes at some pressure below the initial reservoir pressure, the recovery factor will be improved for the wells low on structure. Example 3.2 shows the calculation of the initial gas reserve of a 160-acre unit by volumetric depletion, partial water drive, and complete water drive.

Example 3.2. Calculating the initial gas reserve of a 160-acre unit of the Bell Gas Field by volumetric depletion and under partial and complete water drive.

Given:

Average porosity = 22%

Connate water = 23%

Residual gas saturation after water displacement = 34%

$B_{gi} = 0.00533$ cu ft/SCF at $p_i = 3250$ psia

$B_g = 0.00667$ cu ft/SCF at 2500 psia

    = 0.03623 cu ft/SCF at 500 psia

Area = 160 acres

Net productive thickness = 40 ft

SOLUTION:

Pore volume = $43,560 \times 0.22 \times 160 \times 40 = 61.33 \times 10^6$ cu ft

Initial gas in place:

$G_1 = 61.33 \times 10^6 \times (1 - 0.23) \div 0.00533 = 8860$ MM SCF

Gas in place after volumetric depletion to 2500 psia:

$$G_2 = 61.33 \times 10^6 \times (1 - 0.23) \div 0.00667 = 7080 \text{ MM SCF}$$

Gas in place after volumetric depletion to 500 psia:

$$G_3 = 61.33 \times 10^6 \times (1 - 0.23) \div 0.03623 = 1303 \text{ MM SCF}$$

Gas in place after water invasion at 3250 psia:

$$G_4 = 61.33 \times 10^6 \times 0.34 \div 0.00533 = 3912 \text{ MM SCF}$$

Gas in place after water invasion at 2500 psia:

$$G_5 = 61.33 \times 10^6 \times 0.34 \div 0.00667 = 3126 \text{ MM SCF}$$

Initial reserve by depletion to 500 psia:

$$G_1 - G_3 = (8860 - 1303) \times 10^6 = 7557 \text{ MM SCF}$$

Initial reserve by water drive at 3250 psia:

$$G_1 - G_4 = (8860 - 3912) \times 10^6 = 4948 \text{ MM SCF}$$

Initial reserve by water drive at 2500 psia:

$$(G_1 - G_5) = (8860 - 3126) \times 10^6 = 5734 \text{ MM SCF}$$

If there is one undip well, the initial reserve by water drive at 3250 psia is

$$\tfrac{1}{2}(G_1 - G_4) = \tfrac{1}{2}(8860 - 3912) \times 10^6 = 2474 \text{ MM SCF}$$

The recovery factors calculate to be 85%, 65%, and 56% for the cases of no water drive, partial water drive, and full water drive, respectively. These recoveries are fairly typical and can be explained in the following way. As water invades the reservoir, the reservoir pressure is maintained at a higher level than if there were no water encroachment. This leads to higher abandonment pressures for water-drive reservoirs. Because the main mechanism of production in a gas reservoir is that of depletion, or gas expansion, recoveries are lower, as shown in Ex. 3.2.

Agarwal, Al-Hussainy, and Ramey conducted a theoretical study and showed that gas recoveries increased with increasing production rates from water-drive reservoirs.[6] This technique of "outrunning" the water has been attempted in the field and has been found to be successful. Matthes, Jackson, Schuler, and Marudiak showed that ultimate recovery increased from 69 to 74% by increasing the field production rate from 50 to 75 MM SCF/D in the Bierwang Field in West Germany.[7] Lutes, Chiang, Brady, and Rossen reported an 8.5% increase in ultimate recovery with an increased production rate in a strong water-drive Gulf Coast gas reservoir.[8]

A second technique used in the field is the coproduction technique discussed by Arcaro and Bassiouni.[9] The coproduction technique is defined as the simultaneous production of gas and water. In the coproduction process, as downdip wells begin to be watered out, they are converted to high-rate water production wells, while the updip wells are maintained on gas production. This technique enhances the production of gas by several methods. First, the high-rate downdip water wells act as a pressure sink for the water because the water is drawn to these wells. This retards the invasion of water into productive gas zones in the reservoir, therefore prolonging the useful productive life to these zones. Second, the high-rate production of water lowers the average pressure in the reservoir, allowing for more gas expansion and therefore more gas production. Third, when the average reservoir pressure is lowered, immobile gas in the water-swept portion of the reservoir could become mobile. The coproduction technique performs best before the reservoir is totally invaded by water. Arcaro and Bassiouni reported the improvement of gas production from 62% to 83% from the Louisiana Gulf Coast Eugene Island Block 305 Reservoir by using the coproduction technique instead of the conventional production approach. Water-drive reservoirs are discussed in much more detail in Chapter 8.

5. MATERIAL BALANCE

In the previous sections, the initial gas in place was calculated on a unit basis of 1 ac-ft of bulk productive rock from a knowledge of the porosity and connate water. To calculate the initial gas in place on any particular portion of a reservoir, it is necessary to know, in addition, the bulk volume of that portion of the reservoir. If the porosity, connate water, and/or the bulk volumes are not known with any reasonable precision, the methods described cannot be used. In this case, the material balance method may be used to calculate the initial gas in place; however, this method is applicable only to the reservoir as a whole, because of the migration of gas from one portion of the reservoir to another in both volumetric and water-drive reservoirs.

The general material balance equation for a gas reservoir was derived in Chapter 2.

$$G(B_g - B_{gi}) + GB_{gi}\left[\frac{c_w S_{wi} + c_f}{1 - S_{wi}}\right]\Delta\bar{p} + W_e = G_p B_g + B_w W_p \qquad (2.10)$$

Equation (2.10) could have been derived by applying the law of conservation of mass to the reservoir and associated production.

For most gas reservoirs, the gas compressibility term is much greater than the formation and water compressibilities, and the second term on the left-hand side of Eq. (2.10) becomes negligible

$$G(B_g - B_{gi}) + W_e = G_p B_g + B_w W_p \tag{3.15}$$

When reservoir pressures are abnormally high, this term is not negligible and should not be ignored. This situation is discussed in a later section of this chapter.

When there is neither water encroachment into nor water production from a reservoir of interest, the reservoir is said to be *volumetric*. For a volumetric gas reservoir, Eq. (3.15) reduces to:

$$G(B_g - B_{gi}) = G_p B_g \tag{3.16}$$

Using Eq. (1.16) and substituting expressions for B_g and B_{gi} into Eq. (3.16), the following is obtained:

$$G\left(\frac{p_{sc}zT}{T_{sc}p}\right) - G\left(\frac{p_{sc}z_iT_i}{T_{sc}p_i}\right) = G_p\left(\frac{p_{sc}zT}{T_{sc}p}\right) \tag{3.17}$$

Noting that production is essentially an isothermal process (i.e., the reservoir temperature remains constant), then Eq. (3.17) is reduced to:

$$G\left(\frac{z}{p}\right) - G\left(\frac{z_i}{p_i}\right) = G_p\left(\frac{z}{p}\right)$$

Rearranging:

$$\frac{p}{z} = -\frac{p_i}{z_iG}G_p + \frac{p_i}{z_i} \tag{3.18}$$

Because p_i, z_i, and G are constants for a given reservoir, Eq. (3.18) suggests that a plot of p/z as the ordinate versus G_p as the abscissa would yield a straight line with:

$$\text{slope} = -\frac{p_i}{z_iG}$$

$$y \text{ interecept} = \frac{p_i}{z_i}$$

This plot is shown in Fig. 3.6.

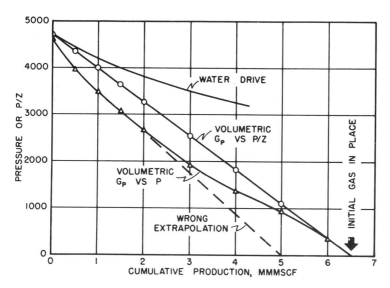

Fig. 3.6. Comparison of theoretical values of p and p/z plotted versus cumulative production from a volumetric gas reservoir.

If p/z is set equal to zero, which would represent the production of all the gas from a reservoir, then the corresponding G_p equals G, the initial gas in place. The plot could also be extrapolated to any abandonment p/z to find the initial reserve. Usually this extrapolation requires at least three years of accurate pressure depletion and gas production data.

Figure 3.6 also contains a plot of cumulative gas production G_p versus pressure. As indicated by Eq. (3.18), this is not linear, and extrapolations from the pressure-production data may be in considerable error. Because the minimum value of the gas deviation factor generally occurs near 2500 psia, the extrapolations will be low for pressures above 2500 psia and high for pressures below 2500 psia. Equation (3.18) may be used graphically as shown in Fig. 3.6 to find the initial gas in place or the reserves at any pressure for any selected abandonment pressure. For example, at 1000 psia (or $p/z = 1220$) abandonment pressure, the initial reserve is 4.85 MMM SCF. At 2500 psia (or $p/z = 3130$), the (remaining) reserve is 4.85 less 2.20—that is, 2.65 MMM SCF.

In water-drive reservoirs, the relation between G_p and p/z is not linear, as can be seen by an inspection of Eq. (3.15) and (3.18). Because of the water influx, the pressure drops less rapidly with production than under volumetric control, as shown in the upper curve of Fig. 3.6. Consequently, the extrapolation technique described for volumetric reservoirs is not applicable. Also, where there is water influx, the initial gas in place calculated at successive stages of depletion, assuming no water influx, takes on successively higher values, whereas with volumetric reservoirs the calculated values of the initial gas should remain substantially constant.

Equation (3.15) may be expressed in terms of the initial pore volume, V_i, by recognizing that $V_i = GB_{gi}$ and using Eq. (1.16) for B_g and B_{gi}:

$$V_i \left[\frac{z_f T \, P_i V_i}{P_f Z_i T} - 1 \right] = \frac{Z_f T \, P_{sc} G_p}{P_f T_{sc}} + B_w W_p - W_e \qquad (3.19)$$

For volumetric reservoirs, this equation can be reduced and rearranged to give:

$$\frac{P_{sc} G_p}{T_{sc}} = \frac{P_i V_i}{z_i T} - \frac{P_f V_i}{z_f T} \qquad (3.20)$$

The following three example problems illustrate the use of the various equations that we have described in gas reservoir calculations.

Example 3.3. Calculating the initial gas in place and the initial reserve of a gas reservoir from pressure-production data for a volumetric reservoir. Note that the base pressure is 15.025 psia.

Given:

$$\text{Initial pressure} = 3250 \text{ psia}$$

$$\text{Reservoir temperature} = 213°F$$

$$\text{Standard pressure} = 15.025 \text{ psia}$$

$$\text{Standard temperature} = 60°F$$

$$\text{Cumulative production} = 1.00 \times 10^9 \text{ SCF}$$

$$\text{Average reservoir pressure} = 2864 \text{ psia}$$

$$\text{Gas deviation factor at 3250 psia} = 0.910$$

$$\text{Gas deviation factor at 2864 psia} = 0.888$$

$$\text{Gas deviation factor at 500 psia} = 0.951$$

SOLUTION: Solve Eq. (3.20) for the reservoir gas pore volume V_i:

$$\frac{15.025 \times 1.00 \times 10^9}{520} = \frac{3250 \times V_i}{0.910 \times 673} - \frac{2864 \, V_i}{0.888 \times 673}$$

$$V_i = 56.17 \text{ MM cu ft}$$

The initial gas in place by the real gas law is:

$$G = \frac{p_i V_i}{z_i T} \times \frac{T_{sc}}{p_{sc}} = \frac{3250 \times 56.17 \times 10^6 \times 520}{0.910 \times 673 \times 15.025}$$

$$= 10.32 \text{ MMM SCF}$$

The gas remaining at 500 psia abandonment pressure is:

$$G_a = \frac{p_a V_i}{z_a T} \times \frac{T_{sc}}{p_{sc}} = \frac{500 \times 56.17 \times 10^6 \times 520}{0.951 \times 673 \times 15.025}$$

$$= 1.52 \text{ MMM SCF}$$

The initial gas reserve based on a 500 psia abandonment pressure is the difference between the initial gas in place and the gas remaining at 500 psia, or:

$$G_r = G - G_a = (10.32 - 1.52) \times 10^9$$

$$= 8.80 \text{ MMM SCF}$$

Example 3.4 illustrates the use of equations to calculate the water influx when the initial gas in place is known. It also shows the method of estimating the residual gas saturation of the portion of the reservoir invaded by water, at which time a reliable estimate of the invaded volume can be made. This is calculated from the isopachous map, the invaded volume being dilineated by those wells that have gone to water production. The residual gas saturation calculated in Ex. 3.4 includes that portion of the lower permeability rock within the invaded area that actually may not have been invaded at all, the wells having been "drowned" by water production from the more permeable beds of the formation. Nevertheless, it is still interpreted as the *average* residual gas saturation, which may be applied to the uninvaded portion of the reservoir.

Example 3.4 Calculating water influx and residual gas saturation in water-drive gas reservoirs.

Given:

Bulk reservoir volume, initial = 415.3 MM cu ft

Average porosity = 0.172

Average connate water = 0.25

Initial pressure = 3200 psia

$B_{gi} = 0.005262$ cu ft/SCF, 14.7 psia and 60°F

Final pressure = 2925 psia

$B_{gf} = 0.005700$ cu ft/SCF, 14.7 psia and 60°F

Cumulative water production = 15,200 bbl (surface)

$B_w = 1.03$ bbl/surface bbl

$G_p = 935.4$ MM SCF at 14.7 psia and 60°F

Bulk volume invaded by water at 2925 psia = 13.04 MM cu ft

SOLUTION:

$$\text{Initial gas in place} = G = \frac{415.3 \times 10^6 \times 0.172 \times (1 - 0.25)}{0.005262}$$

$$= 10,180 \text{ MM SCF at } 14.7 \text{ psia and } 60°F$$

Substitute in Eq. (3.15) to find W_e:

$$W_e = 935.4 \times 10^6 \times 0.005700 - 10,180 \times 10^6$$
$$(0.005700 - 0.005262) + 15,200 \times 1.03 \times 5.615$$
$$= 960,400 \text{ cu ft}$$

This much water has invaded 13.04 MM cu ft of bulk rock that initially contained 25% connate water. Then the final water saturation of the flooded portion of the reservoir is

$$S_w = \frac{\text{Connate water} + \text{Water influx} - \text{Produced water}}{\text{Pore space}}$$

$$= \frac{(13.04 \times 10^6 \times 0.172 \times 0.25) + 960,400 - 15,200 \times 1.03}{13.04 \times 10^6 \times 0.172}$$

$$= 0.67 \text{ or } 67\%$$

Then the residual gas saturation S_{gr} is 33%.

Example 3.5. Using the p/z plot to estimate cumulative gas production. A dry gas reservoir contains gas of the following composition:

	Mole Fraction
Methane	0.75
Ethane	0.20
n-Hexane	0.05

The initial reservoir pressure was 4200 psia, with a temperature of 180°F. The reservoir has been producing for some time. Two pressure surveys have been made at different times.

p/z (psia)	G_p (MMM SCF)
4600	0
3700	1
2800	2

(a) What will be the cumulative gas produced when the average reservoir pressure has dropped to 2000 psia?

(b) Assuming the reservoir rock has a porosity of 12%, the water saturation is 30%, and the reservoir thickness is 15 ft, how many acres does the reservoir cover?

SOLUTION:

		P_c	T_c	YP_c	YT_c
Methane	0.75	673.1	343.2	504.8	257.4
Ethane	0.20	708.3	504.8	141.7	110.0
n-Hexane	0.05	440.1	914.2	22.0	45.7
Totals				668.5	413.1

(a) To get G_p at 2000 psia, calculate z and the p/z. Use pseudocritical properties.

$$p_r = \frac{2000}{668.5} = 2.99$$

$$T_r = \frac{640}{413.1} = 1.55$$

$$z = 0.8$$

$$p/z = \frac{2000}{0.8} = 2500$$

A linear regression of the data plotted in Fig. 3.7 yields the following equation for the best straight line through the data:

$$p/z = -9(10)^{-7}G_p + 4600$$

Substituting a value of $p/z = 2500$ in this equation yields:

$$2500 = -9(10)^{-7}G_p + 4600$$

$$G_p = 2.33(10)^9 \text{ SCF} \quad \text{or} \quad 2.33 \text{ MMM SCF}$$

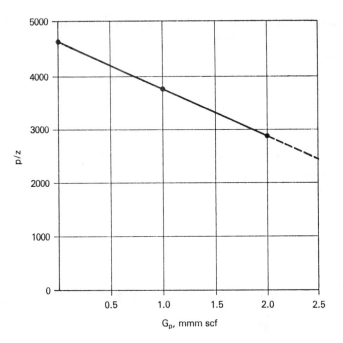

Fig. 3.7. p/z versus G_p for Ex. 3.5.

(b) Substituting a value of $p/z = 0$ into the straight-line equation, would yield the amount of produced gas if all of the initial gas were produced; therefore, the G_p at this p/z is equal to the initial gas in place.

$$0 = -9(10)^{-7}G_p + 4600$$

$$G_p \ (p/z = 0) = G = 5.11(10)^9 \text{ SCF} \qquad \text{or} \qquad 5.11 \text{ MMM SCF}$$

Recognizing that $V_i = GB_{gi}$ and that $B_{gi} = 0.02829 \ (z_i/p_i)T$:

$$V_i = GB_{gi} = 5.11(10)^9 \left[\frac{0.02829(180 + 460)}{4600} \right] = 20.1(10)^6 \text{ cu ft}$$

also

$$V_i = Ah\phi(1 - S_{wi})$$

$$A = \frac{20.1(10)^6}{15(0.12)(1 - 0.30)} = 15.95(10)^6 \text{ ft}^2 \qquad \text{or} \qquad 366 \text{ acres}$$

6. THE GAS EQUIVALENT OF PRODUCED CONDENSATE AND WATER

In the study of gas reservoirs in the preceding section, it was implicitly assumed that the fluid in the reservoir at all pressures as well as on the surface was in a *single* (gas) phase. Most gas reservoirs, however, produce some hydrocarbon liquid, commonly called condensate, in the range of a few to a hundred or more barrels per million standard cubic feet. So long as the *reservoir* fluid remains in a single (gas) phase, the calculations of the previous sections may be used, provided the cumulative gas production G_p is modified to include the condensate liquid production. On the other hand, if a hydrocarbon liquid phase develops in the reservoir, the methods of the previous sections are not applicable, and these *retrograde,* gas-condensate reservoirs must be treated specially, as described in Chapter 4.

The reservoir gas production G_p used in the previous sections must include the separator gas production, the stock tank gas production, and the stock tank liquid production converted to its gas equivalent, symbol *GE*. Figure 3.8 illustrates two common separation schemes. Figure 3.8(a) shows a three-stage separation system with a primary separator, a secondary separator, and a stock tank. The well fluid is introduced into the primary separator where most of the produced gas is obtained. The liquid from the primary separator is then sent to the secondary separator where an additional amount of gas is obtained. The liquid from the secondary separator is then flashed into the stock tank. The liquid from the stock tank is N_p and any gas from the stock tank is added to the primary and secondary gas to arrive at the total produced surface gas, $G_{p(\text{surf})}$. Figure 3.8(b) shows a two-stage separation process similar to the one shown in Fig. 3.8(a) without the secondary separator.

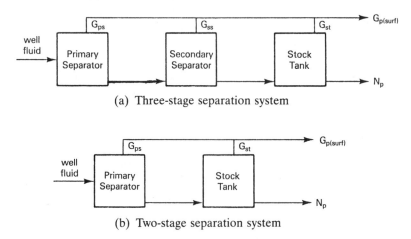

(a) Three-stage separation system

(b) Two-stage separation system

Fig. 3.8. Schematic representation of surface separation systems.

The produced hydrocarbon liquid is converted to its gas equivalent, assuming it behaves as an ideal gas when vaporized in the produced gas. Taking 14.7 psia and 60°F as standard conditions, the gas equivalent of one stock tank barrel of condensate liquid is:

$$GE = V = \frac{nR'T_{sc}}{p_{sc}} = \frac{350.5\gamma_o(10.73)(520)}{M_{wo}(14.7)} = 133{,}000\frac{\gamma_o}{M_{wo}} \qquad (3.21)$$

The gas equivalent of one barrel of condensate of specific gravity of 0.780 (water = 1.00) and molecular weight 138 is 752 SCF. The specific gravity may be calculated from the API gravity. If the molecular weight of the condensate is not measured, as by the freezing point depression method, it may be estimated using Eq. (3.22).

$$M_{wo} = \frac{5954}{\rho_{o,API} - 8.811} = \frac{42.43\gamma_o}{1.008 - \gamma_o} \qquad (3.22)$$

The total gas equivalent for N_p STB of condensate production is $GE(N_p)$. The total reservoir gas production, G_p, is given by Eq. (3.23) for a three-stage separation system and by Eq. (3.24) for a two-stage separation system:

$$G_p = G_{p(surf)} + GE(N_p) = G_{ps} + G_{ss} + G_{st} + GE(N_p) \qquad (3.23)$$

$$G_p = G_{p(surf)} + GE(N_p) = G_{ps} + G_{st} + GE(N_p) \qquad (3.24)$$

If the gas volumes from the low-pressure separators and the stock tank are not measured, then the correlations found in Figs. 3.9 and 3.10 can be used to estimate the vapor equivalent of these gas volumes plus the gas equivalent of the liquid condensate. Figure 3.9 is for a three-stage separator system, and Fig. 3.10 is for a two-stage separator system. The correlations are based on production parameters that are routinely measured (i.e., the primary separator pressure, temperature, and gas gravity), the secondary separator temperature, and the stock tank liquid gravity. The total reservoir gas production, G_p, is given by Eq. (3.25) when using the correlations of Figs. 3.9 and 3.10.

$$G_p = G_{ps} + V_{eq}(N_p) \qquad (3.25)$$

When water is produced on the surface as a condensate from the gas phase in the reservoir, it is fresh water and should be converted to a gas equivalent and added to the gas production. Since the specific gravity of water is 1.00 and its molecular weight is 18, its gas equivalent is

$$GE_w = \frac{nR'T_{sc}}{p_{sc}} = \frac{350.5 \times 1.00}{18} \times \frac{10.73 \times 520}{14.7}$$

$$= 7390 \text{ SCF/surface barrel}$$

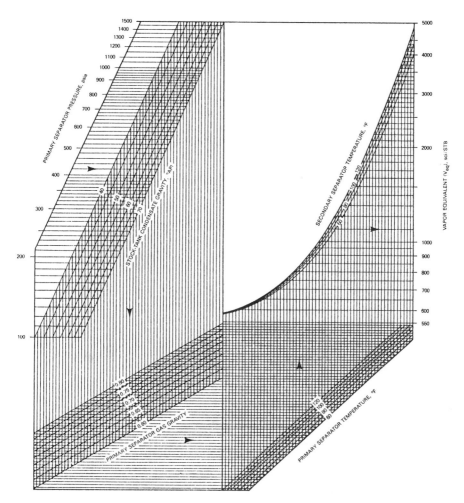

Fig. 3.9. Vapor-Equivalent Nomograph for Three-Stage Separation Systems. (*After* Gold, McCain, and Jennings.[10])

Studies by McCarthy, Boyd, and Reid indicate that the water vapor content of reservoir gases at usual reservoir temperatures and usual *initial* reservoir pressures is in the range of a fraction of one barrel per million standard cubic feet of gas.[11] Production data from a Gulf Coast gas reservoir show a production of 0.64 barrel of water per million standard cubic feet compared with a reservoir *content* of about 1.00 bbl/MM SCF using the data of McCarthy, Boyd, and Reid. The difference is presumably that water remaining in the vapor state at separator temperature and pressure, most of which must be removed by dehydration to a level of about six *pounds* per million standard

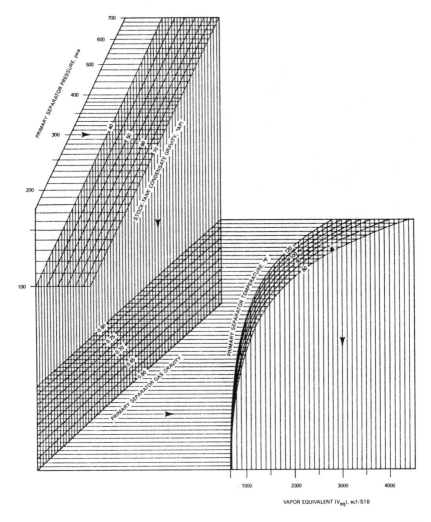

Fig. 3.10. Vapor-Equivalent Nomograph for Two-Stage Separation Systems. (*After* Gold, McCain, and Jennings.[10])

cubic feet. As reservoir pressure declines, the water content increases to as much as three barrels per million standard cubic feet. Since this additional content has come from vaporization of the connate water, it would appear that any *fresh* water produced in excess of the *initial* content should be treated as produced water and taken care of in the W_p term rather than the G_p term. If the water is saline, it definitely is produced water; however, it includes the fraction of a barrel per million cubic feet obtained from the gas phase. If the produced gas is based on the dehydrated gas volume, the gas volume should be increased by the gas equivalent of the water content at the *initial* reservoir pressure and temperature regardless of the subsequent decline in reservoir pressure, and the water production should be diminished by the water content. This amounts to about a 0.05% increase in the produced gas volumes.

7. GAS RESERVOIRS AS STORAGE RESERVOIRS

The demand for natural gas is seasonal. During winter months, there is a much greater demand for natural gas than during the warmer summer months. To meet this variable demand, several means of storing natural gas are used in the industry. One of the best methods of storing natural gas is with the use of depleted gas reservoirs. Gas is injected during the warm summer months when there is an overabundance and produced during the winter months when there is a shortage of supply. Katz and Tek have presented a good overview of this subject.[12]

Katz and Tek listed three primary objectives in the design and operation of a gas storage reservoir: (1) verification of inventory; (2) retention against migration; and (3) assurance of deliverability. Verification of inventory simply means knowing the storage capacity of the reservoir as a function of pressure. This suggests that a p/z plot or some other measure of material balance be known for the reservoir of interest. Retention against migration refers to a monitoring system capable of ascertaining if the injected gas remains in the storage reservoir. Obviously, leaks in casing and so on would be detrimental to the storage process. The operator needs to be assured that the reservoir can be produced during peak demand times in order to provide the proper deliverability. A major concern with the deliverability is that water encroachment not interfere with the gas production. With these design considerations in mind, it is apparent that a good candidate for a storage reservoir would be a depleted volumetric gas reservoir. With a depleted volumetric reservoir, the p/z versus G_p curve is usually known and water influx is not a problem.

Ikoku defines three types of gas involved in a gas storage reservoir.[13] The first is the base gas, or cushion gas, that remains when the base pressure is reached. The base pressure is the pressure at which production is stopped and injection is begun. The second type of gas is the working gas, or working storage, that is produced and injected during the cycle process. The third type is the unused gas that essentially is the unused capacity of the reservoir. Figure 3.11 defines these three types of gas on a p/z plot.

The base pressure, and therefore the amount of base gas, is defined by deliverability needs. Sufficient pressure must be maintained in the reservoir for reservoir gas to be delivered to transporting pipelines. Economics dictates the pressure at which injection of gas during the summer months ends. Compression costs must be balanced with the projected supply and demand of the winter months. In theory, for a volumetric reservoir, the cycles of injection

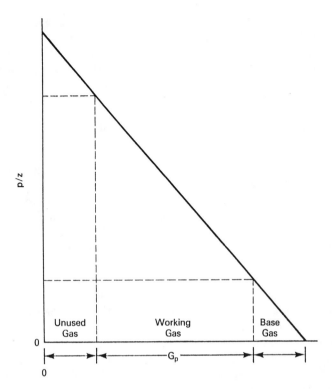

Fig. 3.11. p/z plot showing different types of gas in a gas storage reservoir.

and production simply run up and down the p/z versus G_p curve between the pressure limits just discussed.

In certain applications, the use of the delta pressure concept may be advantageous.[12] The *delta pressure* is defined as the pressure at maximum storage minus the initial reservoir pressure. Under the right conditions, an amount of gas larger than the initial gas in place can be achieved. This again is dictated by the economics of the given situation.

Hollis presented an interesting case history of the considerations involved in changing the Rough Gas Field in the North Sea over to a storage reservoir.[14] Considerations in the design of storage and deliverability rates included the probability of a severe winter occurring in the demand area. A severe winter was given a probability of 1 in 50 of happening. Hollis concluded that the differences between offshore and onshore storage facilities are owing mainly to economics and the integrated planning that must take place in offshore development.

Storage is a useful application of gas reservoirs. We encourage you to pursue the references for more detailed information if it becomes necessary.

8. ABNORMALLY PRESSURED GAS RESERVOIRS

Normal pressure gradients observed in gas reservoirs are in the range of 0.4 to 0.5 psia per foot of depth. Reservoirs with abnormal pressures may have gradients as high as 0.7 to 1.0 pisa per foot of depth.[15,16,17,18] Bernard has reported that over 300 gas reservoirs have been discovered in the offshore Gulf Coast alone with initial gradients in excess of 0.65 psia per foot of depth in formations over 10,000 feet deep.[18]

When the water and formation compressibility term in the material balance equation can be ignored, the normal p/z behavior for a volumetric gas reservoir plots a straight line versus cumulative gas produced (Fig. 3.6, Sect. 5). This is not the case for an abnormally pressured gas reservoir, as can be seen in Figure 3.12, which illustrates the p/z behavior for this type of reservoir.

For an abnormally pressured volumetric reservoir, the p/z plot is a straight line during the early life of production, but then it usually curves downward during the later stages of production. If the early data are used to extrapolate for G or for an abandonment G_p, the extrapolation can yield significant errors.

To explain the curvature in the p/z plot for abnormally pressured reservoirs, Harville and Hawkins postulated a "rock collapse" theory that used a high rock compressibility at abnormally high pressures and a reduced rock compressibility at normal reservoir pressures.[15] However, working with rock

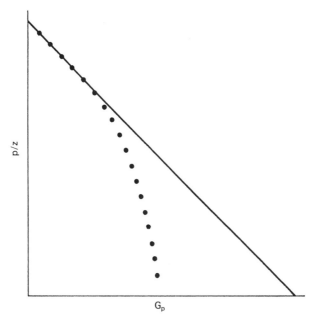

Fig. 3.12. p/z plot illustrating nonlinear behavior of abnormally pressured reservoir.

samples taken from abnormally pressured reservoirs, Jogi, Gray, Ashman, and Thompson and Sinha, Holland, Borshcel, and Schatz reported rock compressibilities measured at high pressures in the order of 2 to 5 $(10)^{-6}$ psi^{-1}.[19, 20] These values are representative of typical values at low pressures and suggest that rock compressibilities do not change with pressure. Ramagost and Farshad showed that in some cases the p/z data could be adjusted to yield straight-line behavior by including the water and formation compressibility term.[21] Bourgoyne, Hawkins, Lavaquial, and Wickenhauser suggested that the non-linear behavior could be due to water influx from shales.[22]

Bernard has proposed a method of analyzing the p/z curve for abnormally pressured reservoirs to determine initial gas in place and gas reserve as a function of abandonment p/z.[18] The method uses two approaches. The first involves the early production life when the p/z plot exhibits linear behavior. Bernard developed a correlation for actual gas in place as a function of apparent gas in place as shown in Fig. 3.13.

The apparent gas in place is obtained by extrapolating the early, linear p/z data. Then, by entering Figure 3.13 with the apparent value of the gas in place, the ratio of actual gas in place to apparent gas in place can be obtained. The correlation appears to be reasonably accurate for the reservoirs that Bernard studied. For later production times, when the p/z data exhibit non-linear behavior, Bernard defined a constant C' according to Eq. (3.26)

$$\frac{p}{z}(1 - C'\Delta p) = \frac{p_i}{z_i} - \frac{p_i}{z_i G}G_p \qquad (3.26)$$

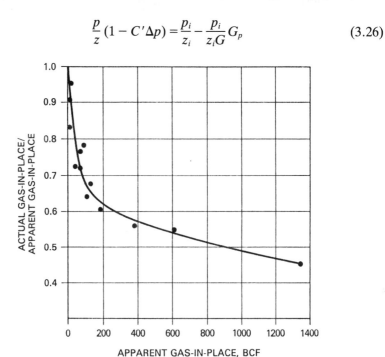

Fig. 3.13. Correction for initial gas in place. (*After* Bernard.[18])

where

C' = constant

Δp = total pressure drop in the reservoir, $p_i - p$

In Eq. (3.26), the only unknowns are C' and G. Bernard suggested that they can be found by the following procedure:

First, calculate A' and B' from:

$$A' = \frac{\left(\dfrac{p}{z} - \dfrac{p_i}{z_i}\right)}{\Delta p(p/z)} \quad \text{and} \quad B' = \frac{\left(\dfrac{p_i}{z_i}\right)G_p}{\Delta p(p/z)}$$

If there are n data points for p/z and G_p, then C' and G can be calculated from the following equations:

$$G = \frac{\Sigma B' \Sigma B'/n - \Sigma (B'^2)}{\Sigma A'B' - \Sigma A' \Sigma B'/n}$$

$$C' = \Sigma A'/n + \left(\frac{1}{G}\right)\Sigma B'/n$$

9. LIMITATIONS OF EQUATIONS AND ERRORS

The precision of the reserve calculations by the volumetric method depends on the accuracy of the data that enter the computations. The precision of the initial gas in place depends on the probable errors in the averages of the porosity, connate water, pressure, and gas deviation factor, and in the error in the determination of the bulk productive volume. With the best of core and log data in rather uniform reservoirs, it is doubtful that the initial gas in place can be calculated more accurately than about 5%, and the figure will range upward to 100% or higher depending on the uniformity of the reservoir and the quantity and quality of the data available.

The reserve is the product of the gas in place and the recovery factor. For volumetric reservoirs, the reserve of the reservoir as a whole, for any selected abandonment pressure, should be known to about the same precision as the initial gas in place. Water-drive reservoirs require, in addition, the estimate of the volume of the reservoir invaded at abandonment and the average residual gas saturation. When the reservoir exhibits permeability stratification, the difficulties are increased, and the accuracy is therefore reduced. In general, reserve calculations are more accurate for volumetric than for water-drive reservoirs. When the reserves are placed on a well or lease basis, the accuracy may be reduced further because of lease drainage, which occurs in both volumetric and water-drive reservoirs.

The use of the material balance equation to calculate gas in place involves the terms of the gas volume factor. The precision of the calculations is, of course, a function of the probable error in these terms. The error in gas production G_p arises from error in gas metering, in the estimate of lease use and leakage, and in the estimate of the low-pressure separator or stock tank gases. Sometimes underground leakage occurs—from the failure in casing cementing, from casing corrosion, or, in the case of dual completions, from leakage between the two zones. When gas is commingled from two reservoirs at the surface prior to metering, the division of the total between the two reservoirs depends on periodic well tests, which may introduce additional inaccuracies. Meters are usually calibrated to an accuracy of 1%, and therefore it is doubtful that the gas production under the best of circumstances is known closer than 2%. Average accuracies are in the range of a few to several percentage points.

Pressure errors are a result of gauge errors and the difficulties in averaging, particularly when there are large pressure differences throughout the reservoir. When reservoir pressures are estimated from measured wellhead pressures, the errors of this technique enter the calculations. When the field is not fully developed, the average pressure is, of course, of the developed portion, which is lower than that of the reservoir as a whole. Water production with gas wells is frequently unreported when the amount is small; when it is appreciable, it is often estimated from periodic well tests.

Under the best of circumstances, the material balance estimates of the gas in place are seldom more accurate than 5% and may range much higher. The estimate of reserves is, of course, one step removed.

PROBLEMS

3.1 A volumetric gas field has an initial pressure of 4200 psia, a porosity of 17.2%, and connate water of 23%. The gas volume factor at 4200 psia is 0.003425 cu ft/SCF and at 750 psia is 0.01852 cu ft/SCF.

   (a) Calculate the initial in-place gas in standard cubic feet on a unit basis.

   (b) Calculate the initial gas reserve in standard cubic feet on a unit basis, assuming an abandonment pressure of 750 psia.

   (c) Explain why the calculated initial reserve depends on the abandonment pressure selected.

   (d) Calculate the initial reserve of a 640-acre unit whose average net productive formation thickness is 34 ft, assuming an abandonment pressure of 750 psia.

   (e) Calculate the recovery factor based on an abandonment pressure of 750 psia.

3.2 The discovery well No. 1 and wells No. 2 and No. 4 produce gas in the 7500-ft reservoir of the Echo Lake Field (Fig. 3.14). Wells Nos. 3 and 7 were dry in the 7500-ft reservoir; however, together with their electric logs and the one from well No. 1, the fault that seals the northeast side of the reservoir was established. The logs of wells Nos. 1, 2, 4, 5, and 6 were used to construct the map of Fig. 3.14, which was used to locate the gas-water contact and to determine the

average net sand thickness. The reservoir had been producing for 18 months
when well No. 6 was drilled at the gas-water contact. The static wellhead pressures of the production wells showed virtually no decline during the 18-month
period before drilling well No. 6 and averaged near 3400 psia. The following data
were available from electric logs, core analysis, and the like.

Average well depth = 7500 ft

Average static wellhead pressure = 3400 psia

Reservoir temperature = 175°F

Gas specific gravity = 0.700

Average porosity = 27%

Average connate water = 22%

Standard conditions = 14.7 psia and 60°F

Bulk volume of productive reservoir rock at the time No. 6 was
drilled = 22,500 ac-ft

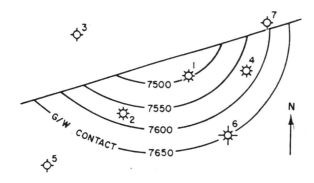

Fig. 3.14. Echo Lake Field, subsurface map, 7500 ft reservoir.

(a) Calculate the reservoir pressure.

(b) Estimate the gas deviation factor and the gas volume factor.

(c) Calculate the reserve at the time well No. 6 was drilled, assuming a residual
gas saturation of 30%.

(d) Discuss the location of well No. 1 with regard to the overall gas recovery.

(e) Discuss the effect of sand uniformity on overall recovery—for example, a
uniform permeable sand versus a sand in two beds of equal thickness, one
of which has a permeability of 500 md and the other, 100 md.

3.3 The M Sand is a small gas reservoir with an initial bottom-hole pressure of 3200
psia and bottom-hole temperature of 220°F. It is desired to inventory the gas in
place at three production intervals. The pressure-production history and gas
volume factors are as follows:

Due
2/15

Pressure psia	Cumulative Gas Production MM SCF	Gas Volume Factor cu ft/SCF
3200	0	0.0052622
2925	79	0.0057004
2525	221	0.0065311
2125	452	0.0077360

(a) Calculate the initial gas in place using production data at the end of each of the production intervals, assuming volumetric behavior.

(b) Explain why the calculations of part (a) indicate a water drive.

(c) Show that a water drive exists by plotting the cumulative production versus p/z.

(d) Based on electric log and core data, volumetric calculations on the M Sand showed that the initial volume of gas in place is 1018 MM SCF. If the sand is under a partial water drive, what is the volume of water encroached at the end of each of the periods? There was no appreciable water production.

3.4 When the Sabine Gas Field was brought in, it had a reservoir pressure of 1700 psia, and a temperature of 160°F. After 5.00 MMM SCF was produced, the pressure had fallen to 1550 psia. If the reservoir is assumed to be under volumetric control, using the deviation factors of Prob. 1.10, calculate the following:

(a) The hydrocarbon pore volume of the reservoir.

(b) The SCF produced when the pressure falls to 1550, 1400, 1100, 500, and 200 psia. Plot cumulative recovery in SCF versus p/z.

(c) The SCF of gas initially in place.

(d) From your graph, find how much gas can be obtained without the use of compressors for delivery into a pipeline operating at 750 psia.

(e) What is the approximate pressure drop per MMM SCF of production?

(f) Calculate the minimum value of the initial reserve if the produced gas measurement is accurate to ± 5% and if the average pressures are accurate to ± 12 psi when 5.00 MMM SCF have been produced and the reservoir pressure has dropped to 1550 psia.

3.5 If, however, during the production of 5.00 MMM SCF of gas in the preceding problem, 4.00 MM bbl of water had encroached into the reservoir and still the pressure had dropped to 1550 psia, calculate the initial in-place gas. How does this compare with Prob. 3.4(c)?

3.6 (a) The gas cap of the St. John Oil Field had a bulk volume of 17,000 ac-ft when the reservoir pressure had declined to 634 psig. Core analysis shows an average porosity of 18% and an average interstitial water of 24%. It is desired to increase the recovery of oil from the field by repressuring the gas cap to 1100 psig. Assuming that no additional gas dissolves in the oil during repressuring, calculate the SCF required. The deviation factors for both the reservoir gas and the injected gas are 0.86 at 634 psig and 0.78 at 1100 psig, both at 130°F.

(b) If the injected gas has a deviation factor of 0.94 at 634 psig and 0.88 at 1100 psig, and the reservoir gas deviation factors are as above, recalculate the injected gas required.

(c) Is the assumption that no additional solution gas will enter the reservoir oil a valid one?

(d) Considering the possibility of some additional solution gas and the production of oil during the time of injection, will the figure of part (a) be maximum or minimum? Explain.

(e) Explain why the gas deviation factors are higher (closer to unity) for the injected gas in part (b) than for the reservoir gas.

3.7 The following production data are available from a gas reservoir produced under volumetric control:

Pressure (psia)	Cumulative Gas Production (MMM SCF)
5000	200
4000	420

The initial reservoir temperature was 237°F and the reservoir gas gravity is 0.7.

(a) What will be the cumulative gas production at 2500 psia?

(b) What fraction of the initial reservoir gas will be produced at 2500 psia?

(c) What was the initial reservoir pressure?

3.8 **(a)** A well drilled into a gas cap for gas recycling purposes is found to be in an isolated fault block. After 50 MM SCF was injected, the pressure increased from 2500 to 3500 psia. Deviation factors for the gas are 0.90 at 3500 and 0.80 at 2500 psia and the bottom-hole temperature is 160°F. What is the cubic feet of gas storage space in the fault block?

(b) If the average porosity is 16%, average connate water is 24%, and average sand thickness is 12 ft, what is the areal extent of the fault block?

3.9 The initial volume of gas in place in the P Sand reservoir of the Holden Field is calculated from electric log and core data to be 200 MMM SCF underlying 2250 productive acres, at an initial pressure of 3500 psia and 140°F. The pressure-production history is

Pressure, psia	Production MMM SCF	Gas Deviation Factor at 140°F
3500 (initial)	0.0	0.85
2500	75.0	0.82

(a) What is the initial volume of gas in place as calculated from the pressure-production history, assuming no water influx?

(b) Assuming uniform sand thickness, porosity, and connate water, if the volume of gas in place from pressure-production data is believed to be correct, how many acres of extension to the present limits of the P Sand are predicted?

(c) If, on the other hand, the gas in place calculated from the log and core data is believed to be correct, how much water influx must have occurred during the 75 MMM SCF of production to make the two figures agree?

3.10 Explain why initial calculations of gas in place are likely to be in greater error during the early life of depletion reservoirs. Will these factors make the predictions high or low? Explain.

3.11 A gas reservoir under partial water drive produced 12.0 MMM SCF when the average reservoir pressure had dropped from 3000 psia to 2200 psia. During the same interval, an estimated 5.20 MM bbl of water entered the reservoir based on the volume of the invaded area. If the gas deviation factor at 3000 psia and bottom-hole temperature of 170°F is 0.88 and at 2200 psia is 0.78, what is the initial volume of gas in place measured at 14.7 psia and 60°F?

3.12 A gas-producing formation has uniform thickness of 32 ft, a porosity of 19%, and connate water saturation of 26%. The gas deviation factor is 0.83 at the initial reservoir pressure of 4450 psia and reservoir temperature of 175°F.

(a) Calculate the initial in-place gas per acre-foot of bulk reservoir rock.

(b) How many years will it take a well to deplete by 50% a 640 ac unit at the rate of 3 MM SCF/day?

(c) If the reservoir is under an active water drive so that the decline in reservoir pressure is negligible, and during the production of 50.4 MMM SCF of gas water invades 1280 acres, what is the percentage of recovery by water drive?

(d) What is the gas saturation as a percentage of total pore space in the portion of the reservoir invaded by water?

3.13 Fifty billion standard cubic feet of gas has been produced from a dry gas reservoir since its discovery. The reservoir pressure during this production has dropped to 3600 psia. Your company, which operates the field, has contracted to use the reservoir as a gas storage reservoir. A gas with a gravity of 0.75 is to be injected until the average pressure reaches 4800 psia. Assume that the reservoir behaves volumetrically, and determine the amount of SCF of gas that must be injected to raise the reservoir pressure from 3600 to 4800 psia. The initial pressure and temperature of the reservoir were 6200 psia and 280°F, respectively, and the specific gravity of the reservoir gas is 0.75.

3.14 The production data for a gas field are given below. Assume volumetric behavior and calculate the following:

Due
2/18

(a) Determine the initial gas in place.

(b) What percentage of the initial gas in place will be recovered at a p/z of 1000?

(c) The field is to be used as a gas storage reservoir into which gas is injected during summer months and produced during the peak demand months of the winter. What is the minimum p/z value that the reservoir needs to be brought back up to if a supply of 50 MMM SCF of gas is required and the abandonment p/z is 1000?

p/z *(psia)*	G_p *(MMM SCF)*
6553	0.393
6468	1.642
6393	3.226
6329	4.260
6246	5.504
6136	7.538
6080	8.749

3.15 Calculate the daily gas production including the condensate and water gas equivalents for a reservoir with the following daily production:

Due
2/19

Separator gas production = 6 MM SCF

Condensate production = 100 STB

Stock tank gas production = 21 M SCF

Fresh water production = 10 bbl

Initial reservoir pressure = 6000 psia

Current reservoir pressure = 2000 psia

Reservoir temperature = 225°F

Water vapor content of 6000 psia and 225°F = 0.86 bbl/MM SCF

Condensate gravity = 50°API

REFERENCES

[1] Harold Vance, *Elements of Petroleum Subsurface Engineering* St. Louis, MO: Educational Publishers, 1950.

[2] L. W. LeRoy, *Subsurface Geologic Methods,* 2nd ed. Golden, CO: Colorado School of Mines, 1950.

[3] W. A. Bruce and H. J. Welge, "The Restored-State Method for Determination of Oil in Place and Connate Water," *Drilling and Production Practice,* API (1947), 166–173.

[4] H. H. Kaveler, "Engineering Features of the Schuler Field and Unit Operation," *Trans.* AIME (1944), **155,** 73.

[5] T. M. Geffen, D. R. Parrish, G. W. Haynes, and R. A. Morse, "Efficiency of Gas Displacement from Porous Media by Liquid Flooding," *Trans.* AIME (1952), **195,** 37.

[6] R. Agarwal, R. Al-Hussainy, and H. J. Ramey, "The Importance of Water Influx in Gas Reservoirs," *Jour. of Petroleum Technology* (Nov. 1965), 1336.

[7] G. Matthes, R. F. Jackson, S. Schüler, and O. P. Marudiak, "Reservoir Evaluation and Deliverability Study, Bierwang Field, West Germany," *Jour. of Petroleum Technology* (Jan. 1973), 23.

[8] T. L. Lutes, C. P. Chiang, M. M. Brady, and R. H. Rossen, "Accelerated Blowdown of a Strong Water Drive Gas Reservoir," SPE 6166. Presented at the 51st Annual Fall Meeting of the Society of Petroleum Engineers of AIME, New Orleans, Oct. 3–6, 1976.

[9] D. P. Arcano and Z. Bassiouni, "The Technical and Economic Feasibility of Enhanced Gas Recovery in the Eugene Island Field by Use of the Coproduction Technique," *Jour. of Petroleum Technology* (May 1987), 585.

[10] D. K. Gold, W. D. McCain, J. W. Jennings, "An Improved Method for the Determination of the Reservoir Gas Specific Gravity for Retrograde Gases," *Jour. of Petroleum Technology* (July 1989).

[11] Eugene L. McCarthy, William L. Boyd, and Lawrence S. Reid, "The Water Vapor Content of Essentially Nitrogen-Free Natural Gas Saturated at Various Conditions of Temperature and Pressure," *Trans.* AIME (1950), **189**, 241–242.

[12] D. I. Katz and M. R. Tek, "Overview on Underground Storage of Natural Gas," *Jour. of Petroleum Technology* (June 1981), 943.

[13] Chi U. Ikoku, *Natural Gas Reservoir Engineering*. New York: Wiley, 1984.

[14] A. P. Hollis, "Some Petroleum Engineering Considerations in the Changeover of the Rough Gas Field to the Storage Mode," *Jour. of Petroleum Technology* (May 1984), 797.

[15] D. W. Harville and M. F. Hawkins, Jr., "Rock Compressibility and Failure as Reservoir Mechanisms in Geopressured Gas Reservoirs," *Jour. of Petroleum Technology* (Dec. 1969), 1528.

[16] I. Fatt, "Compressibility of Sandstones at Low to Moderate Pressures," Bulletin, AAPG (1954), No. 8, 1924.

[17] J. O. Duggan, "The Anderson 'L'—An Abnormally Pressured Gas Reservoir in South Texas," *Jour. of Petroleum Technology* (Feb. 1972), 132.

[18] W. J. Bernard, "Reserves Estimation and Performance Prediction For Geopressured Reservoirs," *Jour. of Petroleum Science and Engineering* (1987), **1**, 15.

[19] P. N. Jogi, K. E. Gray, T. R. Ashman, and T. W. Thompson, "Compaction Measurements of Cores from the Pleasant Bayou Wells." Proc. 5th Conf. Geopressured-Geothermal Energy, Oct. 13–15, 1981, Baton Rouge, LA.

[20] K. P. Sinha, M. T. Holland, T. F. Borschel, and J. P. Schatz, "Mechanical and Geological Characteristics of Rock Samples from Sweezy No. 1 Well at Parcperdue Geopressured/Geothermal Site." Report of Terra Tek Inc. to Dow Chemical Co., U.S. Department of Energy, 1981.

[21] B. P. Ramagost and F. F. Farshad, "*P/Z* Abnormally Pressured Gas Reservoirs." Paper SPE 10125 presented at the SPE Fall Meeting, San Antonio, Oct. 5–7, 1981.

[22] A. T. Bourgoyne, M. F. Hawkins, F. P. Lavaquial, and T. L. Wickenhauser, "Shale Water as a Pressure Support Mechanism in Superpressure Reservoirs." Paper SPE 3851, 1981.

Chapter 4

Gas-Condensate Reservoirs

1. INTRODUCTION

Gas-condensate production may be thought of as being intermediate between oil and gas. Oil reservoirs have a dissolved gas content in the range of zero (dead oil) to a few thousand cubic feet per barrel, whereas in gas reservoirs 1 bbl of liquid (condensate) is vaporized in 100,000 SCF of gas or more, and from which, therefore, a small or negligible amount of hydrocarbon liquid is obtained in surface separators. Gas-condensate production is predominantly *gas* from which more or less liquid is *condensed* in the surface separators, hence the name gas-condensate. the liquid is sometimes called by an older name, *distillate,* and also sometimes simply oil because it is an oil. Gas-condensate reservoirs may be approximately defined as those that produce light-colored or colorless stock tank liquids with gravities above 45° API at gas-oil ratios in the range of 5000 to 100,000 SCF/bbl. Allen has pointed out the inadequacy of classifying wells and the reservoirs from which they produce entirely on the basis of surface gas-oil ratios, for the classification of reservoirs, as discussed in Chapter 1, properly depends on (a) the composition of the hydrocarbon accumulation and (b) the temperature and pressure of the accumulation in the earth.[*,1]

* References throughout the text are given at the end of each chapter.

Table 4.1 presents the mole compositions and some additional properties of five single-phase reservoir fluids. The volatile oil is intermediate between the gas-condensate and the black, or heavy, oil types. Production with gas-oil ratios greater than 100,000 SCF/bbl is commonly called *lean* or *dry gas,* although there is no generally recognized dividing line between the two categories. In some legal work, statutory gas wells are those with gas-oil ratios in excess of 100,000 SCF/bbl (i.e., 10 bbl/M MCF). The term *wet gas* is sometimes used as being more or less equivalent to gas-condensate. In the gas-oil ratios, general trends are noticeable in the methane and heptanes-plus content of the fluids and the color of the tank liquids. Although there is good correlation between the molecular weight of the heptanes-plus and the gravity of the stock tank liquid, there is virtually no correlation between the gas-oil ratios and the gravities of the stock tank liquids, except that most black oil reservoirs have gas-oil ratios below 1000 SCF/bbl and stock tank liquid gravities below 45° API. The gas-oil ratios are a good indication of the overall composition of the fluid, high gas-oil ratios being associated with low concentrations of pentanes and heavier, and vice versa.

The gas-oil ratios given in Table 4.1 are for the initial production of the one-phase reservoir fluids producing through one or more surface separators operating at various temperatures and pressures, which may vary considerably among the several types of production. The gas-oil ratios and consequently the API gravity of the produced liquid vary with the number, pressures, and temperatures of the separators so that one operator may report a somewhat different gas-oil ratio from another, although both produce the same reservoir fluid. Also, as pressure declines in the black oil, volatile oil, and some gas-

TABLE 4.1.

Mole composition and other properties of typical single-phase reservoir fluids

Component	Black Oil	Volatile Oil	Gas-Condensate	Dry Gas	Gas
C_1	48.83	64.36	87.07	95.85	86.67
C_2	2.75	7.52	4.39	2.67	7.77
C_3	1.93	4.74	2.29	0.34	2.95
C_4	1.60	4.12	1.74	0.52	1.73
C_5	1.15	2.97	0.83	0.08	0.88
C_6	1.59	1.38	0.60	0.12	
C_7^+	42.15	14.91	3.80	0.42	
	100.00	100.00	100.00	100.00	100.00
Mol. wt. C_7^+	225	181	112	157	
GOR, SCF/bbl	625	2000	18,200	105,000	Inf.
Tank gravity, °API	34.3	50.1	60.8	54.7	
Liquid color	Greenish black	Medium orange	Light straw	Water white	

condensate reservoirs, there is generally a considerable increase in the gas-oil ratio owing to the reservoir mechanisms that control the relative flow of oil and gas to the well bores. The separator efficiencies also generally decline as flowing wellhead pressures decline, which also contributes to increased gas-oil ratios.

What has been said previously applies to reservoirs initially in a single phase. The initial gas-oil ratios of production from wells completed either in the gas cap or in the oil zone of two-phase reservoirs depends, as discussed previously, on the compositions of the gas cap hydrocarbons and the oil zone hydrocarbons as well as the reservoir temperature and pressure. The gas cap may contain gas-condensate or dry gas, whereas the oil zone may contain black oil or volatile oil. Naturally, if a well is completed in both the gas and oil zones, the production will be a mixture of the two. Sometimes this is unavoidable, as when the gas and oil zones (columns) are only a few feet in thickness. Even when a well is completed in the oil zone only, the downward coning of gas from the overlying gas cap may occur to increase the gas-oil ratio of the production.

When deeper drilling began in many areas, the trend to discoveries was toward reservoirs of the gas and gas-condensate types. Figure 4.1 based on well test data reported in *Ira Rinehart's Yearbooks* shows the discovery trend for 17 parishes in southwest Louisiana for 1952–1956 inclusive.[2] The reservoirs were separated into oil and gas or gas-condensate types on the basis of well test

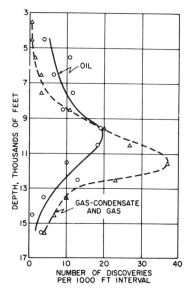

Fig. 4.1. Discovery frequency of oil and gas or gas-condensate reservoirs versus depth. For 17 parishes in southwest Louisiana, 1952–1956, inclusive. (Data from *Ira Rinehart's Yearbooks.*[2])

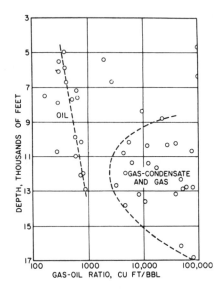

Fig. 4.2. Plot showing trend of increase of gas-oil ratio versus depth. For 17 parishes in southwest Louisiana during 1955. (Data from *Ira Rinehart's Yearbooks.*[2])

gas-oil ratios and the API gravity of the produced liquid. Oil discoveries predominated at depths less than 8000 ft, but gas and gas-condensate discoveries predominated below 10,000 ft. The decline in discoveries below 12,000 ft was due to the fewer number of wells drilled below that depth rather than to a drop in the occurrence of hydrocarbons. Figure 4.2 shows the same data for the year 1955 with the gas-oil ratio plotted versus depth. The dashed line marked "oil" indicates the general trend to increased solution gas in oil with increasing pressure (depth), and the envelop to the lower right encloses those discoveries that were of the gas or gas-condensate types.

Muskat, Standing, Thornton, and Eilerts, have discussed the properties and behavior of gas-condensate reservoirs.[3-6] Table 4.2, taken from Eilerts, shows the distribution of the gas-oil ratio and the API gravity among 172 gas and gas-condensate fields in Texas, Louisiana, and Mississippi.[6] These authors found no correlation between the gas-oil ratio and the API gravity of the tank liquid for these fields.

2. CALCULATION OF INITIAL GAS AND OIL

The initial gas and oil (condensate) for gas-condensate reservoirs, both retrograde and nonretrograde, may be calculated from generally available field data by recombining the produced gas and oil in the correct ratio to find the

TABLE 4.2.

Range of gas-oil ratios and tank oil gravities for 172 gas and gas-condensate fields in Texas, Louisiana, and Mississippi

Range of Liquid-Gas Ratio, GPM[a]	Range of Gas-Oil Ratio M SCF/bbl	Number of Fields				Per Cent of Total
		Texas	Louisiana	Mississippi	Total	
<0.4	>105	38	12	7	57	33.1
0.4 to 0.8	52.5 to 105	33	18	4	55	32.0
0.8 to 1.2	35.0 to 52.5	12	15	5	32	18.6
1.2 to 1.6	26.2 to 35.0	1	8	1	10	5.8
1.6 to 2.0	21.0 to 26.2	1	3	1	5	2.9
>2.0	<21.0	2	5	6	13	7.6
		87	61	24	172	100.0
	Range of Stock Tank Gravity, °API					
	<40	2	1	0	3	1.8
	40–45	4	2	0	6	3.6
	45–50	12	12	0	24	14.6
	50–55	23	17	7	47	28.5
	55–60	24	13	12	49	29.7
	60–65	19	8	3	30	18.2
	>65	3	1	2	6	3.6
		87	54	24	165	100.0

[a] Gallons per 1000 SCF.

average specific gravity (air = 1.00) of the total well fluid, which is presumably being produced initially from a one-phase reservoir. Consider the two-stage separation system shown in Fig. 3.8. The average specific gravity of the total well fluid is given by Eq. (4.1):

$$\gamma_w = \frac{R_1\gamma_1 + 4602.\gamma_o + R_3\gamma_3}{R_1 + \dfrac{133{,}316\gamma_o}{M_{wo}} + R_3} \tag{4.1}$$

where,

R_1, R_3—producing gas-oil ratios from the separator (1) and stock tank (3)

γ_1, γ_3—specific gravities of separator and stock tank gases

γ_o—specific gravity of the stock tank oil (water = 1.00). This is given by:

$$\gamma_o = \frac{141.5}{\rho_{o,\text{API}} + 131.5} \tag{4.2}$$

M_{wo}—molecular weight of the stock tank oil that is given by Eq. (3.22):

$$M_{wo} = \frac{5954}{\rho_{o,\text{API}} - 8.811} = \frac{42.43\gamma_o}{1.008 - \gamma_o} \tag{3.22}$$

Example 4.1 shows the use of Eq. (4.1) to calculate the initial gas and oil in place per acre-foot of a gas-condensate reservoir from the usual production data. The three example problems in this chapter represent the type of calculations that an engineer would perform on data generated from laboratory tests on reservoir fluid samples from gas-condensate systems. Sample reports containing additional example calculations may be obtained from commercial laboratories which conduct PVT studies. The engineer dealing with gas-condensate reservoirs should obtain these sample reports to supplement the material in this chapter. The gas deviation factor at initial reservoir temperature and pressure is estimated from the gas gravity of the recombined oil and gas as shown in Chapter 1. From the estimated gas deviation factor and the reservoir temperature, pressure, porosity, and connate water, the moles of hydrocarbons per acre-foot can be calculated, and from this the initial gas and oil in place.

Example 4.1. Calculate the initial oil and gas in place per acre-foot for a gas-condensate reservoir.

Given:

Initial pressure	2740 psia
Reservoir temperature	215°F = 675°R
Average porosity	25%
Average connate water	30%
Daily tank oil	242 STB
Oil gravity, 60°F	48.0°API
Daily separator gas	3100 MCF
Separator gas gravity	0.650
Daily tank gas	120 MCF
Tank gas gravity	1.20

SOLUTION:

$$\gamma_o = \frac{141.5}{48.0 + 131.5} = 0.788$$
$$\underset{API}{\uparrow}$$

$$M_{wo} = \frac{5954}{\rho_{o,API} - 8.811} = \frac{5954}{48.0 - 8.811} = 151.9$$

$$R_1 = \frac{3,100,000}{242} = 12,810$$

$$\underset{\text{Tank oil}}{\curvearrowleft}$$

$$R_2 = \frac{120,000}{242} = 496$$

$\frac{133,316}{4602} = 28.97 = MW_{air}$

$$\gamma_w = \frac{12,180(0.650) + 4602.(0.788) + 496(1.20)}{12,810 + \dfrac{133,316(0.788)}{151.9} + 496} = 0.896$$

From Fig. 1.4, $T_c = 423°R$ and $p_c = 637$ psia. Then $T_r = 1.60$ and $p_r = 4.30$, from which, using Fig. 1.5, the gas deviation factor is 0.825 at the initial conditions. Then the total initial gas in place per acre-foot of bulk reservoir is

SCF/lbmole   SCF/lbmole p ft^3/ac ϕ S_w

$$G = \frac{379.4 \, pV}{zR'T} = \frac{379.4(2740)(43560)(0.25)(1 - 0.30)}{0.825(10.73)(675)} = 1326 \; MCF/ac\text{-}ft$$

$\underset{z}{} \quad \underset{R'}{} \quad \underset{°R}{}$

Because the volume fraction equals the mole fraction in the gas state, the fraction of the total produced on the surface as gas is

$$f_g = \frac{n_g}{n_g + n_o} = \frac{\dfrac{R_1}{379.4} + \dfrac{R_3}{379.4}}{\dfrac{R_1}{379.4} + \dfrac{R_3}{379.4} + \dfrac{350\gamma_o}{M_{wo}}}$$

(4.3)

$\left(62.37 \% \dfrac{lb\,oil}{cu\,ft\,oil}\right)\left(5.615 \dfrac{cu\,ft\,oil}{STB}\right)$

$$f_g = \frac{\dfrac{12,810}{379.4} + \dfrac{496}{379.4}}{\dfrac{12,810}{379.4} + \dfrac{496}{379.4} + \dfrac{350(0.788)}{151.9}} = 0.951$$

See McCain
p 202

Then

$$f_g \quad G$$

$$\text{Initial gas in place} = 0.951(1326) = 1261 \; MCF/ac\text{-}ft$$

$$\text{Initial oil in place} = \frac{1261(10^3)}{12{,}810 + 496} = 94.8 \; STB/ac\text{-}ft$$

Because the gas production is 95.1% of the total moles produced, the total daily gas-condensate production in MCF is

$$\Delta G_p = \frac{\text{daily gas}}{0.951} = \frac{3100 + 120}{0.951} = 3386 \; MCF/day$$

The total daily reservoir voidage by the gas law is

$$\Delta V = 3{,}386{,}000 \overset{\Delta G_p}{} \left(\frac{\overset{T_R \; P_{sc} \quad Z}{675(14.7)(0.825)}}{\underset{T_{sc} \quad P_R}{520(2740)}} \right) = 19{,}450 \; cu\;ft/day$$

The gas deviation factor of the total well fluid at reservoir temperature and pressure can also be calculated from its composition. The composition of the total well fluid is calculated from the analyses of the produced gas(es) and liquid by recombining them in the ratio in which they are produced. When the composition of the stock tank liquid is known, a unit of this liquid must be combined with the proper amounts of gas(es) from the separator(s) and the stock tank, each of which has its own composition. When the compositions of the gas and liquid in the first or high-pressure separator are known, the shrinkage the separator liquid undergoes in passing to the stock tank must be measured or calculated in order to know the proper proportions in which the separator gas and liquid must be combined. For example, if the volume factor of the separator liquid is 1.20 separator bbl per stock tank barrel and the measured gas-oil ratio is 20,000 SCF of high-pressure gas per bbl of stock tank liquid, then the separator gas and liquid samples should be recombined in the proportions of 20,000 SCF of gas to 1.20 bbl of separator liquid, since 1.20 bbl of separator liquid shrinks to 1.00 bbl in the stock tank.

Example 4.2 shows the calculation of initial gas and oil in place for a gas-condensate reservoir from the analyses of the high pressure gas and liquid, assuming the well fluid to be the same as the reservoir fluid. The calculation is the same as that shown in Ex. 4.1 except that the gas deviation factor of the reservoir fluid is found from the pseudoreduced temperature and pressure,

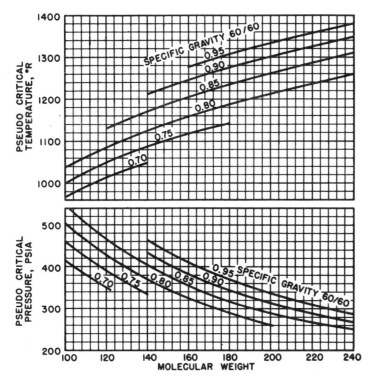

Fig. 4.3. Correlation charts for estimation of the pseudocritical temperature and pressure of heptanes plus fractions from molecular weight and specific gravity. (*After Mathews, Roland, and Katz,*[7] *Proc. NGAA.*)

which are determined from the composition of the total well fluid rather than from its specific gravity. Figure 4.3 presents charts for estimating the pseudocritical temperature and pressure of the heptanes-plus fraction from its molecular weight and specific gravity.

Example 4.2. To calculate the initial gas and oil in place from the compositions of the gas and liquid from the high-pressure separator.

Given:

Reservoir pressure	4350 psia
Reservoir temperature	217°F
Hydrocarbon porosity	17.4 per cent

Std. cond.	15.025 psia, 60°F
Separator gas	842,600 SCF/day
Stock tank oil	31.1 STB/day
Mol. wt. C_7^+ in separator liquid	185.0
Sp. gr. C_7^+ in separator liquid	0.8343
Sp. gr. separator liquid at 880 psig and 60°F	0.7675
Separator liquid volume factor	1.235 bbl at 880 psia/STB, both at 60°F
Compositions of high-pressure gas and liquid	Cols. (2) and (3), Table 4.3
Molar volume at 15.025 psia and 60°F	371.2 cu ft/mole.

SOLUTION: (column numbers refer to Table 4.3):

1. Calculate the mole proportions in which to recombine the separator gas and liquid. Multiply the mole fraction of each component in the liquid, Col. (3), by its molecular weight, Col. (4), and enter the products in Col. (5). The sum of Col. (5) is the molecular weight of the separator liquid, 127.48. Because the specific gravity of the separator liquid is 0.7675 at 880 psig and 60°F, the moles per barrel is

$$\frac{0.7675 \times 350 \text{ lb/bbl}}{127.48 \text{ lb/mole}} = 2.107 \text{ moles/bbl for the separator liquid}$$

TABLE 4.3.
Calculations for Example 4.2 on gas-condensate fluid

(1)			(4)	(5)	(6)	(7)	(8)
					Table 1.1		*Moles*
	Mole				*G/lbmole/42gal*	*(3) × (6)*	*of Each*
	Composition					*bbl of*	*Component*
	of Separator				*Liquid bbl*	*Each*	*in 59.11*
	Fluids				*bbl/mole*	*Component*	*Moles of*
	(2)	*(3)*	*Mol.*	*(3) × (4)*	*for Each*	*per Mole of*	*Gas*
	Gas	*Liquid*	*wt.*	*lb/mole*	*Component*	*sep. liq.*	*(2) × 59.11*
CO_2	0.0120	0.0000					0.709
C_1	0.9404	0.2024	16.04	3.247	0.1317	0.02666	55.587
C_2	0.0305	0.0484	30.07	1.455	0.1771	0.00857	1.803
C_3	0.0095	0.0312	44.09	1.376	0.2480	0.00774	0.562
$i{-}C_4$	0.0024	0.0113	58.12	0.657	0.2948	0.00333	0.142
$n{-}C_4$	0.0023	0.0196	58.12	1.139	0.2840	0.00557	0.136
$i{-}C_5$	0.0006	0.0159	72.15	1.147	0.3298	0.00524	0.035
$n{-}C_6$	0.0003	0.0170	72.15	1.227	0.3264	0.00555	0.018
C_6	0.0013	0.0384	86.17	3.309	0.3706	0.01423	0.077
C_7^+	0.0007	0.6158	185.0	113.923	0.6336[a]	0.39017	0.041
	1.0000	1.0000		127.480		0.46706	59.110

[a] 185 lb/mole ÷ (0.8343 × 350 lb/bbl) = 0.6336 bbl/mole.

The separator liquid rate is 31.1 STB/day × 1.235 sep. bbl/STB so that the separator gas-oil ratio is

$$\frac{842,600}{31.1 \times 1.235} = 21,940 \text{ SCF sep. gas/bbl sep. liquid}$$

$SCF/mole$

Because the 21,940 SCF is 21,940/371.2, or 59.11 moles, the separator gas and liquid must be recombined in the ratio of 59.11 moles of gas to 2.107 moles of liquid.

If the specific gravity of the separator liquid is not available, the mole per barrel figure may be calculated as follows. Multiply the mole fraction of each component in the liquid, Col. (3), by its barrel per mole figure, Col. (6), obtained from data in Table 1.1 and enter the product in Col. (7). The sum of Col. (7), 0.46706 is the number of barrels of separator liquid per mole of separator liquid, and the reciprocal is 2.141 moles/bbl (versus 2.107 measured).

2. Recombine 59.11 moles of gas and 2.107 moles of liquid. Multiply the mole fraction of each component in the gas, Col. (2), by 59.11 moles, and enter in Col. (8). Multiply the mole fraction of each component in the liquid, Col. (3), by 2.107 moles, and enter in Col. (9). Enter the sum of the moles of

(9) Moles of Each Component in 2.107 Moles of Liquid (3) × 2.107	(10) Moles of Each Component in 61.217 Moles of Gas and Liquid (8) + (9)	(11) Mole Composition of Total Well Fluid (10) ÷ 61.217	(12) Critical Pressure, psia	(13) Partial Critical Pressure, psia (11) × (12)	(14) Critical Temp. °R	(15) Partial Critical Temp. °R (11) × (14)
0.0000	0.7090	0.0116	1070	12.41	548	6.36
0.4265	56.0135	0.9150	673	615.80	343	313.85
0.1020	1.9050	0.0311	708	22.02	550	17.11
0.0657	0.6277	0.0102	617	6.29	666	6.79
0.0238	0.1658	0.0027	529	1.43	735	1.98
0.0413	0.1773	0.0029	550	1.60	766	2.22
0.0335	0.0685	0.0011	484	0.53	830	0.91
0.0358	0.0538	0.0009	490	0.44	846	0.76
0.0809	0.1579	0.0026	440	1.14	914	2.38
1.2975	1.3385	0.0219	300[6]	6.57	1227[b]	26.87
2.1070	61.2170	1.0000		668.23		379.23

[b] From Fig. 4.3, *after* Mathews, Roland, and Katz for Mol. wt. $C_7^+ = 185$ and sp. gr. = 0.8342.[7]

each component in the gas and liquid, Col. (8) plus Col. (9), in Col. (10). Divide each figure in Col. (10) by the sum of Col. (10), 61.217, and enter the quotients in Col. (11), which is the mole composition of the total well fluid. Calculate the pseudocritical temperature, 379.23°R and pressure 668.23 psia from the composition. From the pseudocriticals, find the pseudoreduced criticals, and then the deviation factor at 4350 psia and 217°F, which is 0.963.

3. Find the gas and oil (condensate) in place per acre-foot of net reservoir rock. From the gas law, the initial moles per acre-foot at 17.4% hydrocarbon porosity is:

$$\frac{pV}{zRT} = \frac{4350 \times (43{,}560 \times 0.174)}{0.963 \times 10.73 \times 677} = 4713 \text{ moles/ac-ft}$$

$$\text{Gas mole fraction} = \frac{59.11}{59.11 + 2.107} = 0.966$$

$$\text{Initial gas in place} = \frac{0.966 \times 4713 \times 371.2}{1000} = 1690 \text{ MCF/ac-ft}$$

$$\text{Initial oil in place} = \frac{(1 - 0.966) \times 4713}{2.107 \times 1.235} = 61.6 \text{ STB/ac-ft}$$

Because the high-pressure gas is 96.6% of the total mole production, the daily gas-condensate production expressed in standard cubic feet is

$$\Delta G_p = \frac{\text{Daily hp gas}}{0.966} = \frac{842{,}600}{0.966} = 872{,}200 \text{ SCF/day}$$

The daily reservoir voidage at 4350 psia is

$$\Delta V = 872{,}200 \times \frac{677}{520} \times \frac{15.025}{4350} \times 0.963 = 3777 \text{ cu ft/day}$$

3. THE PERFORMANCE OF VOLUMETRIC RESERVOIRS

The behavior of single-phase gas and gas-condensate reservoirs has been treated in Chapter 3. Since no liquid phase develops within the reservoir, where the temperature is above the cricondentherm, the calculations are simplified. When the reservoir temperature is below the cricondentherm, however, a liquid phase develops within the reservoir when pressure declines below the dew point owing to retrograde condensation, and the treatment is considerably more complex, even for volumetric reservoirs.

One solution is to duplicate closely the reservoir depletion by laboratory studies on a representative sample of the initial, single-phase reservoir fluid. The sample is placed in a high-pressure cell at reservoir temperature and initial reservoir pressure. During the depletion, the volume of the cell is held constant to duplicate a volumetric reservoir, and care is taken to remove only gas-phase hydrocarbons from the cell because for most reservoirs, the retrograde condensate liquid that forms is trapped as an immobile liquid phase within the pore spaces of the reservoir.

Laboratory experiments have shown that with most rocks the oil phase is essentially immobile until it builds up to a saturation in the range of 10 to 20% of the pore space, depending on the nature of the rock pore spaces and the connate water. Because the liquid saturations for most retrograde fluids seldom exceed 10%, this is a reasonable assumption for most retrograde condensate reservoirs. In this same connection, it should be pointed out that in the vicinity of the well bore retrograde liquid saturations often build up to higher values so that there is two-phase flow, both gas and retrograde liquid. This buildup of liquid occurs as the one-phase gas suffers a pressure drop as it approaches the well bore. Continued flow increases the retrograde liquid saturation until there is liquid flow. Although this phenomenon does not affect the overall performance seriously or enter into the present performance predictions, it can (a) reduce, sometimes seriously, the flow rate of gas-condensate wells and (b) affect the accuracy of well samples taken assuming one-phase flow into the well bore.

The continuous depletion of the gas phase (only) of the cell at constant volume can be closely duplicated by the following more convenient technique. The content of the cell is expanded from the initial volume to a larger volume at a pressure a few hundred psi below the initial pressure by withdrawing mercury from the bottom of the cell, or otherwise increasing the volume. Time is allowed for equilibrium to be established between the gas phase and the retrograde liquid phase that has formed, and for the liquid to drain to the bottom of the cell so that only gas-phase hydrocarbons are produced from the top of the cell. Mercury is injected into the bottom of the cell and gas is removed at the top at such a rate as to maintain constant pressure in the cell. Thus the volume of gas removed, measured at this lower pressure and cell (reservoir) temperature, equals the volume of mercury injected when the hydrocarbon volume, now two-phase, is returned to the initial cell volume. The volume of retrograde liquid is measured, and the cycle—expansion to a next lower pressure followed by the removal of a second increment of gas—is repeated down to any selected abandonment pressure. Each increment of gas removed is analyzed to find its composition, and the volume of each increment of produced gas is measured at subatmospheric pressure to determine the standard volume, using the ideal gas law. From this, the gas deviation factor at cell pressure and temperature may be calculated using the real gas law. Alternatively, the gas deviation factor at cell pressure and temperature may be calculated from the composition of the increment.

Figure 4.4 and Table 4.4 give the composition of a retrograde gas-condensate reservoir fluid at initial pressure and the composition of the gas

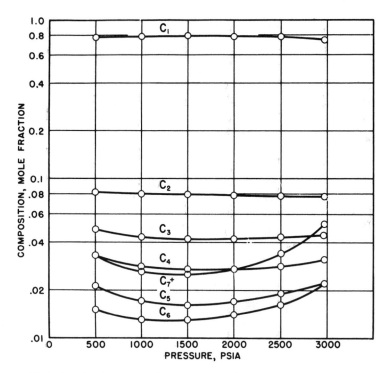

Fig. 4.4. Variations in the composition of the produced gas phase material of a retrograde gas-condensate fluid with pressure decline. (Data from Table 4.4.)

removed from a PVT cell in each of five increments, as previously described. Table 4.4 also gives the volume of retrograde liquid in the cell at each pressure and the gas deviation factor and volume of the produced gas increments at cell pressure and temperature.

The liquid recovery from the gas increments produced from the cell may be measured by passing the gas through small-scale separators, or it may be calculated from the composition for usual field separation methods, or for gasoline plant methods.[8,9,10] Liquid recovery of the pentanes-plus is somewhat greater in gasoline plants than is obtained by field separation, and much greater for the propanes and butanes, commonly called liquified petroleum gas (LPG). For simplicity, the liquid recovery from the gas increments of Table 4.4 is calculated in Ex. 4.3 assuming 25% of the butanes, 50% of the pentanes, 75% of the hexane, and 100% of the heptanes-plus is recovered as liquid.

Example 4.3. Calculating the volumetric depletion performance of a retrograde gas-condensate reservoir based on the laboratory tests given in Table 4.4.

TABLE 4.4.
Volume, composition, and gas deviation factors for a retrograde condensate fluid

(1)	(2)	(3)	(4)	(5)	(6)	(7)	(8)	(9)	(10)	(11)	(12)
	Composition of Produced Gas Increments, Mole Fraction							Produced Gas, cu cm at 195°F and Cell Pressure	Retrograde Liquid Volume, cu cm Cell Volume, 947.5 cu cm	Retrograde Volume, Per Cent of Hydrocarbon Volume	Gas Deviation Factor at 195°F and Cell Pressure
Pressure psia	C_1	C_2	C_3	C_4	C_5	C_6	C_7^+				
2960	0.752	0.077	0.044	0.031	0.022	0.022	0.052	0.0	0.0	0.0	0.771
2500	0.783	0.077	0.043	0.028	0.019	0.016	0.034	175.3	62.5	6.6	0.794
2000	0.795	0.078	0.042	0.027	0.017	0.014	0.027	227.0	77.7	8.2	0.805
1500	0.798	0.079	0.042	0.027	0.016	0.013	0.025	340.4	75.0	7.9	0.835
1000	0.793	0.080	0.043	0.028	0.017	0.013	0.026	544.7	67.2	7.1	0.875
500	0.768	0.082	0.048	0.033	0.021	0.015	0.033	1080.7	56.9	6.0	0.945

$\Delta \forall_{wb}$

Given:

Initial pressure (dew point)	2960 psia
Abandonment pressure	500 psia
Reservoir temperature	195°F
Connate water	30%
Porosity	25%
Standard conditions	14.7 psia and 60°F
Initial cell volume	947.5 cu cm
Mol. wt. of C_7^+ in Initial Fluid	114 lb/lb-mole
Sp. gr. of C_7^+ in Initial Fluid	0.755 at 60°F

Compositions, volumes, and deviation factors given in Table 4.4.

Assume: The same mol. wt. and sp. gr. for the C_7^+ content for all produced gas.

Assume: Liquid recovery from the gas is 25% of the butanes, 50% of the pentane, 75% of the hexane, and 100% of the heptanes and heavier.

SOLUTION: (column numbers refer to Table 4.5):

1. Calculate the increments of gross production in M SCF per ac-ft of net, bulk reservoir rock. Enter in Col. (2):

$$V_{HC} = 43,560 \times 0.25 \times (1 - 0.30) = 7623 \text{ cu ft/ac-ft}$$

For the increment produced from 2960 to 2500 psia, for example,

$$\Delta V = 7623 \times \frac{175.3 \text{ cu cm}}{947.5 \text{ cu cm}} = 1410 \text{ cu ft/ac-ft at 2500 psia and 195°F}$$

$$\Delta G_p = \frac{379.4 \, p\,\Delta V}{1000 \, zRT} = \frac{379.4 \times 2500 \times 1410}{1000 \times 0.794 \times 10.73 \times 655} = 240.1 \text{M SCF}$$

Find the cumulative gross gas production, $G_p = \Sigma \, \Delta G_p$ and enter in Col. (3).

2. Calculate the M SCF of residue gas and the barrels of liquid obtained from each increment of gross gas production. Enter in Col. (4) and Col. (6). Assume that 0.25 C_4, 0.50 C_5, 0.75 C_6, and all C_7^+ is recovered as stock tank liquid. For example, in the 240.1 M SCF produced from 2960 to 2500 psia, the mole fraction recovered as liquid is

$$\Delta n_L = 0.25 \times 0.028 + 0.50 \times 0.019 + 0.75 \times 0.016 + 0.034$$

$$= 0.0070 + 0.0095 + 0.0120 + 0.034 = 0.0625 \text{ mole fraction}$$

TABLE 4.5.
Gas and liquid recoveries in percentage and per acre-foot for Example 4.3

(1)	(2)	(3)	(4)	(5)	(6)	(7)	(8)	(9)	(10)	(11)
Pressure, psia	Increments of Gross Gas Production, M SCF	Cumulative Gross Gas Production, M SCF, Σ(2)	Residue Gas in Each Increment, M SCF	Cumulative Residue Gas Production, M SCF Σ(4)	Liquid in Each Increment, bbl	Cumulative Liquid Production, bbl Σ(6)	Average Gas-Oil Ratio of Each Increment, SCF Residue Gas per bbl, (4) ÷ (6)	Cumulative Gross Gas Recovery, percentage (3) × 100/1580	Cumulative Residue Gas Recovery, percentage (5) × 100/1441	Cumulative Liquid Recovery, percentage (7) × 100/143.2
2960	0	0	0	0	0	0	10,600	0	0	0
2500	240.1	240.1	225.1	225.1	15.3	15.3	14,700	15.2	15.6	10.7
2000	245.2	485.3	232.3	457.4	13.1	28.4	17,730	30.7	31.7	19.8
1500	266.0	751.3	252.8	710.2	13.3	41.7	19,010	47.6	49.3	29.1
1000	270.8	1022.1	256.9	967.1	14.0	55.7	18,350	64.7	67.1	38.9
500	248.7	1270.8	233.0	1200.1	15.9	71.6	14,650	80.4	83.3	50.0

As the mole fraction also equals the volume fraction in gas, the M SCF recovered as liquid from 240.1 M SCF is

$$\Delta G_L = 0.0070 \times 240.1 + 0.0095 \times 240.1 + 0.0120 \times 240.1 + 0.034 \times 240.1$$

$$= 1.681 + 2.281 + 2.881 + 8.163 = 15.006 \text{M SCF}$$

The gas volume can be converted to gallons of liquid using the gal/M SCF figures of Table 1.1 for C_4, C_5 and C_6. The average of the iso and normal compounds is used for C_4 and C_5.

For C_7^+

$$\frac{114 \text{ lb/lb-mole}}{0.3794 \text{M SCF/lb-mole} \times 8.337 \text{ lb/gal} \times 0.755} = 47.71 \text{ gal/M SCF}$$

0.3794 is the molar volume at standard conditions of 14.7 psia and 60°F. Then the total liquid recovered from 240.1M SCF is $1.681 \times 32.04 + 2.281 \times 36.32 + 2.881 \times 41.03 + 8.163 \times 47.71 = 53.9 + 82.8 + 118.2 + 389.5 = 644.4$ gal = 15.3 bbl. The residue gas recovered from the 240.1 M SCF is $240.1 \times (1 - 0.0625) = 225.1$ M SCF. Calculate the cumulative residue gas and stock tank liquid recoveries from Cols. (4) and (6) and enter in Cols. (5) and (7), respectively.

    3. Calculate the gas-oil ratio for each increment of gross production in units of residue gas per barrel of liquid. Enter in Col. (8). For example, the gas-oil ratio of the increment produced from 2960 to 2500 psia is

$$\frac{225.1 \times 1000}{15.3} = 14,700 \text{ SCF/bbl}$$

    4. Calculate the cumulative recovery percentages of gross gas, residue gas, and liquid. Enter in Cols. (9), (10), and (11). The initial gross gas in place is:

$$\frac{379.4 \, pV}{1000 \, zRT} = \frac{379.4 \times 2960 \times 7623}{1000 \times 0.771 \times 10.73 \times 655} = 1580 \text{ M SCF/ac-ft}$$

Of this, the liquid mole fraction is 0.088 and the total liquid recovery is 3.808 gal/M SCF of gross gas, which are calculated from the initial composition in the same manner as shown in Part 2. Then

$$G = (1 - 0.088) \times 1580 = 1441 \text{ M SCF residue gas/ac-ft}$$

$$N = \frac{3.808 \times 1580}{42} = 143.2 \text{ bbl/ac-ft}$$

At 2500 psia, then

$$\text{Gross gas recovery} = \frac{100 \times 240.1}{1580} = 15.2\%$$

$$\text{Residue gas recovery} = \frac{100 \times 225.1}{1441} = 15.6\%$$

$$\text{Liquid recovery} = \frac{100 \times 15.3}{143.2} = 10.7\%$$

The results of the laboratory tests and calculations of Ex. 4.3 are plotted versus pressure in Fig. 4.5. The gas-oil ratio rises sharply from 10,060 SCF/bbl to about 19,000 SCF/bbl near 1600 psia. Maximum retrograde liquid and maximum gas-oil ratios do not occur at the same pressure because, as pointed out previously, the retrograde liquid volume is much larger than its equivalent obtainable stock tank volume, and there is more stock tank liquid in 6.0% retrograde liquid volume at 500 psia than in 7.9% at 1500 psia. Revaporization below 1600 psia helps to reduce the gas-oil ratio; however, there is some doubt that revaporization equilibrium is reached in the reservoir, the field gas-oil ratios generally remain higher than that predicted. Part of this is probably a result of the lower separator efficiency of liquid recovery at the lower pressure and higher separator temperatures. Lower separator temperatures occur at higher wellhead pressures owing to the greater cooling of the gas by free expansion in flowing through the choke. Although the overall recovery at 500 psia abandonment pressure is 80.4%, the liquid recovery is only 50.0%, owing to retrograde condensation. The cumulative production plots are slightly curved because of the variation in the gas deviation factor with pressure and with the composition of the reservoir fluid.

The volumetric depletion performance of a retrograde condensate fluid, such as given in Example 4.3, may also be calculated from the initial composition of the single-phase reservoir fluid, using *equilibrium* ratios. An equilibrium ratio (K) is the ratio of the mole fraction (y) of any component in the vapor phase to the mole fraction (x) of the same component in the liquid phase, or $K = y/x$. These ratios depend on the temperature and pressure, and, unfortunately, on the composition of the system. If a set of equilibrium ratios can be found that are applicable to a given condensate system, then it is possible to calculate the mole distribution between the liquid and vapor phases at any pressure, and reservoir temperature, and also, the composition of the separate vapor and liquid phases, as shown in Fig. 4.4. From the composition and total moles in each phase, it is also possible to calculate with reasonable accuracy the liquid and vapor volumes at any pressure.

The prediction of retrograde condensate performance using equilibrium ratios is a specialized technique. Standing and Rodgers, Harrison, and Regier give methods for adjusting published equilibrium ratio data for condensate

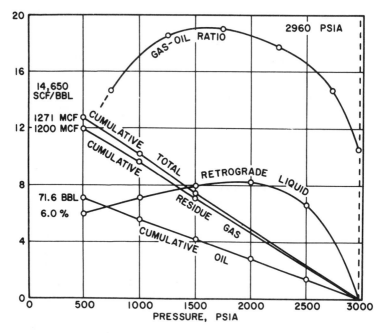

Fig. 4.5. Gas-oil ratios, retrograde liquid volumes, and recoveries for the depletion performance of a retrograde gas-condensate reservoir. (Data from Tables 4.4 and 4.5.)

systems to apply to systems of different compositions.[4,8,11,12,13,14] They also give step-by-step calculation methods for volumetric performance that consist of starting with a unit volume of the initial reservoir vapor of known composition. An increment of vapor phase material is assumed to be removed from the initial volume at constant pressure, and the remaining fluid is expanded to the initial volume. The final pressure, the division in volume between the vapor and retrograde liquid phases, and the individual compositions of the vapor and liquid phases are then calculated using the adjusted equilibrium ratios. A second increment of vapor is removed at the lower pressure, and the pressure, volumes, and compositions are calculated as before. An account is kept of the produced moles of each component so that the total moles of any component remaining at any pressure is known by subtraction from the initial amount. This calculation may be continued down to any abandonment pressure, just as in the laboratory technique.

Standing points out that the prediction of condensate reservoir performance from equilibrium ratios alone is likely to be in considerable error and that some laboratory data should be available to check the accuracy of the adjusted equilibrium ratios.[4] Actually, the equilibrium ratios are changing because the composition of the system remaining in the reservoir or cell changes. The changes in the composition of the heptanes-plus particularly affect the calculations. Rodgers, Harrison, and Regier point out the need for improved procedures for developing the equilibrium ratios for the heavier hydrocarbons to improve the overall accuracy of the calculation.[8]

Jacoby, Koeller, and Berry studied the phase behavior of eight mixtures of separator oil and gas from a lean gas-condensate reservoir at recombined ratios in the range of 2000 to 25,000 SCF/bbl and at several temperatures in the range of 100 to 250°F.[15] The results are useful in predicting the depletion performance of gas-condensate reservoirs for which laboratory studies are not available. They show that there is a gradual change in the surface production performance from the volatile oil to the rich gas-condensate type of reservoir and that a laboratory examination is necessary to distinguish between the dew-point and bubble-point reservoirs in the range of 2000 to 6000 SCF/bbl gas-oil ratios.

4. USE OF MATERIAL BALANCE

The laboratory test on the retrograde condensate fluid in Ex. 4.3 is itself a material balance study of the *volumetric* performance of the reservoir from which the sample was taken. The application of the basic data and the calculated data of Ex. 4.3 to a volumetric reservoir is straightforward. For example, suppose the reservoir had produced 12.05 MMM SCF of gross well fluid when the average reservoir pressure declined from 2960 psia initial to 2500 psia. According to Table 4.5 the recovery at 2500 psia under volumetric depletion is 15.2% of the initial gross gas in place, and therefore the initial gross gas in place is

$$G = \frac{12.05 \times 10^9}{0.152} = 79.28 \text{MMM SCF}$$

Because Table 4.5 shows a recovery of 80.4% down to an abandonment pressure of 500 psia, the initial *recoverable* gross gas or the initial *reserve* is

Initial reserve $= 79.28 \times 10^9 \times 0.804 = 63.74$ MMM SCF

Since 12.05MMM SCF has already been recovered, the *reserve* at 2500 psia is

Reserve at 2500 psia $= 63.74 - 12.05 = 51.69$ MMM SCF

The accuracy of these calculations depends, among other things, on the sampling accuracy and the degree of which the laboratory test represents the volumetric performance. Generally there are pressure gradients throughout a reservoir that indicate that the various portions of the reservoir are in varying stages of depletion. This is due to greater withdrawals in some portions and/or to lower reserves in some portions, because of lower porosities and/or lower net productive thicknesses. As a consequence the gas-oil ratios of the wells differ, and the average composition of the total reservoir production at any prevailing average reservoir pressure does not exactly equal the composition of the total cell production at the same pressure.

Although the gross gas production history of a volumetric reservoir follows the laboratory tests more or less closely, the division of the production into residue gas and liquid follows with less accuracy. This is due to the differences in the stage of depletion of various portions of the reservoir, as explained in the preceding paragraph, and also to the differences between the calculated liquid recoveries in the laboratory tests and the actual efficiency of separators in recovering liquid from the fluid in the field.

The previous remarks apply only to volumetric single-phase gas-condensate reservoirs. Unfortunately, most retrograde gas-condensate reservoirs that have been discovered are initially at their dew-point pressures, rather than above it. This indicates the presence of an oil zone in contact with the gas-condensate cap. The oil zone may be negligibly small or commensurate with the size of the cap; or it may be much larger. The presence of a small oil zone affects the accuracy of the calculations based on the single-phase study and is more serious for a larger oil zone. When the oil zone is of a size at all commensurate with the gas cap, the two must be treated together as a two-phase reservoir, as explained in Chapter 6.

Many gas-condensate reservoirs are produced under a partial or total water drive. When the reservoir pressure stabilizes or stops declining, as occurs in many reservoirs, recovery depends on the value of the pressure at stabilization and the efficiency with which the invading water displaces the gas phase from the rock. The liquid recovery is lower for the greater retrograde condensation, because the retrograde liquid is generally immobile and is trapped together with some gas behind the invading water front. This situation is aggravated by permeability variations because the wells become "drowned" and are forced off production before the less permeable strata are depleted. In many cases, the recovery by water drive is less than by volumetric performance, as explained in Chapter 3, Sect. 4.

When an oil zone is absent or negligible, the material balance Eq. (3.15) may be applied to retrograde reservoirs under both volumetric and water-drive performance, just as for the single-phase (nonretrograde) gas reservoirs for which it was developed:

$$G(B_g - B_{gi}) + W_e = G_p B_g + B_w W_p \qquad (3.15)$$

This equation may be used to find either the water influx W_e or the initial gas in place G. The equation contains the gas deviation factor z at the lower pressure. It is included in the gas volume factor B_g in Eq. 3.15. Because this deviation factor applies to the gas-condensate fluid remaining in the reservoir, when the pressure is below the dew-point pressure in retrograde reservoirs, it is a *two-phase* gas deviation factor. The actual volume in Eq. (1.6) includes the volume of both the gas and liquid phases, and the ideal volume is calculated from the total moles of gas and liquid, assuming ideal gas behavior. For volumetric performance, this two-phase deviation factor may be obtained from such laboratory data as obtained in Example 4.3. For example, from the data of Table 4.6 the cumulative gross gas production down to 2000 psia is 485.3M SCF/ac-ft out of an initial content of 1580.0M SCF/ac-ft. Since the initial hydrocarbon pore volume is 7623 cu ft/ac-ft, Ex. 4.3, the two-phase volume factor for the fluid remaining in the reservoir at 2000 psia and 195°F as calculated using the gas law is

$$z = \frac{379.4 \times pV}{(G - G_p)R'T} = \frac{379.4 \times 2000 \times 7623}{(1580.0 - 485.3)10^3 \times 10.73 \times 655} = 0.752$$

Table 4.6 gives the two-phase gas deviation factors for the fluid remaining in the reservoir at pressures down to 500 psia, calculated as before for the gas-condensate fluid of Ex. 4.3. These data are not strictly applicable when there is some water influx because they are based on cell performance in which vapor equilibrium is maintained between all the gas and liquid remaining in the cell, whereas in the reservoir some of the gas and retrograde liquid are enveloped by the invading water and are prevented from entering into equilibrium with the hydrocarbons in the rest of the reservoir. The deviation factors of Col. (4), Table 4.6, may be used with volumetric reservoirs and, with some reduction in accuracy, with water-drive reservoirs.

When laboratory data such as given in Ex. 4.3 have not been obtained, the gas deviation factors of the initial reservoir gas may be used to approxi-

TABLE 4.6.
Two-phase and single-phase gas deviation factors for the retrograde gas-condensate fluid of Example 4.3

(1)	*(2)*	*(3)*	*(4)*	*(5)*	*(6)*
				Gas Deviation Factors	
Pressure,	$G_p{}^a$	$(G - G_p)^a$		*Initial*	*Produced*
psia	*M SCF/ac-ft*	*M SCF/ac-ft*	*Two-phase*[b]	*Gas*[c]	*Gas*[a]
2960	0.0	1580.0	0.771	0.780	0.771
2500	240.1	1339.9	0.768	0.755	0.794
2000	485.3	1094.7	0.752	0.755	0.805
1500	751.3	828.7	0.745	0.790	0.835
1000	1022.1	557.9	0.738	0.845	0.875
500	1270.8	309.2	0.666	0.920	0.945

[a] Data from Table 4.5 and Example 4.3.

[b] Calculated from the data of Table 4.5 and Example 4.3.

[c] Calculated from initial gas composition using correlation charts.

mate those of the remaining reservoir fluid. These are best measured in the laboratory but may be estimated from the initial gas gravity or well-stream composition using the pseudoreduced correlations. Although the measured deviation factors for the initial gas of Ex. 4.3 are not available, it is believed that they are closer to the two-phase factors in Col. (4) than those given in Col. (5) of Table 4.6, which are calculated using the pseudoreduced correlations, since the latter method presumes single-phase gases. The deviation factors of the produced gas phase are given in Col. (6) for comparison.

5. COMPARISON BETWEEN THE PREDICTED AND ACTUAL PRODUCTION HISTORIES OF VOLUMETRIC RESERVOIRS

Allen and Roe have reported the performance of a retrograde condensate reservoir that produces from the Bacon Lime Zone of a field located in East Texas.[13] The production history of this reservoir is shown in Figs. 4.6 and 4.7. The reservoir occurs in the lower Glen Rose Formation of Cretaceous age at a depth of 7600 ft (7200 ft subsea) and comprises some 3100 acres. It is composed of approximately 50 ft of dense, crystalline, fossiliferous dolomite with an average permeability of 30 to 40 millidarcys in the more permeable stringers, and an estimated average porosity of about 10%. Interstitial water

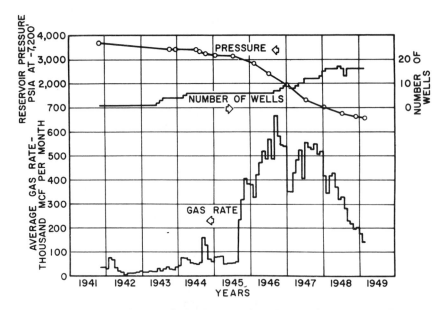

Fig. 4.6. Production history of the Bacon Lime Zone of an eastern Texas gas-condensate reservoir. (*After* Allen and Roe,[13] *Trans.* AIME.)

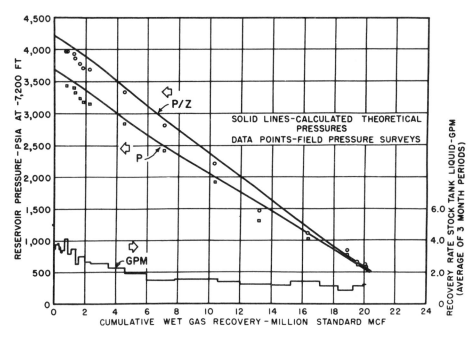

Fig. 4.7. Calculated and measured pressure and p/z values versus cumulative gross gas recovery from the Bacon Lime Zone of an eastern Texas gas-condensate reservoir. (*After* Allen and Roe,[13] *Trans.* AIME.)

is approximately 30%. The reservoir temperature is 220°F, and the initial pressure was 3691 psia at 7200 ft subsea. Because the reservoir was very heterogeneous regarding porosity and permeability, and, because very poor communication between wells was observed, cycling (Sect. 4.6) was not considered feasible. The reservoir was therefore produced by pressure depletion, using three-stage separation to recover the condensate. The recovery at 600 psia was 20,500 MM SCF and 830,000 bbl of condensate, or a cumulative (average) gas-oil ratio of 24,700 SCF/bbl, or 1.70 GPM (gallons per MCF). Since the initial gas-oil ratios were about 12,000 SCF/bbl (3.50 GPM), the condensate recovery of 600 psia was $100 \times 1.7/3.5$, or 48.6% of the liquid originally contained in the produced gas. Theoretical calculations based on equilibrium ratios predicted a recovery of only 1.54 GPM (27,300 gas-oil ratio), or 44% recovery, which is about 10% lower.

The difference between the actual and predicted recoveries may have been due to sampling errors. The initial well samples may have been deficient in the heavier hydrocarbons owing to retrograde condensation of liquid from the flowing fluid as it approached the well bore (Sect. 4.3). Another possibility suggested by Allen and Roe is the omission of nitrogen as a constituent of the gas from the calculations. A small amount of nitrogen, always below 1 mole

%, was found in several of the samples during the life of the reservoir. Finally, they suggested the possibility of retrograde liquid flow in the reservoir to account for a liquid recovery higher than that predicted by their theoretical calculations, which presume the immobility of the retrograde liquid phase. Considering the many variables that influence both the calculated recovery using equilibrium ratios and the field performance, the agreement between the two appears good.

Figure 4.8 shows good general agreement between the butanes-plus content calculated from the composition of the production from two wells and the content calculated from the study based on equilibrium ratios. The liquid content expressed in butanes-plus is higher than the stock tank GPM (Fig. 4.7) because not all the butanes, or, for that matter, all the pentanes-plus are recovered in the field separators. The higher actual butanes-plus content down to 1600 psia is undoubtedly a result of the same causes given in the preceding paragraph to explain why the actual overall recovery of stock tank liquid exceeded the recovery based on equilibrium ratios. The stock tank GPM in Fig. 4.7 shows no revaporization; however, the well-stream compositions below 1600 psia in Fig. 4.8 clearly show revaporization of the butanes-plus, and therefore certainly of the pentanes-plus, which make up the majority of the separator liquid. The revaporization of the retrograde liquid in the reservoir below 1600 psia is evidently just about offset by the decrease in separator efficiency at lower pressures.

Figure 4.8 also shows a comparison between the calculated reservoir behavior based on the *differential process* and the *flash process*. In the *differential process*, only the gas is produced and is therefore removed from contact

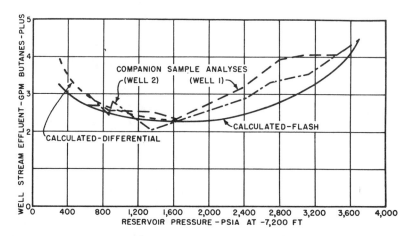

Fig. 4.8. Calculated and measured butanes-plus in the well streams of the Bacon Lime Zone of an eastern Texas gas-condensate reservoir. (*After* Allen and Roe,[13] *Trans.* AIME.)

with the liquid phase in the reservoir. In the *flash process,* all the gas remains in contact with the retrograde liquid, and for this to be so the volume of the system must increase as the pressure declines. Thus the differential process is one of constant volume and changing composition, and the flash process is one of constant composition and changing volume. Laboratory work and calculations based on equilibrium ratios are simpler with the flash process, where the overall composition of the system remains constant; however, the reservoir mechanism for the volumetric depletion of retrograde condensate reservoirs is essentially a differential process. The laboratory work and the use of equilibrium ratios discussed in Sect. 4.3 approaches the differential process by a series of step-by-step flash processes. Figure 4.8 shows the close agreement between the flash and differential calculations down to 1600 psia. Below 1600 psia, the well performance is closer to the differential calculations because the reservoir mechanism largely follows the differential process, provided that only gas phase materials are produced from the reservoir (i.e., the retrograde liquid is immobile).

Figure 4.9 shows the good agreement between the reservoir field data and the laboratory data for a small (one well), noncommercial, gas-condensate accumulation in the Paradox limestone formation at a depth of 5775 ft in San

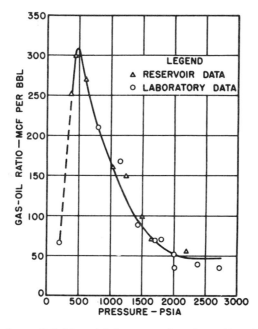

Fig. 4.9. Comparison of field and laboratory data for a Paradox limestone gas-condensate reservoir in Utah. (*After* Rodgers, Harrison, and Regier,[8] courtesy AIME.)

Juan County, Utah. This afforded Rodgers, Harrison, and Regier a unique opportunity to compare laboratory PVT studies and studies based on equilibrium ratios with actual field depletion under closely controlled and observed conditions.[8] In the laboratory, a 4000 cu cm cell was charged with representative well samples at reservoir temperature and initial reservoir pressure. The cell was pressure depleted so that only the gas phase was removed, and the produced gas was passed through miniature three-stage separators, which were operated at optimum field pressures and temperatures. The calculated performance was also obtained, as explained previously, from equations involving equilibrium ratios, assuming the differential process. Rodgers et al. concluded that the model laboratory study can adequately reproduce and predict the behavior of condensate reservoirs. Also, they found that the performance can be calculated from the composition of the initial reservoir fluid, provided representative equilibrium ratios are available.

Table 4.7 shows a comparison between the initial compositions of the Bacon Lime and Paradox limestone formation fluids. The lower gas-oil ratios for the Bacon Lime are consistent with the much larger concentration of the pentanes and heavier in the Bacon Lime fluid.

6. LEAN GAS CYCLING AND WATER DRIVE

Because the liquid content of many condensate reservoirs is a valuable and important part of the accumulation, and because through retrograde condensation a large fraction of this liquid may be left in the reservoir at abandonment, the practice of lean gas cycling has been adopted in many condensate reservoirs. In gas cycling, the condensate liquid is removed from the produced

TABLE 4.7.

Comparison of the compositions of the initial fluids in the Bacon Lime and Paradox formations

	Bacon Lime	*Paradox Formation*
Nitrogen	?	0.0099
Carbon dioxide	0.0135	0.0000
Methane	0.7690	0.7741
Ethane	0.0770	0.1148
Propane	0.0335	0.0531
Butane	0.0350	0.0230
Pentane	0.0210	0.0097
Hexane	0.0150	0.0054
Heptane-plus	0.0360	0.0100
	1.0000	1.0000
Mol. wt. C_7^+	130	116.4
Sp. gr. C_7^+ (60°F)	0.7615	0.7443

(wet) gas, usually in a gasoline plant, and the residue, or dry gas, is returned to the reservoir through injection wells. The injected gas maintains reservoir pressure and retards retrograde condensation. At the same time, it drives the wet gas toward the producing wells. Because the removed liquids represent part of the wet gas volume, unless additional dry (make-up) gas is injected, reservoir pressure will decline slowly. At the conclusion of cycling (i.e., when the producing wells have been invaded by the dry gas), the reservoir is then pressure depleted (blown down) to recover the gas plus some of the remaining liquids from the portions not swept.

Although lean gas cycling appears to be an ideal solution to the retrograde condensate problem, a number of practical considerations make it less attractive. First, there is the deferred income from the sale of the gas, which may not be produced for 10 to 20 years. Second, cycling requires additional expenditures, usually some more wells, a gas compression and distribution system to the injection wells, and a liquid recovery plant. Third, it must be realized that even when reservoir pressure is maintained above the dew point, the liquid recovery by cycling may be considerably less than 100%.

Cycling recoveries may be broken down into three separate recovery factors, or efficiencies. When dry gas displaces wet gas within the pores of the reservoir rock, the microscopic *displacement* efficiency is in the range of 70 to 90%. Then, owing to the location and flow rates of the production and injection wells, there are areas of the reservoir that are not swept by dry gas at the time the producing wells have been invaded by dry gas, resulting in *sweep* efficiencies in the range of 50 to 90% (i.e., 50 to 90% of the initial pore volume is invaded by dry gas). Finally, many reservoirs are stratified in such a way that some stringers are much more permeable than others, so that the dry gas sweeps through them quite rapidly. Although much wet gas remains in the lower permeability (tighter) stringers, dry gas will have entered the producing wells in the more permeable stringers, eventually reducing the liquid content of the gas to the plant to an unprofitable level.

Now suppose a particular gas-condensate reservoir has a displacement efficiency of 80%, a sweep efficiency of 80%, and a permeability stratification factor of 80%. The product of these separate factors given an overall condensate recovery by cycling of 51.2%. Under these conditions, cycling may not be particularly attrative, because retrograde condensate losses by depletion performance seldom exceed 50%. However, during pressure depletion (blow down) of the reservoir following cycling, some additional liquid may be recovered from both the swept and unswept portions of the reservoir. Also, liquid recoveries of propane and butane in gasoline plants are much higher than those from stage separation of low-temperature separation, which would be used if cycling is not adopted. From what has been said, it is evident that whether to cycle is a problem whose many aspects must be carefully studied before a proper decision can be reached.

Cycling is also adopted in nonretrograde gas caps overlying oil zones, particularly when the oil is itself underlain by an active body of water. If the gas cap is produced concurrently with the oil, as the water drives the oil zone into the shrinking gas cap zone, unrecoverable oil remains not only in the original oil zone but also in that portion of the gas cap invaded by the oil. On the other hand, if the gas cap is cycled at essentially initial pressure, the active water drive displaces the oil into the producing oil wells with maximum recovery. In the meantime, some of the valuable liquids from the gas cap may be recovered by cycling. Additional benefit will accrue, of course, if the gas cap is retrograde. Even when water drive is absent, the concurrent depletion of the gas cap and the oil zone results in lowered oil recoveries; and increased oil recovery is had by depleting the oil zone first and allowing the gas cap to expand and sweep through the oil zone.

When gas-condensate reservoirs are produced under an active water drive such that reservoir pressure declines very little below the initial pressure, there is little or no retrograde condensation and the gas-oil ratio of the production remains substantially constant. The recovery is the same as in nonretrograde gas reservoirs under the same conditions and depends on (a) the initial connate water, S_{wi}, (b) the residual gas saturation, S_{gr}, in the portion of the reservoir invaded by water, and (c) the fraction, F, of the initial reservoir volume invaded by water. The gas volume factor B_{gi} in cu ft/SCF remains substantially constant because reservoir pressure does not decline, so the fractional recovery is

$$\text{Recovery} = \frac{V_i \phi (1 - S_{wi} - S_{gr}) B_{gi} F}{V_i \phi (1 - S_{wi}) B_{gi}} = \frac{(1 - S_{wi} - S_{gr}) F}{(1 - S_{wi})} \qquad (4.4)$$

where V_i is the initial gross reservoir volume, S_{gr} is the residual gas saturation in the flooded area, S_{wi} is the initial connate water saturation, and F is the fraction of the total volume invaded. Table 3.3 shows that residual gas saturations lie in range of 20 to 50% following water displacement. The fraction of the total volume invaded at any time or at abandonment depends primarily on well location and the effect of permeability stratification in edge-water drives, and primarily on well spacing and the degree of water coning in bottom-water drives.

Table 4.8 shows the recovery factors calculated from Eq. (4.4) assuming a reasonable range of values for the connate water, residual gas saturation, and the fractional invasion by water at abandonment. The recovery factors apply equally to gas and gas-condensate reservoirs because under active water drive there is no retrograde loss.

Table 4.9 shows a comparison of gas-condensate recovery for the reservoir of Ex. 4.3 by (a) volumetric depletion, (b) water drive at initial pressure of 2960 psia, and (c) partial water drive where the pressure stabilizes at 2000 psia. The initial gross fluid, gas, and condensate, and the recoveries by depletion performance at an assumed abandonment pressure of 500 psia are

TABLE 4.8.
Recovery factors for complete water-drive reservoirs based on Eq. (4.4)

S_{gr}	S_w	$F = 40$	$F = 60$	$F = 80$	$F = 90$	$F = 100$
20	10	31.1	46.7	62.2	70.0	77.8
	20	30.0	45.0	60.0	67.5	75.0
	30	28.6	42.8	57.1	64.3	71.4
	40	26.7	40.0	53.4	60.0	66.7
30	10	26.7	40.0	53.4	60.0	66.7
	20	25.0	37.5	50.0	56.3	62.5
	30	22.8	34.3	45.7	51.4	57.1
	40	20.0	30.0	40.0	45.0	50.0
40	10	22.2	33.3	44.4	50.0	55.6
	20	20.0	30.0	40.0	45.0	50.0
	30	17.1	25.7	34.2	38.5	42.8
	40	13.3	20.0	26.6	30.0	33.3
50	10	17.7	26.6	35.5	40.0	44.4
	20	15.0	22.5	30.0	33.8	37.5
	30	11.4	17.1	22.8	25.7	28.5
	40	6.7	10.0	13.6	15.0	16.7

TABLE 4.9.
Comparison of gas-condensate recovery by volumetric performance, complete water drive, and partial water drive (based on the data of Tables 4.4 and 4.5 and Ex. 4.3. $S_w = 30\%$; $S_{gr} = S_{or} + S_{gr} = 20\%$; $F = 80\%$)

Recovery Mechanism	Condensate Recovery		Gas Recovery		Gross Recovery	
	bbl/ac-ft	Per-centage	MCF/ac-ft	Per-centage	MCF/ac-ft	Per-centage
Initial in-place	143.2	100.0	1441	100.0	1580	100.0
Depletion to 500 psia	71.6	50.0	1200	83.3	1271	80.4
Water drive at 2960 psia	81.8	57.1	823	57.1	902	57.1
(a) Depletion to 2000 psia	28.4	19.8	457	31.7	485	30.7
(b) Water drive at 2000 psia	31.2	21.8	553	38.4	584	37.0
Total by partial water drive, (a) + (b)	59.6	41.6	1010	70.1	1069	67.7

obtained from Ex. 4.3 and Tables 4.4 and 4.5. Under complete water drive, the recovery is 57.1% for a residual gas saturation of 20%, a connate water of 30%, and a fractional invasion of 80% at abandonment, as may be found by Eq. (4.4) or Table 4.8. Because there is no retrograde loss, this figure applies equally to the gross gas, gas, and condensate recovery.

When a partial water drive exists and the reservoir pressure stabilizes at some pressure, here 2000 psia, the recovery is approximately the sum of the recovery by pressure depletion down to the stabilization pressure, plus the recovery of the remaining fluid by complete water drive at the stabilization pressure. Because the retrograde liquid at the stabilization pressure is immobile, it is enveloped by the invading water, and the residual hydrocarbon saturation (gas plus retrograde liquid) is about the same as for gas alone, or 20% for this example. The recovery figures of Table 4.9 by depletion down to 2000 psia are obtained from Table 4.5. The additional recovery by water drive at 2000 psia may be explained using the figures of Table 4.10. At 2000 psia the retrograde condensate volume is 625 cu ft/ac-ft, or 8.2% of the initial hydrocarbon pore volume of 7623 cu ft/ac-ft, 8.2% being found from the PVT data given in Table 4.4. If the residual hydrocarbon (both gas and condensate) saturation after water invasion is assumed to be 20%, as previously assumed for the residual gas saturation by complete water drive, the water volume after water drive is 80% of 10,890, or 8712 cu ft/ac-ft. Of the remainder (i.e., 2178 cu ft/ac-ft) since the pressure is assumed to stabilize at 2000 psia, 8.2%, or 625 cu ft/ac-ft, will be condensate liquid and 1553 free gas. The reservoir vapor at 2000 psia prior to water drive is

$$\frac{2000 \times 6998 \times 379.4}{1000 \times 0.805 \times 10.73 \times 655} = 938.5 \text{ M SCF/ac-ft}$$

The fractional recovery of this vapor phase by complete water drive at 2000 psia is

$$\frac{6998 - 1553}{6998} = 0.778 \text{ or } 77.8\%$$

If $F = 0.80$—or only 80% of each acre-foot, on the average, is invaded by water at abandonment—the overall recovery reduces to 0.80×0.778, or 62.2% of the vapor content at 2000 psia, or 584M SCF/ac-ft. Table 4.5 indicates that at 2000 psia the ratio of gross gas to residue gas after separation is

TABLE 4.10.
Volumes of water, gas, and condensate in one acre-foot of bulk rock for the reservoir of Example 4.3

	Initial Reservoir Volumes, cu ft/ac-ft	Volumes After Depletion to 2000 psia cu ft/ac-ft	Volumes After Water Drive at 2000 psia cu ft/ac-ft
Water	3267	3267	8712
Gas	7623	6998	1553
Condensate	. . . .	625	625
Total	10,890	10,890	10,890

245.2 to 232.3, and that the gas-oil ratio on a residue gas basis is 17,730 SCF/bbl. Then 584 M SCF of gross gas contains residue gas in the amount of

$$584 \times \frac{232.3}{245.2} = 553 \text{ M SCF/ac-ft}$$

and tank or surface condensate liquid in the amount of

$$\frac{553 \times 1000}{17,730} = 31.2 \text{ bbl/ac-ft}$$

Table 4.9 indicates that for the gas-condensate reservoir of Example 4.3, using the assumed values for F and S_{gr}, best overall recovery is obtained by straight depletion performance. Best condensate recovery is by active water drive because no retrograde liquid forms. The value of the products obtained depends, of course, on the relative unit prices at which the gas and condensate are sold.

7. USE OF NITROGEN FOR PRESSURE MAINTENANCE

One of the major disadvantages associated with the use of lean gas in gas cycling applications is that the income that would be derived from the sale of the lean gas is deferred for several years. For this reason, the use of nitrogen has been suggested as a replacement for the lean gas.[16] However, one might expect the phase behavior of nitrogen and a wet gas to exhibit different characteristics from that of lean gas and the same wet gas. Researchers have found that mixing nitrogen and a typical wet gas causes the dew point of the resulting mixture to be higher than the dew point of the original wet gas.[17,18] This is also true for lean gas, but the dew point is raised higher with nitrogen.[17] If, in a reservoir situation, the reservoir pressure is not maintained higher than this new dew point, then retrograde condensation will occur. This condensation may be as much or more than what would occur if the reservoir was not cycled with gas. Studies have shown, however, that very little mixing occurs between an injected gas and the reservoir gas in the reservoir.[17,18] Mixing occurs as a result of molecular diffusion and dispersion forces, and the resulting mixing zone width is usually only a few feet.[19,20] The dew point may be raised in this local area of mixing, but this will be a very small volume and, as a result, only a small amount of condensate may drop out. Vogel and Yarborough have also shown that, under certain conditions, nitrogen revaporizes the condensate.[18] The conclusion from these studies indicates that nitrogen can be used as a replacement for lean gas in cycling operations with the potential for some condensate formation that should be minimal in most applications.

Kleinsteiber, Wendschlag, and Calvin conducted a study to determine the optimum plan of depletion for the Anschutz Ranch East Unit, which is located in Summit County, Utah and Uinta County, Wyoming.[21] The Anschutz Ranch East Field, discovered in 1979, is one of the largest hydrocarbon accumulations found in the Western Overthrust Belt. Tests have indicated that the original in-place hydrocarbon content was over 800 million bbl of oil equivalent. Laboratory experiments conducted on several surface-recombined samples indicated that the reservoir fluid was a rich gas condensate. The fluid had a dew point only 150 to 300 psia below the original reservoir pressure of 5310 psia. The dew-point pressure was a function of the structural position in the reservoir, with fluid near the water-oil contact having a dew point about 300 psia lower than the original pressure and field near the crest having a dew point only about 150 psia lower. The liquid saturation, observed in constant composition expansion tests, accumulated very rapidly below the dew point, suggesting that depletion of the reservoir and the subsequent drop in reservoir pressure could cause the loss of significant amounts of condensate. Because of this potential loss of valuable hydrocarbons, a project was undertaken to determine the optimum method of production.[21]

To begin the study, a modified Redlich-Kwong equation of state was calibrated with the laboratory phase behavior data that had been obtained.[17, 22] The equation of state was then used in a compositional reservoir simulator. Several depletion schemes were considered, including primary depletion and partial or full pressure maintenance. Wet hydrocarbon gas, dry hydrocarbon gas, carbon dioxide, combustion flue gas, and nitrogen were all considered as potential gases to inject. Carbon dioxide and flue gas were eliminated due to lack of availability and high cost. The results of the study led to the conclusion that full pressure maintenance should be used. Liquid recoveries were found to be better with dry hydrocarbon gas than with nitrogen. However, when nitrogen injection was preceded by a 10 to 20% buffer of dry hydrocarbon gas, the liquid recoveries were nearly the same. When an economic analysis was coupled with the simulation study, the decision was to conduct a full-pressure maintenance program with nitrogen as the injected gas. A 10% pore volume buffer, consisting of 35% nitrogen and 65% wet hydrocarbon gas, was to be injected before the nitrogen to improve the recovery of liquid condensate.

The approach taken in the study by Kleinsteiber, Wendschlag, and Calvin would be appropriate for the evaluation of any gas-condensate reservoir. The conclusions regarding which injected material is best or whether a buffer would be necessary may be different for a reservoir gas of different composition.

PROBLEMS

4.1 A gas-condensate reservoir initially contains 1300M SCF of residue (dry or sales gas) per acre-foot and 115 STB of condensate. Gas recovery is calculated to be 85% and condensate recovery 58% by depletion performance. Calculate the value of the initial gas and condensate reserves per acre-foot if the condensate sells for $20.00/bbl, and the gas sells for $1.90 per 1000 std cu ft.

4.2 A well produces 45.3 STB of condensate and 742 M SCF of sales gas daily. The condensate has a molecular weight of 121.2 and a gravity of 52.0°API at 60°F.

(a) What is the gas-oil ratio on a dry gas basis?

(b) What is the liquid content expressed in barrels per million standard cubic feet on a dry gas basis?

(c) What is the liquid content expressed in GPM on a dry gas basis?

(d) Repeat parts (a), (b), and (c) expressing the figures on a wet, or gross, gas basis.

4.3 The initial daily production from a gas-condensate reservoir is 186 STB of condensate, 3750M SCF of high-pressure gas, and 95 M SCF of stock tank gas. The tank oil has a gravity of 51.2°API at 60°F. The specific gravity of the separator gas is 0.712 and of the stock tank gas, 1.30. The initial reservoir pressure is 3480 psia, and reservoir temperature is 220°F. Average hydrocarbon porosity is 17.2%. Assume standard conditions of 14.7 psia and 60°F.

(a) What is the average gravity of the produced gases?

(b) What is the initial gas-oil ratio?

(c) Estimate the molecular weight of the condensate.

(d) Calculate the specific gravity (air $= 1.00$) of the total well production.

(e) Calculate the gas deviation factor of the initial reservoir fluid (vapor) at initial reservoir pressure.

(f) Calculate the initial moles in place per acre-foot.

(g) Calculate the mole fraction that is gas in the initial reservoir fluid.

(h) Calculate the initial (sales) gas and condensate in place per acre-foot.

4.4 (a) Calculate the gas deviation factor for the gas-condensate fluid the composition of which is given in Table 4.1 at 5820 psia and 265°F. Use the critical values of C_8 for the C_{7+} fraction.

(b) If half the butanes and all the pentanes and heavier gases are recovered as liquids, calculate the gas-oil ratio of the initial production. Compare with the measured gas-oil ratio.

4.5 Calculate the composition of the reservoir retrograde liquid at 2500 psia for the data of Tables 4.4 and 4.5 and Ex. 4.3. Assume the molecular weight of the heptanes-plus fraction to be the same as for the initial reservoir fluid.

4.6 Estimate the gas and condensate recovery for the reservoir of Ex. 4.3 under partial water drive if reservoir pressure stabilizes at 2500 psia. Assume a residual hydrocarbon saturation of 20% and $F = 52.5\%$.

4.7 Calculate the recovery factor by cycling in a condensate reservoir if the displacement efficiency is 85%, the sweep efficiency is 65%, and the permeability stratification factor is 60%.

4.8 The following data are taken from a study on a recombined sample of separator gas and separator condensate in a PVT cell with an initial hydrocarbon volume of 3958.14 cu cm. The wet gas GPM and the residue gas-oil ratios were calculated using equilibrium ratios for production through a separator operating at

300 psia and 70°F. The initial reservoir pressure was 4000 psia, which was also close to the dew-point pressure, and reservoir temperature was 186°F.

(a) On the basis of an initial reservoir content of 1.00 MM SCF of wet gas, calculate the wet gas, residue gas, and condensate recovery by pressure depletion for each pressure interval.

(b) Calculate the dry gas and condensate initially in place in 1.00 MM SCF of wet gas.

(c) Calculate the cumulative recovery and the percentage of recovery of wet gas, residue gas, and condensate by depletion performance at each pressure.

(d) Place the recoveries at an abandonment pressure of 605 psia on an acre-foot basis for a porosity of 10% and a connate water of 20%.

| | Composition in Mole Percentages | | | | | |
Pressure, psia	4000	3500	2900	2100	1300	605
CO_2	0.18	0.18	0.18	0.18	0.19	0.21
N_2	0.13	0.13	0.14	0.15	0.15	0.14
C_1	67.72	63.10	65.21	69.79	70.77	66.59
C_2	14.10	14.27	14.10	14.12	14.63	16.06
C_3	8.37	8.25	8.10	7.57	7.73	9.11
$i\text{-}C_4$	0.98	0.91	0.95	0.81	0.79	1.01
$n\text{-}C_4$	3.45	3.40	3.16	2.71	2.59	3.31
$i\text{-}C_5$	0.91	0.86	0.84	0.67	0.55	0.68
$n\text{-}C_5$	1.52	1.40	1.39	0.97	0.81	1.02
C_6	1.79	1.60	1.52	1.03	0.73	0.80
C_{7+}	6.85	5.90	4.41	2.00	1.06	1.07
Mol. wt. C_{7+}	143	138	128	116	111	110
Gas deviation factor for wet gas at 186°F	0.867	0.799	0.748	0.762	0.819	0.902
Wet gas production, cu cm at cell P and T	0	224.0	474.0	1303	2600	5198
Wet gas GPM (calculated)	5.254	4.578	3.347	1.553	0.835	0.895
Residue gas-oil ratio	7,127	8,283	11,621	26,051	49,312	45,872
Retrograde liquid, percentage of cell volume	0	3.32	19.36	23.91	22.46	18.07

4.9 If the retrograde liquid for the reservoir of Prob. 4.8 becomes mobile at 15% retrograde liquid saturation, what effect will this have on the condensate recovery?

4.10 If the initial pressure of the reservoir of Prob. 4.8 had been 5713 psia with the dew point at 4000 psia, calculate the additional recovery of wet gas, residue gas, and condensate per acre-foot. The gas deviation factor at 5713 psia is 1.107, and the GPM and GOR between 5713 and 4000 psia are the same as at 4000 psia.

4.11 Calculate the value of the products by each mechanism in Table 4.9 assuming (a) $17.50 per STB for condensate and $1.50 per M SCF for gas; (b) $20.00 per STB and $1.90 per M SCF; and (c) $20.00 per STB and $2.50 per M SCF.

4.12 In a PVT study of a gas-condensate fluid, 17.5 cu cm of wet gas (vapor), measured at cell pressure of 2500 psia and temperature of 195°F, was displaced into an evacuated low-pressure receiver of 5000 cu cm volume that was maintained at 250°F to ensure that no liquid phase developed in the expansion. If the pressure of the receiver rises to 620 mm Hg, what will be the deviation factor of the gas in the cell at 2500 psia and 195°F, assuming the gas in the receiver behaves ideally?

4.13 Using the assumptions of Ex. 4.3 and the data of Table 4.4, show that the condensate recovery between 2000 and 1500 psia is 13.3 STB/ac-ft and the residue gas-oil ratio is 19,010 SCF/bbl.

4.14 A stock tank barrel of condensate has a gravity of 55°API. Estimate the volume in ft³ occupied by this condensate as a single-phase gas in a reservoir at 2740 psia and 215°F. The reservoir wet gas has a gravity of 0.76.

4.15 A gas-condensate reservoir has an areal extent of 200 acres, an average thickness of 15 ft, an average porosity of 0.18, and an initial water saturation of 0.23. A PVT cell is used to simulate the production from the reservoir, and the following data are collected:

Pressure (psia)	Wet Gas Produced (cc)	z Wet Gas	Condensate Produced from Separator (moles)
4000 (dew point)	0	0.75	0
3700	400	0.77	0.0003
3300	450	0.81	0.0002

The initial cell volume was 1850 cc, and the initial gas contained 0.002 moles of condensate. The initial pressure is 4000 psia, and the reservoir temperature is 200°F. Calculate the amount of dry gas (SCF) and condensate (STB) recovered at 3300 psia from the reservoir. The molecular weight and specific gravity of the condensate are 145 and 0.8, respectively.

4.16 Production from a gas-condensate reservoir is listed below. The molecular weight and the specific gravity of the condensate are 150 and 0.8, respectively. The initial wet gas in place was 35 MMM SCF, and the initial condensate was 2 MM STB. Assume a volumetric reservoir and that the recoveries of condensate and water are identical, and determine the following:

(a) What is the percentage of recovery of residue gas at 3300 psia?

(b) Can a PVT cell experiment be used to simulate the production from this reservoir? Why or why not?

Pressure, psia	4000	3500	3300
Compressibility of wet gas, z	0.85	0.80	0.83
Wet gas produced during pressure increment, SCF	0	2.4 MMM	2.2 MMM
Liquid condensate produced during pressure increment, STB	0	80,000	70,000
Water produced during pressure increment, STB	0	5000	4375

4.17 A PVT cell is used to simulate a gas-condensate reservoir. The initial cell volume is 1500 cc, and the initial reservoir temperature is 175°F. Show by calculations that the PVT cell will or will not adequately simulate the reservoir behavior. The data generated by the PVT experiments as well as the actual production history are as follows:

Pressure, psia	4000	3600	3000
Wet gas produced in pressure increment, cc	0	300	700
Compressibility of produced gas	0.70	0.73	0.77
Actual Production history, M SCF	0	1000	2300

REFERENCES

[1] J. C. Allen, "Factors Affecting the Classification of Oil and Gas Wells," API *Drilling and Production Practice* (1952), p. 118.

[2] *Ira Rinehart's Yearbooks.* Dallas, TX: Rinehart Oil News, 1953–1957, Vol. No. 2.

[3] M. Muskat, *Physical Principles of Oil Production,* (New York: McGraw-Hill, *Book Company, Inc,* 1949), Chap. 15.

[4] M. B. Standing, *Volumetric and Phase Behavior of Oil Field Hydrocarbon Systems.* New York: Reinhold Publishing, 1952, Chap. 6.

[5] O. F. Thornton, "Gas-Condensate Reservoirs—a Review," API *Drilling and Production Practice* (1946), p. 150.

[6] C. K. Eilerts, *Phase Relations of Gas-Condensate Fluids,* Monograph 10, Bureau of Mines. New York: American Gas Association, 1957, Vol. I.

[7] T. A. Mathews, C. H. Roland, and D. L. Katz, "High Pressure Gas Measurement," *Proc.* NGAA (1942), p. 41.

[8] J. K. Rodgers, N. H. Harrison, and S. Regier, "Comparison Between the Predicted and Actual Production History of a Condensate Reservoir." Paper No. 883–G presented before the AIME Meeting, Dallas, TX, Oct. 1957.

[9] W. E. Portman and J. M. Campbell, "Effect of Pressure, Temperature, and Wellstream Composition on the Quantity of Stabilized Separator Fluid," *Trans.* AIME (1956), **207,** 308.

[10] R. L. Huntington, *Natural Gas and Natural Gasoline.* New York: McGraw-Hill, 1950, Chap. 7.

[11] *Natural Gasoline Supply Men's Association Engineering Data Book,* 7th ed. Tulsa, OK: Natural Gasoline Supply Men's Association, 1957), p. 161.

[12] A. E. Hoffmann, J. S. Crump, and C. R. Hocott, "Equilibrium Constants for a Gas-Condensate System," *Trans.* AIME (1953), **198,** 1.

[13] F. H. Allen and R. P. Roe, "Performance Characteristics of a Volumetric Condensate Reservoir," *Trans.* AIME (1950), **189,** 83.

[14] J. E. Berryman, "The Predicted Performance of a Gas-Condensate System, Washington Field, Louisiana," *Trans.* AIME (1957), **210,** 102.

[15] R. H. Jacoby, R. C. Koeller, and V. J. Berry, Jr., "Effect of Composition and

Temperature on Phase Behavior and Depletion Performance of Gas-Condensate Systems." Paper presented before SPE of AIME, Houston, TX, Oct. 5–8, 1958.

[16] C. W. Donohoe and R. D. Buchanan, "Economic Evaluation of Cycling Gas-Condensate Reservoirs with Nitrogen," *Jour. of Petroleum Technology* (Feb. 1981), 263.

[17] P. L. Moses and K. Wilson, "Phase Equilibrium Considerations in Using Nitrogen for Improved Recovery From Retrograde Condensate Reservoirs," *Jour. of Petroleum Technology* (Feb. 1981), 256.

[18] J. L. Vogel and L. Yarborough, "The Effect of Nitrogen on the Phase Behavior and Physical Properties of Reservoir Fluids." Paper SPE 8815 presented at the First Joint SPE/DOE Symposium on Enhanced Oil Recovery, Tulsa, OK, April, 1980.

[19] P. M. Sigmund, "Prediction of Molecular Diffusion at Reservoir Conditions. Part I—Measurement and Prediction of Binary Dense Gas Diffusion Coefficients," *Jour. of Canadian Petroleum Technology* (April–June 1976), 48.

[20] P. M. Sigmund, "Prediction of Molecular Diffusion of Reservoir Conditions. Part II—Estimating the Effects of Molecular Diffusion and Convective Mixing in Multicomponent Systems," *Jour. of Canadian Petroleum Technology* (July–Sept. 1976), 53.

[21] S. W. Kleinsteiber, D. D. Wendschlag, and J. W. Calvin, "A Study for Development of a Plan of Depletion in a Rich Gas Condensate Reservoir: Anschutz Ranch East Unit, Summit County, Utah, Uinta County, Wyoming." Paper SPE 12042 presented at 58th Annual Conference of SPE of AIME, San Francisco, Oct. 1983.

[22] L. Yarborough, "Application of a Generalized Equation of State to Petroleum Reservoir Fluids," in *Equations of State in Engineering, Advances in Chemical Series,* K. C. Chao and R. L. Robinson, ed., American Chemical Society, 1979, 385.

Chapter 5

Undersaturated Oil Reservoirs

1. INTRODUCTION

Oil reservoir fluids are mainly complex mixtures of the hydrocarbon compounds, which frequently contain impurities such as nitrogen, carbon dioxide, and hydrogen sulfide. The composition in mole percentages of several typical reservoir liquids is given in Table 5.1, together with the tank gravity of the crude oil, the gas-oil ratio of the reservoir mixture, and other characteristics of the fluids.[*,1] The composition of the tank oils obtained from the reservoir fluids are quite different from the composition of the reservoir fluids, owing mainly to the release of most of the methane and ethane from solution and the vaporization of sizeable fractions of the propane, butanes, and pentanes as pressure is reduced in passing from the reservoir to the stock tank. The table shows a good correlation between the gas-oil ratios of the fluids and the percentages of methane and ethane they contain over a range of gas-oil ratios from only 22 SCF/STB up to 4053 SCF/STB.

Several methods are available for collecting samples of reservoir fluids. The samples may be taken with subsurface sampling equipment lowered into the well on a wire line, or samples of the gas and oil may be collected at the

---

[*] References throughout the text are given at the end of each chapter.

TABLE 5.1.
Reservoir fluid compositions and properties (*After* Kennery,[1] courtesy Core
Laboratories, Inc.)

Component or Property	Calif.	Wyo.	South Texas	North Texas	West Texas	South La.
Methane	22.62	1.08	48.04	25.63	28.63	65.01
Ethane	1.69	2.41	3.36	5.26	10.75	7.84
Propane	0.81	2.86	1.94	10.36	9.95	6.42
iso-Butane	0.51	0.86	0.43	1.84	4.36	2.14
n-Butane	0.38	2.83	0.75	5.67	4.16	2.91
iso-Pentane	0.19	1.68	0.78	3.14	2.03	1.65
n-Pentane	0.19	2.17	0.73	1.91	3.83	0.83
Hexanes	0.62	4.51	2.79	4.26	2.35	1.19
Heptanes-plus	72.99	81.60	41.18	41.93	33.94	12.01
Density heptanes-plus: g/cc	0.957	0.920	0.860	0.843	0.792	0.814
Mol. wt. heptanes-plus	360	289	198	231	177	177
Sampling depth, feet	2980	3160	8010	4520	12,400	10,600
Reservoir temperature, °F	141	108	210	140	202	241
Saturation pressure, psig	1217	95	3660	1205	1822	4730
GOR, SCF/STB	105	22	750	480	895	4053
Formation volume factor, bbl/STB	1.065	1.031	1.428	1.305	1.659	3.610
Tank oil gravity, °API	16.3	25.1	34.8	40.6	50.8	43.5
Gas gravity (air = 1.00)	0.669	...	0.715	1.032	1.151	0.880

surface and later recombined in proportion to the gas-oil ratio measured at
the time of sampling. Samples should be obtained as early as possible in the
life of the reservoir, preferably at the completion of the discovery well, so that
the sample approaches as nearly as possible the original reservoir fluid. The
type of fluid collected in a sampler is dependent on the well history prior
to sampling. Unless the well has been properly conditioned before sampling,
it is impossible to collect representative samples of the reservoir fluid. A
complete well-conditioning procedure has been described by Kennerly
and Reudelhuber.[1,2] The information obtained from the usual fluid sample
analysis includes the following properties:

1. Solution and evolved gas-oil ratios and liquid phase volumes.
2. Formation volume factors, tank oil gravities, and separator and stock tank
   gas-oil ratios for various separator pressures.
3. Bubble-point pressure of the reservoir fluid.
4. Compressibility of the saturated reservoir oil.

5. Viscosity of the reservoir oil as a function of pressure.

6. Fractional analysis of a casing head gas sample and of the saturated reservoir fluid.

If laboratory data are not available, satisfactory estimations for a preliminary analysis can often be made from empirical correlations based on data that are usually available. These data include the gravity of the tank oil, the specific gravity of the produced gas, the initial producing gas-oil ratio, the viscosity of the tank oil, reservoir temperature, and initial reservoir pressure.

In most reservoirs, the variations in the reservoir fluid properties among samples taken from different portions of the reservoir are not large, and they lie within the variations inherent in the techniques of fluid sampling and analysis. In some reservoirs, on the other hand, particularly those with large closures, there are large variations in the fluid properties. For example, in the Elk Basin Field, Wyoming and Montana, under initial reservoir conditions there was 490 SCF of gas in solution per barrel of oil in a sample taken near the crest of the structure but only 134 SCF/bbl in a sample taken on the flanks of the field, 1762 ft lower in elevation.[3] This is a solution gas gradient of 20 SCF/bbl per 100 ft of elevation. Because the quantity of solution gas has a large effect on the other fluid properties, large variations also occur in the fluid viscosity, the formation volume factor, and the like. Similar variations have been reported for the Weber sandstone reservoir of the Rangely Field, Colorado, and the Scurry Reef Field, Texas, where the solution gas gradients were 25 and 46 SCF/bbl per 100 ft of elevation, respectively.[4,5] These variations in fluid properties may be explained by a combination of (a) temperature gradients, (b) gravitational segregation, and (c) the lack of equilibrium between the oil and the solution gas. Cook, Spencer, Bobrowski, and Chin, and McCord have presented methods for handling calculations when there are significant variations in the fluid properties.[5,6]

2. CALCULATION OF INITIAL OIL IN PLACE BY THE VOLUMETRIC METHOD AND ESTIMATION OF OIL RECOVERIES

One of the important functions of the reservoir engineer is the periodic calculation of the reservoir oil (and gas) in place and the recovery anticipated under the prevailing reservoir mechanism(s). In some companies, this work is done by a group that periodically renders an accounting of the company's reserves together with the rates at which they can be recovered in the future. The company's financial position depends primarily on its reserves, the rate at which it increases or loses them, and the rates at which they can be recovered. A knowledge of the reserves and rates of recovery is also important in the sale or exchange of oil properties. The calculation of reserves of new discoveries is particularly important because it serves as a guide to sound development programs. Likewise, an accurate knowledge of the initial contents of reservoirs is invaluable to the reservoir engineer who studies the reservoir behavior with the aim of calculating and/or improving primary recoveries, for it eliminates one of the unknown quantities in equations.

Oil reserves are usually obtained by applying recovery factors to the oil in place. They are also estimated from decline curve studies and by applying appropriate barrel-per-acre-foot recovery figures obtained from experience or statistical studies. The oil in place is calculated either (a) by the *volumetric method* or (b) by *material balance studies,* both of which were presented for gas reservoirs in Chapter 3 and which will be given for oil reservoirs in this and following chapters. The recovery factors are determined from (a) displacement efficiency studies and (b) correlations based on statistical studies of particular types of reservoir mechanisms.

The *volumetric method* for estimating oil in place is based on log and core analysis data to determine the bulk volume, the porosity, and the fluid saturations, and on fluid analysis to determine the oil volume factor. Under initial conditions 1 ac-ft of bulk oil productive rock contains

Interstitial water	$7758 \times \phi \times S_w$
Reservoir oil	$7758 \times \phi \times (1 - S_w)$
Stock tank oil	$\dfrac{7758 \times \phi \times (1 - S_w)}{B_{oi}}$

where 7758 barrels is the equivalent of 1 ac-ft, ϕ is the porosity as a fraction of the bulk volume, S_w is the interstitial water as a fraction of the pore volume, and B_{oi} is the initial formation volume factor of the reservoir oil. Using somewhat average values, $\phi = 0.20$, $S_w = 0.20$, and $B_{oi} = 1.24$, the initial stock tank oil in place per acre-foot is on the order of 1000 STB/ac-ft, or

$$\text{Stock tank oil} = \frac{7758 \times 0.20 \times (1 - 0.20)}{1.24}$$

$$= 1000 \text{ STB/ac-ft}$$

For oil reservoirs under *volumetric control,* there is no water influx to replace the produced oil, so it must be replaced by gas the saturation of which increases as the oil saturation decreases. If S_g is the gas saturation and B_o the oil volume factor at abandonment, then at abandonment conditions 1 ac-ft of bulk rock contains

Interstitial water	$7758 \times \phi \times S_w$
Reservoir gas	$7758 \times \phi \times S_g$
Reservoir oil	$7758 \times \phi \times (1 - S_w - S_g)$
Stock tank oil	$\dfrac{7758 \times \phi \times (1 - S_w - S_g)}{B_o}$

Then the recovery in stock tank barrels per acre-foot is

$$\text{Recovery} = 7758 \times \phi \left[\frac{(1 - S_w)}{B_{oi}} - \frac{(1 - S_w - S_g)}{B_o} \right] \qquad (5.1)$$

and the fractional recovery in terms of stock tank barrels is

$$\text{Recovery} = 1 - \frac{(1 - S_w - S_g)}{(1 - S_w)} \times \frac{B_{oi}}{B_o} \qquad (5.2)$$

The total free gas saturation to be expected at abandonment can be estimated from the oil and water saturations as reported in core analysis.[7] This expectation is based on the assumption that, while being removed from the well, the core is subjected to fluid removal by the expansion of the gas liberated from the residual oil, and that this process is somewhat similar to the depletion process in the reservoir. In a study of the well spacing problem, Craze and Buckley collected a large amount of statistical data on 103 oil reservoirs, 27 of which were considered to be producing under volumetric control.[8] The final gas saturation in most of these reservoirs ranged from 20 to 40% of the pore space, with an average saturation of 30.4%. Recoveries may also be calculated for depletion performance from a knowledge of the properties of the reservoir rock and fluids.

In the case of reservoirs under *hydraulic control,* where there is no appreciable decline in reservoir pressure, water influx is either inward and parallel to the bedding planes as found in thin, relatively steep dipping beds (edge-water drive), or upward where the producing oil zone (column) is underlain by water (bottom-water drive). The oil remaining at abandonment in those portions of the reservoir invaded by water, in barrels per acre-foot, is:

Reservoir oil $7758 \times \phi \times S_{or}$

Stock tank oil $\dfrac{7758 \times \phi \times S_{or}}{B_{oi}}$

where S_{or} is the residual oil saturation remaining after water displacement. Since it was assumed that the reservoir pressure was maintained at its initial value by the water influx, no free gas saturation develops in the oil zone and the oil volume factor at abandonment remains B_{oi}. The recovery by active water drive then is

$$\text{Recovery} = \frac{7758 \times \phi (1 - S_w - S_{or})}{B_{oi}} \text{ STB/ac-ft} \qquad (5.3)$$

and the recovery factor is

$$\text{Recovery factor} = \frac{(1 - S_w - S_{or})}{(1 - S_w)} \qquad (5.4)$$

It is generally believed that the oil content of cores, reported from the analysis of cores taken with a water-base drilling fluid, is a reasonable estimation of the unrecoverable oil, because the core has been subjected to a partial water displacement (by the mud filtrate) during coring and to displacement by the expansion of the solution gas as the pressure on the core is reduced to atmospheric pressure.[10] If this figure is used for the resident oil saturation in Eqs. (5.3) and (5.4), it should be increased by the formation volume factor. For example, a residual oil saturation of 20% from core analysis indicates a residual reservoir saturation of 30% for an oil volume factor of 1.50 bbl/STB. The residual oil saturation may also be estimated using the data of Table 3.3, which should be applicable to residual oil saturations as well as gas saturations (i.e., in the range of 25 to 40% for the consolidated sandstones studied).

In the reservoir analysis made by Craze and Buckley, some 70 of the 103 fields analyzed produced wholly or partially under water-drive conditions, and the residual oil saturations ranged from 17.9 to 60.9% of the pore space.[8] According to Arps, the data apparently relate according to the reservoir oil viscosity and permeability.[7] The average correlation between oil viscosity and residual oil saturation, both under reservoir conditions, is shown in Table 5.2. Also included in Table 5.2 is the deviation of this trend against average formation permeability. For example, the residual oil saturation under reser-

TABLE 5.2.
Correlation between reservoir oil viscosity, average reservoir permeability, and residual oil saturation (*After* Craze and Buckley[8] and Arps[7])

Reservoir Oil Viscosity (in cp)	*Residual Oil Saturation (percentage of pore space)*
0.2	30
0.5	32
1.0	34.5
2.0	37
5.0	40.5
10.0	43.5
20.0	64.5

Average Reservoir Permeability (in md)	*Deviation of Residual Oil Saturation from Viscosity Trend (percentage of pore space)*
50	+12
100	+ 9
200	+ 6
500	+ 2
1000	− 1
2000	− 4.5
5000	− 8.5

voir conditions for a formation containing 2 cp oil and having an average permeability of 500 md is estimated at 37 + 2, or 39% of the pore space.

Because Craze and Buckley's data were arrived at by comparing recoveries from the reservoir as a whole with the estimated initial content, the residual oil calculated by this method includes a sweep efficiency as well as the residual oil saturation; that is, the figures are higher than the residual oil saturations in those portions of the reservoir invaded by water at abandonment. This sweep efficiency reflects the effect of well location, the bypassing of some of the oil in the less permeable strata, and the abandonment of some leases before the flooding action in all zones is complete, owing to excessive water-oil ratios, both in edge-water and bottom-water drives.

In a statistical study of Craze and Buckley's *water-drive* recovery data, Guthrie and Greenberger, using multiple correlation analysis methods, found the following correlation between water-drive recovery and five variables that affect recovery in *sandstone* reservoirs.[11]

$$RF = 0.114 + 0.272 \log k + 0.256 \, S_w - 0.136 \log \mu_o$$
$$- 1.538\phi - 0.00035 \, h \qquad (5.5)$$

For $k = 1000$ md, $S_w = 0.25$, $\mu_o = 2.0$ cp, $\phi = 0.20$, and $h = 10$ ft,

$$RF = 0.114 + 0.272 \times \log 1000 + 0.256 \times 0.25 - 0.136$$
$$\log 2 - 1.538 \times 0.20 - 0.00035 \times 10$$
$$0.642 \text{ or } 64.2\% \text{ (of initial stock tank oil)}$$

where,

RF = recovery factor

A test of the equation showed that 50% of the fields had recoveries within ± 6.2 recovery % of that predicted by Eq. (5.5), 75% were within ± 9.0 recovery %, and 100% were within ± 19.0 recovery %. For instance, it is 75% probable that the recovery from the foregoing example is $64.2 \pm 9.0\%$.

Although it is usually possible to determine a reasonably accurate recovery factor for a reservoir as a whole, the figure may be wholly unrealistic when applied to a particular lease or portion of a reservoir, owing to the problem of fluid migration in the reservoir, also referred to as lease drainage. For example, a flank lease in a water-drive reservoir may have 50,000 STB of recoverable stock tank oil in place but will divide its reserve with all updip wells in line with it. The degree to which migration may affect the ultimate recoveries from various leases is illustrated in Fig. 5.1.[12] If the wells are located on 40-acre units, if each well has the same daily allowable, if there is uniform permeability, and if the reservoir is under an active water drive so that the

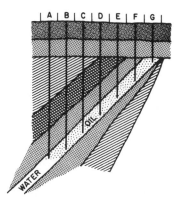

Fig. 5.1. Effect of water drive on oil migration. (*After* Buckley,[12] AIME.)

water advances along a horizontal surface, then the recovery from lease *A* is only one-seventh of the recoverable oil in place, whereas lease *G* recovers one-seventh of the recoverable oil under lease *A*, one-sixth under lease *B*, one-fifth under lease *C*, and so on. Lease drainage is generally less severe with other reservoir mechanisms, but it occurs to some extent in all reservoirs.

3. MATERIAL BALANCE IN UNDERSATURATED RESERVOIRS

The material balance equation for undersaturated reservoirs was developed in Chapter 2 and is

$$N(B_t - B_{ti}) + N B_{ti}\left[\frac{c_w S_{wi} + c_f}{1 - S_{wi}}\right]\Delta \bar{p} + W_e = N_p[B_t + (R_p - R_{soi}) B_g] + B_w W_p \quad (2.8)$$

Neglecting the change in porosity of rocks with the change of internal fluid pressure, which is treated later, reservoirs with zero or negligible water influx are constant volume or volumetric reservoirs. If the reservoir oil is initially undersaturated, then initially it contains only connate water and oil, with their solution gas. The solubility of gas in reservoir waters is generally quite low and is considered negligible for the present discussion. Because the water production from volumetric reservoirs is generally small or negligible, it will be considered as zero. From initial reservoir pressure down to the bubble point, then, the reservoir oil volume remains a constant, and oil, is produced by liquid expansion. Incorporating these assumptions into Eq. (2.8), we get

$$N(B_t - B_{ti}) = N_p[B_t + (R_p - R_{soi})B_g] \quad (5.6)$$

While the reservoir pressure is maintained above the bubble-point pressure and the oil remains undersaturated, only liquid will exist in the reservoir. Any gas that is produced on the surface will be gas coming out of solution as the oil moves up through the wellbore and through the surface facilities. All this gas will be gas that was in solution at reservoir conditions. Therefore, during this period, R_p will equal R_{so} and R_{so} will equal R_{soi} since the solution gas-oil ratio remains constant (see Chapter 1). The material balance equation becomes

$$N(B_t - B_{ti}) = N_p B_t \qquad (5.7)$$

This can be rearranged to yield fractional recovery, RF, as

$$RF = \frac{N_p}{N} = \frac{B_t - B_{ti}}{B_t} \qquad (5.8)$$

The fractional recovery is generally expressed as a fraction of the initial stock tank oil in place. The PVT data for the 3–A–2 reservoir of a field is given in Fig. 5.2.

The formation volume factor plotted in Fig. 5.2 is the single-phase formation volume factor, B_o. The material balance equation has been derived using the two-phase formation volume factor, B_t. B_o and B_t are related by Eq. (1.28)

$$B_t = B_o + B_g(R_{soi} - R_{so}) \qquad (1.28)$$

It should be apparent that $B_t = B_o$ above the bubble-point pressure because R_{so} is constant and equal to R_{soi}.

The reservoir fluid has an oil volume factor of 1.572 bbl/STB at the initial pressure 4400 psia and 1.600 bbl/STB at the bubble-point pressure of 3550 psia. Then by volumetric depletion, the fractional recovery of the stock tank oil at 3550 psia by Eq. (5.8) is

$$RF = \frac{1.600 - 1.572}{1.600} = 0.0175 \text{ or } 1.75\%$$

If the reservoir produced 680,000 STB when the pressure dropped at 3550 psia, then the initial oil in place by Eq. (5.7) is

$$N = \frac{\overset{B_t}{1.600} \times \overset{N_p}{680,000}}{\underset{B_t}{1.600} - \underset{B_{ti}}{1.572}} = 38.8\text{MM STB}$$

Below 3550 psia a *free gas* phase develops; and for a volumetric, undersaturated reservoir with no water production, the hydrocarbon pore volume remains constant, or

$$V_{oi} = V_o + V_g \qquad (5.9)$$

Figure 5.3 shows schematically the changes that occur between initial reservoir pressure and some pressure below the bubble point. The free-gas phase does not necessarily rise to form an *artificial* gas cap, and the equations are the same

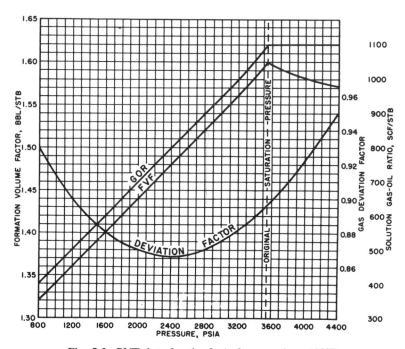

Fig. 5.2. PVT data for the 3–A–2 reservoir at 190°F.

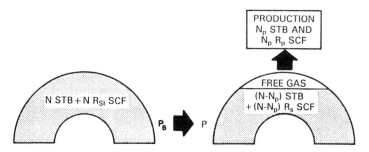

Fig. 5.3. Diagram showing the formation of a free-gas phase in a volumetric reservoir below the bubble point.

if the *free gas* remains distributed throughout the reservoir as isolated bubbles. Equation (5.6) can be rearranged to solve for N and the fractional recovery, RF, for any undersaturated reservoir below the bubble point.

$$N = \frac{N_p[B_t + (R_p - R_{soi})B_g]}{(B_t - B_{ti})} \tag{5.10}$$

$$RF = \frac{N_p}{N} = \frac{(B_t - B_{ti})}{[B_t + (R_p - R_{soi})B_g]} \tag{5.11}$$

The *net cumulative produced gas-oil ratio* R_p is the quotient of all the gas produced from the reservoir G_p and all the oil produced N_p. In some reservoirs, some of the produced gas is returned to the *same* reservoir, so that the *net* produced gas is only that which is not returned to the reservoir. When all the produced gas is returned to the reservoir, R_p is zero.

An inspection of Eq. (5.11) indicates that all the terms except the produced gas-oil ratio, R_p, are functions of pressure only and are the properties of the reservoir fluid. Because the nature of the fluid is fixed, it follows that the fractional recovery RF is fixed by the PVT properties of the reservoir fluid and the produced gas-oil ratio. Since the produced gas-oil ratio occurs in the denominator of Eq. (5.11), large gas-oil ratios give low recoveries and vice versa.

Example 5.1. Calculations to show the effect of the produced gas-oil ratio R_p on fractional recovery in volumetric, undersaturated reservoirs.

Given:

The PVT data for the 3–A–2 reservoir, Fig. 5.1.2
Cumulative GOR at 2800 psia = 3300 SCF/STB.
Reservoir temperature = 190°F = 650°R.
Standard conditions = 14.7 psia and 60°F.

SOLUTION: R_{soi} = 1100 SCF/STB; B_{oi} = 1.572 bbl/STB; R_{so} at 2800 psia = 900 SCF/STB; B_o at 2800 psia = 1.520 bbl/STB; R_p = 3300 SCF/STB; and B_g at 2800 psia calculated as

$$B_g = \frac{znR'T}{5.615\,p} = \frac{0.870 \times 10.73 \times 650}{5.615 \times 379.4 \times 2800} = 0.00102 \text{ bbl/SCF}$$

$$B_t = B_o + B_g(R_{soi} - R_{so})$$

$$B_t = 1.520 + 0.00102(1100 - 900) = 1.724 \text{ bbl/STB}$$

Then at 2800 psia $\left(using\ 5.11 \right)$

$$RF = \frac{1.724 - 1.572}{1.724 + 0.00102(3300 - 1100)}$$

$$= 0.0383, \text{ or } 3.83\%$$

If two-thirds of the produced gas had been returned to the reservoir, at the same pressure (i.e., 2800 psia), the fractional recovery would have been

$$RF = \frac{1.724 - 1.572}{1.724 + 0.00102(1100 - 1100)}$$

$$= 0.088, \text{ or } 8.8\%$$

Equation (5.10) may be used to find the initial oil in place. For example, if 1.486 MM STB had been produced down to 2800 psia, for $R_p = 3300$ SCF/STB, the initial oil in place is

$$N = \frac{1.486 \times 10^6 [1.724 + 0.00102(3300 - 1100)]}{1.724 - 1.572}$$

$$= 38.8 \text{ MM STB}$$

The calculations of Ex. 5.1 for the 3–A–2 reservoir show that for $R_p = 3300$ SCF/STB the recovery at 2800 psia is 3.83%, and that if R_p had been only 1100 SCF/STB, the recovery would have been 8.80%. Neglecting in each case the 1.75% recovery by liquid expansion down to the bubble-point pressure, the effect of reducing the gas-oil ratio by one-third is approximately to triple the recovery. The produced gas-oil ratio can be controlled by working over high gas-oil ratio wells, shutting in or reducing the producing rates of high ratio wells, and/or by returning some or all of the produced gas to the reservoir. If gravitational segregation occurs during production so that a gas cap forms, as shown in Fig. 5.3 and if the producing wells are completed low in the formation, their gas-oil ratios will be lower and recovery will be improved. Simply from the material balance point of view, by returning all produced gas to the reservoir it is possible to obtain 100% recoveries. From the point of view of flow dynamics, however, a practical limit is reached when the reservoir gas saturation rises to values in the range of 10 to 40% because the reservoir becomes so permeable to gas that the returned gas moves rapidly from the injection wells to the production wells, displacing with it only a small quantity of oil. Thus although gas-oil ratio control is important in solution gas-drive reservoirs, recoveries are inherently low because the gas is produced faster than the oil. Outside of the energy stored up in the liquid above the bubble point, the energy for producing the oil is stored up in the solution gas. When this gas has been produced, the only remaining natural source of energy is gravity drainage, and there may be a considerable period in which the oil drains downward to the wells from which it is pumped to the surface.

In the next section, we present a method that allows the material balance equation to be used as a predictive tool. The method was used by engineers performing calculations on the Canyon Reef Reservoir in the Kelly-Snyder Field.

4. KELLY-SNYDER FIELD, CANYON REEF RESERVOIR

The Canyon Reef reservoir of the Kelly-Snyder Field, Texas, was discovered in 1948. During the early years of production, there was much concern about the very rapid decline in reservoir pressure; however, reservoir engineers were able to show that this was to be expected of a volumetric undersaturated reservoir with an initial pressure of 3112 psig and a bubble-point pressure of only 1725 psig, both at a datum of 4300 ft subsea.[13] Their calculations further showed that when the bubble-point pressure is reached, the pressure decline should be much less rapid, and that the reservoir could be produced without pressure maintenance for many years thereafter without prejudice to the pressure maintenance program eventually adopted. In the meantime, with additional pressure drop and production, further reservoir studies could evaluate the potentialities of water influx, gravity drainage, and intrareservoir communication. These, together with laboratory studies on cores to determine recovery efficiencies of oil by depletion and by gas and water displacement, should enable the operators to make a more prudent selection of the pressure maintenance program to be used, or demonstrate that a pressure maintenance program would not be successful.

Although additional and revised data have become available in subsequent years, the following calculations, which were made in 1950 by reservoir engineers, are based on data available in 1950. Table 5.3 gives the basic reservoir data for the Canyon Reef reservoir, Geologic and other evidence indicated that the reservoir was volumetric (i.e., that there would be negligible water influx), so the calculations were based on volumetric behavior. If any water entry should occur, the effect would be to make the calculations more optimistic, that is, there would be more recovery at any reservoir pressure. The reservoir was undersaturated, so the recovery from initial pressure to bubble-point pressure is by liquid expansion, and the fractional recovery at the bubble point is

$$RF = \frac{B_t - B_{ti}}{B_t} = \frac{1.4509 - 1.4235}{1.4509} = 0.0189 \text{ or } 1.89\%$$

Based on an initial content of 1.4235 reservoir barrels or 1.00 STB, this is recovery of 0.0189 STB. Because the solution gas remains at 885 SCF/STB

TABLE 5.3.

Reservoir rock and fluid properties for the Canyon Reef Reservoir of the Kelly-Snyder Field, Texas *(Courtesy, The Oil and Gas Journal* [14]*)*

Initial reservoir pressure	3112 psig (at 4300 ft subsea)
Bubble-point pressure	1725 psig (at 4300 ft subsea)
Average reservoir temperature	125°F
Average porosity	7.7%
Average connate water	20%
Critical gas saturation (estimated)	10%

Differential Liberation Analyses of a Bottom-hole Sample from the Standard Oil Company of Texas No. 2–1, J. W. Brown, at 125°F

Pressure psig	B_o bbl/STB	B_g bbl/SCF	Solution GOR SCF/STB	B_t bbl/STB
3112	1.4235	...	885	1.4235
2800	1.4290	...	885	1.4290
2400	1.4370	...	885	1.4370
2000	1.4446	...	885	1.4446
1725	1.4509	...	885	1.4509
1700	1.4468	0.00141	876	1.4595
1600	1.4303	0.00151	842	1.4952
1500	1.4139	0.00162	807	1.5403
1400	1.3978	0.00174	772	1.5944

down to 1725 psig, the producing gas-oil ratio and the cumulative produced gas-oil ratio should remain near 885 SCF/STB during this pressure decline.

Below 1725 psig, a free gas phase develops in the reservoir. As long as this gas phase remains immobile, it can neither flow to the well bores nor migrate upward to develop a gas cap but must remain distributed throughout the reservoir, increasing in size as the pressure declines. Because pressure changes much less rapidly with reservoir voidage for gases than for liquids, the reservoir pressure declines at a much lower rate below the bubble point. It was estimated that the gas in the Canyon Reef reservoir would remain immobile until the gas saturation reached a value near 10% of the pore volume. When the free gas begins to flow, the calculations become quite complex (see Chapter 9); but as long as the free gas is immobile, calculations may be made assuming that the producing gas-oil ratio R at any pressure will equal the solution gas-oil ratio R_{so} at the pressure, since the only gas that reaches the well bore is that in solution, the free gas being immobile. Then the average producing (daily) gas-oil ratio between any two pressures p_1 and p_2 is approximately

$$R_{avg} = \frac{R_{so1} + R_{so2}}{2} \qquad (5.12)$$

and the cumulative gas-oil ratio at any pressure is

$$R_p = \frac{\Sigma \Delta N_p \times R}{N_p}$$

$$= \frac{N_{pb} \times R_{soi} + (N_{p1} - N_{pb})R_{avg1} + (N_{p2} - N_{p1})R_{avg2} + \text{etc.}}{N_{pb} + (N_{p1} - N_{pb}) + (N_{p2} - N_{p1}) + \text{etc.}} \qquad (5.13)$$

On the basis of 1.00 STB of initial oil, the production at bubble-point pressure N_{pb} is 0.0189 STB. The average producing gas-oil ratio between 1725 and 1600 psig will be

$$R_{avg1} = \frac{885 + 842}{2} = 864 \text{ SCF/STB}$$

The cumulative recovery at 1600 psig N_{p1} is unknown; however the cumulative gas-oil ratio R_p may be expressed by Eq. (5.13) as

$$R_{p1} = \frac{0.0189 \times 885 + (N_{p1} - 0.0189)864}{N_{p1}}$$

This value of R_{p1} may be placed in Eq. (5.11) together with the PVT values at 1600 psig as,

$$N_{p1} = \frac{1.4952 - 1.4235}{1.4952 + 0.00151\left[\dfrac{0.0189 \times 885 + (N_{p1} - 0.0189)864}{N_{p1}} - 885\right]}$$

$$= 0.0486 \text{ STB at 1600 psig}$$

In a similar manner the recovery at 1400 psig may be calculated, the results being valid only if the gas saturation remains below the critical gas saturation, assumed to be 10% for the present calculations.

When N_p stock tank barrels of oil have been produced from a *volumetric undersaturated* reservoir, and the average reservoir pressure is p, the volume of the remaining oil is $(N - N_p)B_o$. Since the initial pore volume of the reservoir V_p is

$$V_p = \frac{NB_{oi}}{(1 - S_{wi})} \qquad (5.14)$$

and since the oil saturation is the oil volume divided by the pore volume,

$$S_o = \frac{(N - N_p)B_o(1 - S_{wi})}{NB_{oi}} \qquad (5.15)$$

On the basis of $N = 1.00$ STB initially, N_p is the fractional recovery RF, or N_p/N, and Eq. (5.15) may be written as

$$S_o = (1 - RF)(1 - S_{wi}) \left(\frac{B_o}{B_{oi}}\right) \tag{5.16}$$

where S_{wi} is the connate water, which is assumed to remain constant for volumetric reservoirs. Then at 1600 psig the oil saturation is

$$S_o = (1 - 0.0486)(1 - 0.20) \left(\frac{1.4303}{1.4235}\right)$$

$$= 0.765$$

The gas saturation is $(1 - S_o - S_{wi})$, or

$$S_g = 1 - 0.765 - 0.200 = 0.035$$

Figure 5.4 shows the calculated performance of the Kelly-Snyder Field down to a pressure of 1400 psig. Calculations were not continued beyond this point because the free gas saturation had reached approximately 10%, the estimated critical gas saturation for the reservoir. The graph shows the rapid pressure decline above the bubble point and the predicted flattening below the

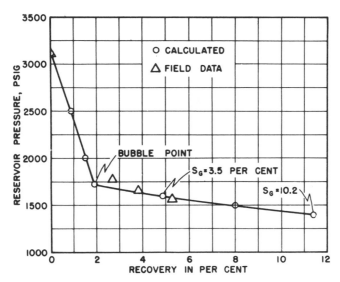

Fig. 5.4. Material balance calculations and performance, Canyon Reef reservoir, Kelly-Snyder Field.

bubble point. The predictions are in good agreement with the field performance, which is calculated in Table 5.4 using field pressures and production
data, and a value of 2.25 MMM STB for the initial oil in place. The producing
gas-oil ratio, Col. (2), increases instead of decreasing, as predicted by the
previous theory. This is due to the more rapid depletion of some portions of
the reservoir—for example, those drilled first, those of low net productive
thickness, and those in the vicinity of the well bores. For the present predictions, it is pointed out that the previous calculations would not be altered
greatly if a constant producing gas-oil ratio of 885 SCF/STB (i.e., the initial
dissolved ratio) had been assumed throughout the entire calculation.

The initial oil under a 40 acre unit of the Canyon Reef reservoir for a net
formation thickness of 200 feet is

$$N = \frac{7758 \times 40 \times 200 \times 0.077 \times (1 - 0.20)}{1.4235}$$

$$= 2.69 \text{MM STB}$$

Then at the average daily well rate of 92 BOPD in 1950, the time to produce
11.35% of the initial oil (i.e., at 1400 psig when the gas saturation is calculated
to be near 10%) is

$$t = \frac{0.1135 \times 2.69 \times 10^6}{92 \times 365} \cong 9.1 \text{ years}$$

By means of this calculation, the reservoir engineers were able to show that
there was no immediate need for a curtailment of production and that there
was plenty of time in which to make further reservoir studies and carefully

TABLE 5.4.
Recovery from Kelly-Snyder Canyon Reef Reservoir based on production
data and measured average reservoir pressures, and assuming an initial oil
content of 2.25 MMM STB

(1)	(2)	(3)	(4)	(5)
				Percentage
Pressure	*Ave. Producing*	*Incremental*	*Cumulative*	*Recovery*
Interval,	*Gas-Oil Ratio,*	*Oil Production,*	*Oil Production,*	*(N = 2.25 MMM*
psig	*SCF/STB*	*MM STB*	*MM STB*	*STB)*
3312 to 1771	896	60.421	60.421	2.69
1771 to 1713	934	11.958	72.379	3.22
1713 to 1662	971	13.320	85.699	3.81
1662 to 1570	1023	20.009	105.708	4.70
1570 to 1561	1045	11.864	117.572	5.23

considered plans for the optimum pressure maintenance program. Following comprehensive and exhaustive studies by engineers, the field was unitized in March, 1953, and placed under the management of an operating committee. This group proceeded to put into operation a pressure maintenance program consisting of (a) water injection into wells located along the longitudinal axis of the field, and (b) shutting in the high gas-oil ratio wells and transferring their allowables to low gas-oil ratio wells. The high-ratio wells were shut in as soon as the field was unitized, and water injection was started in 1954. The operation has gone as planned and approximately 50% of the initial oil in place has been recovered, in contrast to approximately 25% by primary depletion, an increase of approximately 600 MM STB of recoverable oil.[15]

5. THE GLOYD-MITCHELL ZONE OF THE RODESSA FIELD

Many reservoirs are of the volumetric undersaturated type and their production, therefore, is controlled largely by the solution gas-drive mechanism. In many cases, the mechanism is altered to a greater or lesser extent by gravitational segregation of the gas and oil, by small water drives, and by pressure maintenance, all of which improve recovery. The important characteristics of this type of production may be summarized as follows and observed in the graph of Fig. 5.5 for the Gloyd-Mitchell zone of the Rodessa Field. Above the bubble point, the reservoir is produced by *liquid-expansion,* and there is a rapid decline in reservoir pressure that accompanies the recovery of a fraction of 1% to a few percentage points of the initial oil in place. The gas-oil ratios remain low and generally near the value of the initial solution gas-oil ratio. Below the bubble point, a gas phase develops that in most cases is immobile until the gas saturation reaches the critical gas saturation in the range of a few percentage points to 20%. During this period, the reservoir produces by *gas expansion,* which is characterized by a much slower decline in pressure and gas-oil ratios near, or in some cases even below the initial solution gas-oil ratio. After the critical gas saturation is reached, free gas begins to flow. This reduces the oil flow rate and depletes the reservoir of its main source of energy. By the time the gas saturation reaches a value usually in the range of 15 to 30%, the flow of oil is small compared with the gas (high gas-oil ratios), and the reservoir gas is rapidly depleted. At abandonment the recoveries are usually in the range of 10 to 25% by the solution gas-drive mechanism alone, but they may be improved by gravitational segregation and the control of high gas-oil ratio wells.

The production of the Gloyd-Mitchell zone of the Rodessa Field, Louisiana, is a good example of a reservoir which produced during the major portion of its life by the dissolved gas-drive mechanism.[16] Reasonably accurate data on this reservoir relating to oil and gas production, reservoir pressure decline, and sand thickness, and the number of producing wells provide an excellent example of the theoretical features of the dissolved gas-drive mechanism. The Gloyd-Mitchell zone is practically flat and produced an oil of 42.8°API gravity which, under the original bottom-hole pressure of 2700 psig, had a solution gas-oil ratio of 627 SCF/STB. There was no free gas originally present, and there is no evidence of an active water drive. The wells were produced at high rates and had a rapid decline in production. The behavior of the gas-oil ratios, reservoir pressures, and oil production, had the characteristics expected of a dissolved gas drive, although there is some evidence that there was a modification of the recovery mechanism in the later stages of depletion. The ultimate recovery was estimated at 20% of the initial oil in place.

Many unsuccessful attempts were made to decrease the gas-oil ratios by shutting in the wells, by blanking off upper portions of the formation in producing wells, and by perforating only the lowest sand members. The failure to reduce the gas-oil ratios is typical of the dissolved gas-drive mechanism, because when the critical gas saturation is reached, the gas-oil ratio is a function of the decline in reservoir pressure or depletion and is not materially changed by production rate or completion methods. Evidently there was negligible gravitational segregation by which an artificial gas cap develops and causes abnormally high gas-oil ratios in wells completed high on the structure or in the upper portion of the formation.

Table 5.5 gives the number of producing wells, average daily production, average gas-oil ratio, and average pressure for the Gloyd-Mitchell zone. The daily oil production per well, monthly oil production, cumulative oil production, monthly gas production, cumulative gas production, and cumulative gas-oil ratios have been calculated from these figures. The source of data is of interest. The number of producing wells at the end of any period is obtained either from the operators in the field, from the completion records as filed with the state regulatory body, or from the periodic potential tests. The average daily oil production is available from the monthly production reports filed with the state regulatory commission. Accurate values for the average daily gas-oil ratios can be obtained only when all the produced gas is metered. Alternatively, this information is obtained from the potential tests. To obtain the average daily gas-oil ratio from potential tests during any month, the gas-oil ratio for each well is multiplied by the daily oil allowable or daily production rate for the same well, giving the total daily gas production. The average daily gas-oil ratio for any month is the total daily gas production from all producing wells divided by the total daily oil production from all the wells involved. For example, if the gas-oil ratio of well A is 1000 SCF/STB and the daily rate is 100 bbl/day, and the ratio of well B is 4000 SCF/STB and the daily rate is 50 bbl/day, then the average daily gas-oil ratio R of the two wells is

$$R = \frac{1000 \times 100 + 4000 \times 50}{150} = 2000 \text{ SCF/STB}$$

This figure is lower than the arithmetic average ratio of 2500 SCF/STB. The average gas-oil ratio of a large number of wells, then, can be expressed by

$$R_{\text{avg}} = \frac{\Sigma R \times q_o}{\Sigma q_o} \tag{5.17}$$

where R and q_o are the individual gas-oil ratios and stock tank oil production rates.

Figure 5.5 shows, plotted in block diagram, the number of producing wells, the daily gas-oil ratio, and the daily oil production per well. Also in a smooth curve, pressure is plotted against time. The initial increase in daily oil production is due to the increase in the number of producing wells, and not to the improvement in individual well rates. If all the wells had been completed and put on production at the same time, the daily production rate would have been a plateau during the time all the wells could make their allowables, followed by an exponential decline, which is shown beginning at 16 months after the start of production. Since the daily oil allowable and daily production of a well are dependent on the bottom-hole pressure and gas-oil ratio, the oil recovery is larger for wells completed early in the life of a field. Because the controlling factor in this type of mechanism is gas flow in the reservoir, the rate of production has no material effect on the ultimate recovery, unless some gravity drainage occurs. Likewise, well spacing has no proven effect on recovery; however, well spacing and production rate directly affect the economic return.

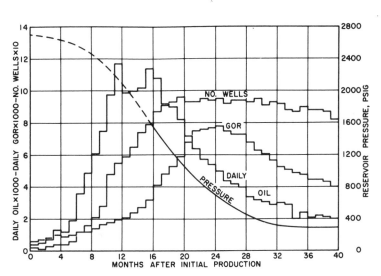

Fig. 5.5. Development, production, and reservoir pressure curves for the Gloyd-Mitchell Zone, Rodessa Field, Louisiana.

TABLE 5.5.
Average monthly production data, Gloyd-Mitchell Zone of the Rodessa Field

(1) Months after Start of Production	(2) No. Wells	(3) Average Daily Oil, Barrels	(4) Average Daily GOR, SCF/STB	(5) Average Pressure, psig	(6) Daily Oil per Well, (3) ÷ (2)	(7) Monthly Oil, Barrels, 30.4 × (3)	(8) Cumulative Oil, Barrels, Sum (7)	(9) Monthly Gas, M SCF, (4) × (7)	(10) Cumulative Gas, M SCF, Sum (9)	(11) Cumulative GOR, SCF/STB, 10 ÷ (8)
1	2	400	625	2700[a]	200	12,160	12,160	7,600	7,600	625
2	1	500	750		500	15,200	27,360	11,400	19,000	694
3	3	700	875		233	21,280	48,640	18,620	37,620	773
4	4	1,300	1,000	2490	325	39,520	88,160	39,520	77,140	875
5	4	1,200	950		300	36,480	124,640	34,656	111,796	897
6	6	1,900	1,000		316	57,760	182,400	57,760	169,556	930
7	12	3,600	1,200	2280	300	109,440	291,840	131,328	300,884	1031
8	16	4,900	1,200		306	148,960	440,800	178,752	479,636	1088
9	21	6,100	1,400		290	185,440	626,240	259,616	739,252	1181
10	28	7,500	1,700	2070	268	228,000	854,240	387,600	1,127M	1319
11	48	9,800	1,800		204	297,920	1,152,160	536,256	1,663M	1443
12	55	11,700	1,900		213	355,680	1,507,840	675,792	2,339M	1551
13	59	9,900	2,100	1860	168	300,960	1,808,800	632,016	2,971M	1643
14	65	10,000	2,400		154	304,000	2,112,800	729,600	3,701M	1752
15	74	10,200	2,750		138	310,080	2,422,880	852,720	4,554M	1880
16	79	11,400	3,200	1650[b]	144	346,560	2,769,440	1,108,992	5,662M	2045
17	87	10,800	4,100		124	328,320	3,097,760	1,346,112	7,008M	2262
18	91	9,200	4,800		101	279,680	3,377,440	1,342,464	8,351M	2473
19	93	9,000	5,300	1250	97	273,600	3,651,040	1,450,080	9,801M	2684

[a] Linear decline assumed from original to first measured BHP.
[b] First measured reservoir pressure. All subsequent pressures are measured.

TABLE 5.5.
(Continued)

(1) Months after Start of Production	(2) No. Wells	(3) Average Daily Oil, Barrels	(4) Average Daily GOR, SCF/STB	(5) Average Pressure, psig	(6) Daily Oil per Well, (3) ÷ (2)	(7) Monthly Oil, Barrels, 30.4 × (3)	(8) Cumulative Oil, Barrels, Sum (7)	(9) Monthly Gas, M SCF, (4) × (7)	(10) Cumulative Gas, M SCF, Sum (9)	(11) Cumulative GOR, SCF/STB, 10 ÷ (8)
20	96	8,300	5,900	1115	86	252,320	3,903,360	1,488,688	11,290M	2892
21	93	7,200	6,800	1000	77	218,880	4,122,240	1,488,384	12,778M	3100
22	93	6,400	7,500	900	69	194,560	4,316,800	1,459,200	14,237M	3298
23	95	5,800	7,600	825	61	176,320	4,493,120	1,340,032	15,577M	3467
24	94	5,400	7,700	740	57	164,160	4,657,280	1,264,032	16,841M	3616
25	95	5,000	7,800	725	53	152,000	4,809,280	1,185,600	18,027M	3748
26	92	4,400	7,500	565	48	133,760	4,943,040	1,003,200	19,030M	3850
27	94	4,200	7,300	530	45	127,680	5,070,720	932,064	19,962M	3937
28	94	4,000	7,300	500	43	121,600	5,192,320	887,680	20,850M	4016
29	93	3,400	6,800	450	37	103,360	5,295,680	702,848	21,553M	4070
30	95	3,200	6,300	405	34	97,280	5,392,960	612,864	22,165M	4110
31	91	3,100	6,100	350	34	94,240	5,487,200	574,864	22,740M	4144
32	93	2,900	5,700	310	31	88,160	5,575,360	502,512	23,243M	4169
33	92	3,000	5,300	390	33	91,200	5,666,560	483,360	23,726M	4187
34	88	2,900	5,100	300	33	88,160	5,754,720	449,616	24,176M	4201
35	87	2,000	4,900	280	23	60,800	5,815,520	297,920	24,474M	4208
36	90	2,400	4,800	310	27	72,960	5,888,480	350,208	24,824M	4216
37	88	2,100	4,500	300	24	63,840	5,952,320	287,280	25,111M	4219
38	88	2,200	4,500	325	25	66,880	6,019,200	300,960	25,412M	4222
39	87	2,100	4,300	300	24	63,840	6,083,040	274,512	25,687M	4223
40	82	2,000	4,000	275	24	60,800	6,143,840	243,200	25,930M	4220
41	85	2,100	3,600	225	25	63,840	6,207,680	229,824	26,160M	4214

The rapid increase in gas-oil ratios in the Rodessa Field led to the enactment of a gas-conservation order. In this order, oil and gas production were allocated partly on a volumetric basis to restrict production from wells with high gas-oil ratios. The basic ratio for oil wells was set at 2000 SCF/bbl. For leases on which the wells produced more than 2000 SCF/STB, the allowable in barrels per day per well, based on acreage and pressure, was multiplied by 2000 and divided by the gas-oil ratio of the well. This cut in production produced a double hump in the daily production curve.

In addition to a graph showing the production history versus *time,* it is usually desirable to have a graph that shows the production history plotted versus the *cumulative produced oil.* Figure 5.6 is such a plot for the Gloyd-Mitchell zone data and is also obtained from Table 5.5. This graph shows some features that do not appear in the time graph. For example, a study of the reservoir pressure curve shows the Gloyd-Mitchell zone was producing by liquid expansion until approximately 200,000 bbl were produced. This was followed by a period of production by gas expansion with a limited amount of free gas flow. When approximately 3 million bbl had been produced, the gas began to flow much more rapidly than the oil, resulting in a rapid increase in the gas-oil ratio. In the course of this trend, the gas-oil ratio curve reached a maximum, then declined as the gas was depleted and the reservoir pressure approached zero. The decline in gas-oil ratio beginning after approximately 4.5 million bbl were produced was due mainly to the expansion of the flowing reservoir gas as pressure declined. Thus the same gas-oil ratio in standard cubic feet per day gives approximately twice the reservoir flow rate at 400 psig

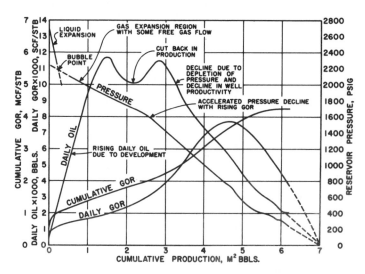

Fig. 5.6. History of the Gloyd-Mitchell Zone of the Rodessa Field plotted versus cumulative recovery.

as at 800 psig; hence, the surface gas-oil ratio may decline and yet the ratio of the rate of flow of gas to the rate of flow of oil *under reservoir conditions* continues to increase. It may also be reduced by the occurrence of some gravitational segregation, and also, from a quite practical point of view, by the failure of operators to measure or report gas production on wells producing fairly low volumes of low-pressure gas.

The results of a differential gas-liberation test on a bottom-hole sample from the Gloyd zone show that the solution gas-oil ratio was 624 SCF/STB, which is in excellent agreement with the initial producing gas-oil ratio of 625 SCF/STB.[17] In the absence of gas-liberation tests on a bottom-hole sample, the initial gas-oil ratio of a properly completed well in either a dissolved gas drive, gas cap drive, or water-drive reservoir, is usually a reliable value to use for the initial solution gas-oil ratio of the reservoir. The extrapolations of the pressure, oil rate, and producing gas-oil ratio curves on the cumulative oil plot all indicate an ultimate recovery of about 7 million bbl. However, no such extrapolation can be made on the time plot. It is also of interest that whereas the daily producing rate is exponential on the time plot, it is close to a straight line on the cumulative oil plot.

The average gas-oil ratio during any production interval and the cumulative gas-oil ratio may be indicated by integrals and shaded areas on a typical daily gas-oil ratio versus cumulative stock tank oil production curve as shown in Fig. 5.7. If R represents the daily gas-oil ratio at any time, and N_p the cumulative stock tank production at the same time, then the production during a short interval of time is dN_p and the total volume of gas produced during that production interval is $R\, dN_p$. The gas produced over a longer period when the gas-oil ratio is changing is given by

$$\Delta G_p = \int_{N_{p1}}^{N_{p2}} R\, dN_p \qquad (5.18)$$

The shaded area between N_{p1} and N_{p2} is proportional to the gas produced during the interval. The average daily gas-oil ratio during the production

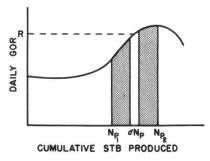

Fig. 5.7. Typical daily gas-oil ratio curve for a dissolved gas drive reservoir.

interval equals the area under the gas-oil ratio curve between N_{p1} and N_{p2} in units given by the coordinate scales, divided by the oil produced in the interval $(N_{p2} - N_{p1})$, and

$$R_{avg} = \frac{\int_{N_{p1}}^{N_{p2}} R \, dN_p}{(N_{p2} - N_{p1})} \qquad (5.19)$$

The cumulative gas-oil ratio, R_p, is the total net gas produced up to any period divided by the total oil produced up to that period, or

$$R_p = \frac{\int_0^{N_p} R \, dN_p}{N_p} \qquad (5.20)$$

The cumulative produced gas-oil ratio was calculated in this manner in Col. (11) of Table 5.5. For example, at the end of the third period,

$$R_p = \frac{625 \times 12{,}160 + 750 \times 15{,}200 + 875 \times 21{,}280}{12{,}160 + 15{,}200 + 21{,}280} = 773 \text{ SCF/STB}$$

6. CALCULATIONS INCLUDING FORMATION AND WATER COMPRESSIBILITIES

In Chapter 1, it was shown that both formation and water compressibilities are functions of pressure. This suggests that there are in fact no volumetric reservoirs—that is, those in which the hydrocarbon pore volume of the reservoir remains constant. Hall showed the magnitude of the effect of formation compressibility on volumetric reservoir calculations.[18] The term *volumetric*, however, is retained to indicate those reservoirs in which there is no water influx but in which volumes change slightly with pressure due to the effects just mentioned.

The effect of compressibilities above the bubble point on calculations for N are examined first. Equation (2.8), with $R_p = R_{soi}$ above the bubble point becomes

$$N(B_t - B_{ti}) + N B_{ti} \left[\frac{c_w S_{wi} + c_f}{1 - S_{wi}} \right] \Delta \bar{p} + W_e = N_p B_t + B_w W_p \qquad (5.21)$$

This equation may be rearranged to solve for N

$$N = \frac{N_p B_t - W_e + B_w W_p}{B_t - B_{ti} + B_{ti} \left[\dfrac{c_w S_{wi} + c_f}{1 - S_{wi}} \right] \Delta \bar{p}} \qquad (5.22)$$

Although this equation is entirely satisfactory, often an oil compressibility, c_o, is introduced with the following defining relationship

$$c_o = \frac{v_o - v_{oi}}{v_{oi}(p_i - \bar{p})} = \frac{B_o - B_{oi}}{B_{oi}\Delta\bar{p}}$$

and

$$B_o = B_{oi} + B_{oi}c_o\Delta\bar{p} \qquad (5.23)$$

The definition of c_o uses the single-phase formation volume factor, but it should be apparent that as long as the calculations are being conducted above the bubble point, $B_o = B_t$. If Eq. (5.23) is substituted into the first term in Eq. (5.21), the result is

$$N(B_{ti} - B_{ti}) + NB_{ti}c_o\Delta\bar{p} + N\,B_{ti}\left[\frac{c_w\,S_{wi} + c_f}{1 - S_{wi}}\right]\Delta\bar{p} = N_pB_t - W_e + B_wW_p \qquad (5.24)$$

Multiplying both the numerator and the denominator of the term containing c_o by S_o and realizing that above the bubble point there is no gas saturation, $S_o = 1 - S_{wi}$, Eq. (5.24) becomes

$$NB_{ti}\left[\frac{c_oS_o\Delta\bar{p}}{1 - S_{wi}}\right] + N\,B_{ti}\left[\frac{c_w\,S_{wi} + c_f}{1 - S_{wi}}\right]\Delta\bar{p} = N_pB_t - W_e + B_wW_p$$

or

$$NB_{ti}\left[\frac{c_oS_o + c_wS_{wi} + c_f}{1 - S_{wi}}\right]\Delta\bar{p} = N_pB_t - W_e + B_wW_p \qquad (5.25)$$

The expression in brackets of Eq. (5.25) is called the *effective fluid compressibility*, c_e, which includes the compressibilities of the oil, the connate water, and the formation, or

$$c_e = \frac{c_oS_o + c_wS_{wi} + c_f}{1 - S_{wi}} \qquad (5.26)$$

Finally, Eq. (5.26) may be written as

$$NB_{ti}c_e\Delta\bar{p} = N_pB_t - W_e + B_wW_p \qquad (5.27)$$

For volumetric reservoirs, $W_e = 0$ and W_p is generally negligible, and Eq. (5.27) can be rearranged to solve for N.

$$N = \frac{N_p}{c_e\Delta\bar{p}}\left(\frac{B_t}{B_{ti}}\right) \qquad (5.28)$$

Finally, if the formation and water compressibilities, c_f and c_w both equal zero, then c_e is simply c_o, and Eq. (5.28) reduces to Eq. (5.8) derived in Sect. 3, for production above the bubble point.

$$\frac{N_p}{N} = \frac{B_t - B_{ti}}{B_{ti}}$$

Example 5.3 shows the use of Eq. (5.22) and (5.28) to find the initial oil in place from the pressure-production data of a reservoir that all geologic evidence indicates is volumetric (i.e., it is bounded on all sides by impermeable rocks). Because the equations are basically identical, they give the same calculation of initial oil, 51.73 MM STB. A calculation is also included to show that an error of 61% is introduced by neglecting the formation and water compressibilities.

Example 5.3. Calculation of initial oil in place in a volumetric, undersaturated reservoir.

Given:

$B_{ti} = 1.35469$ bbl/STB

B_t at 3600 psig $= 1.37500$ bbl/STB

Connate water $= 0.20$

$c_w = 3.6 \ (10)^{-6}$ psi^{-1}

B_w at 3600 psig $= 1.04$ bbl/STB

$c_f = 5.0 \ (10)^{-6}$ psi^{-1}

$p_i = 5000$ psig

$N_p = 1.25$ MM STB

$\Delta \bar{p}$ at 3600 psig $= 1400$ psi

$W_p = 32,000$ STB

$W_e = 0$

SOLUTION: Substituting into Eq. (5.23)

$$N = \frac{1,250,000(1.37500) + 32,000(1.04)}{1.37500 - 1.35469 + 1.35469 \left[\dfrac{3.6(10^{-6})(0.20) + 5.0(10^{-6})}{1 - 0.20} \right](1400)}$$

$$= 51.73 \text{ MM STB}$$

The average compressibility of the reservoir oil is

$$c_o = \frac{B_o - B_{oi}}{B_{oi}\Delta \bar{p}} = \frac{1.375 - 1.35469}{1.35469 \ (5000 - 3600)} = 10.71 \ (10)^{-6} \text{ psi}^{-1}$$

and the effective fluid compressibility by Eq. (5.26) is

$$c_e = \frac{[0.8(10.71) + 0.2(3.6) + 5.0]10^{-6}}{0.8} = 17.86 \ (10)^{-6} \ psi^{-1}$$

Then the initial oil in place by Eq. (5.28) is

$$N = \frac{1{,}250{,}000(1.37500) + 32{,}000(1.04)}{17.86(10)^{-6}(1400)1.35469} = 51.73 \ MM \ STB$$

If the water and formation compressibilities are neglected, $c_e = c_o$, and the initial oil in place is calculated to be

$$N = \frac{1{,}250{,}000(1.37500) + 32{,}000(1.04)}{10.71(10)^{-6}(1400)1.35469} = 86.25 \ MM \ STB$$

As can be seen from the example calculations, the inclusion of the compressibility terms significantly affects the value of N. This is true above the bubble point where the oil-producing mechanism is depletion, or the swelling of reservoir fluids. After the bubble point is reached, the compressibilities have a much smaller effect on the calculations.

When Eq. (2.8) is rearranged and solved for N, we get the following:

$$N = \frac{N_p[B_t + (R_p - R_{soi})B_g] - W_e + B_w W_p}{B_t - B_{ti} + B_{ti}\left[\dfrac{c_w S_{wi} + c_f}{1 - S_{wi}}\right]\Delta\bar{p}} \tag{5.29}$$

This is the general material balance equation written for an undersaturated reservoir below the bubble point. The effect of water and formation compressibilities are accounted for in this equation. Example Problem 5.4 compares the calculations for recovery factor, N_p/N, for an undersaturated reservoir with and without including the effects of the water and formation compressibilities.

Example 5.4 Calculation of N_p/N for an undersaturated reservoir with no water production and neglible water influx. The calculation is performed with and without including the effect of compressibilities. Assume that the critical gas saturation is not reached until after the reservoir pressure drops below 2200 psia.

Given:

$p_i = 4000$ psia	$c_w = 3 \times 10^{-6} \ psi^{-1}$
$p_b = 2500$ psia	$c_f = 5 \times 10^{-6} \ psi^{-1}$
$S_w = 30\%$	$\phi = 10\%$

Pressure psia	R_{so} SCF/STB	B_g bbl/SCF	B_t bbl/SCF
4000	1000	0.00083	1.3000
2500	1000	0.00133	1.3200
2300	920	0.00144	1.3952
2250	900	0.00148	1.4180
2200	880	0.00151	1.4412

SOLUTION: The calculations are performed first by including the effect of compressibilities. Equation (5.22) is used to calculate the recovery at the bubble point.

$$\frac{N_p}{N} = \frac{B_t - B_{ti} + B_{ti}\left[\frac{c_w S_{wi} + c_f}{1 - S_{wi}}\right]\Delta\bar{p}}{B_t}$$

$$\frac{N_p}{N} = \frac{1.32 - 1.30 + 1.30\left[\frac{3(10^{-6})0.3 + 5(10^{-6})}{1 - 0.3}\right]1500}{1.32} = 0.0276$$

Below the bubble point, Eqs. (5.29) and (5.13) are used to calculate the recovery.

$$\frac{N_p}{N} = \frac{B_t - B_{ti} + B_{ti}\left[\frac{c_w S_{wi} + c_f}{1 - S_{wi}}\right]\Delta\bar{p}}{B_t + (R_p - R_{soi})B_g}$$

and

$$R_p = \frac{\Sigma(\Delta N_p)R}{N_p} = \frac{\Sigma(\Delta N_p/N)R}{N_p/N}$$

During the pressure increment 2500 − 2300 psia, the calculations yield

$$\frac{N_p}{N} = \frac{1.3952 - 1.30 + 1.30\left[\frac{3(10^{-6})0.3 + 5(10^{-6})}{1 - 0.3}\right]1700}{1.3952 + (R_p - 1000)0.00144}$$

$$R_p = \frac{0.0276(1000) + (N_p/N - 0.0276)R_{ave1}}{N_p/N}$$

where R_{ave1} equals the average value of the solution GOR during the pressure increment.

$$R_{ave1} = \frac{1000 + 920}{2} = 960$$

Solving these three equations for N_p/N yields

$$\frac{N_p}{N} = 0.08391$$

Repeating the calculations for the pressure increment $2300 - 2250$ psia, the N_p/N is found to be

$$\frac{N_p}{N} = 0.11754$$

Now, the calculations are performed by assuming that the effect of including the compressibility terms is negligible. For this case, at the bubble point, the recovery can be calculated by using Eq. (5.8)

$$\frac{N_p}{N} = \frac{B_t - B_{ti}}{B_t} = \frac{1.32 - 1.30}{1.32} = 0.01515$$

Below the bubble point, Eq. (5.11) and (5.13) are used to calculate N_p/N

$$\frac{N_p}{N} = \frac{B_t - B_{ti}}{B_t + (R_p - R_{soi})B_g}$$

and

$$R_p = \frac{\Sigma(\Delta N_p)R}{N_p} = \frac{\Sigma(\Delta N_p/N)R}{N_p/N}$$

For the pressure increment $2500 - 2300$ psia

$$\frac{N_p}{N} = \frac{1.3952 - 1.30}{1.3952 + (R_p - 1000)0.00144}$$

$$R_p = \frac{0.01515(1000) + (N_p/N - 0.01515)R_{ave1}}{N_p/N}$$

where R_{ave1} is given by

$$R_{ave1} = \frac{1000 + 920}{2} = 960$$

Solving these three equations yields

$$\frac{N_p}{N} = 0.07051$$

Repeating the calculations for the pressure increment 2300 − 2250 psia:

$$\frac{N_p}{N} = 0.08707$$

For the pressure increment 2250 − 2200 psia

$$\frac{N_p}{N} = 0.10377$$

Figure 5.8 is a plot of the results for the two different cases—that is, with and without considering the compressibility term.

The calculations suggest that there is a very significant difference in the results of the two cases down to the bubble point. The difference is the result of the fact that the rock and water compressibilities are on the same order of magnitude as the oil compressibility. By including them, the fractional recovery has been significantly affected. The case that used the rock and water compressibilities comes closer to simulating real production above the bubble point from this type of reservoir. This is because the actual mechanism of oil

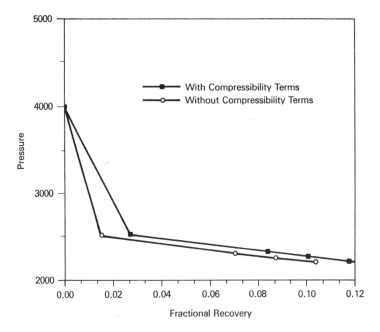

Fig. 5.8. Pressure versus fractional recovery for the calculations of Example Problem 5.4.

production is the expansion of the oil, water, and rock phases; there is no free gas phase.

Below the bubble point, the magnitude of the fractional recoveries calculated by the two schemes still differ by about what the difference was at the bubble point, suggesting that below the bubble point, the compressibility of the gas phase is so large that the water and rock compressibilities do not contribute significantly to the calculated fractional recoveries. This corresponds to the actual mechanism of oil production below the bubble point where gas is coming out of solution and free gas is expanding as the reservoir pressure declines.

The results of the calculations of Ex. 5.4 are meant to help you to understand the fundamental production mechanisms that occur in undersaturated reservoirs. They are not meant to suggest that the calculations can be made easier by ignoring terms in equations for particular reservoir situations. The calculations are relatively easy to perform, whether or not all terms are included. Since nearly all calculations are conducted with the use of a computer, there is no need to neglect terms from the equations.

PROBLEMS

5.1 Using the letter symbols for reservoir engineering, write expressions for the following terms for a volumetric, undersaturated reservoir:

   (a) The initial reservoir oil in place in stock tank barrels.

   (b) The fractional recovery after producing N_p STB.

   (c) The volume occupied by the remaining oil (liquid) after producing N_p STB.

   (d) The SCF of gas produced.

   (e) The SCF of initial gas.

   (f) The SCF of gas in solution in the remaining oil.

   (g) By difference, the SCF of escaped or free gas in the reservoir after producing N_p STB.

   (h) The volume occupied by the escaped, or free, gas.

5.2 The physical characteristics of the 3–A–2 reservoir are given in Fig. 5.2:

   (a) Calculate the percentage of recovery, assuming this reservoir could be produced at a constant cumulative produced gas-oil ratio of 1100 SCF/STB, when the pressure falls to 3550, 2800, 2000, 1200, and 800 psia. Plot the percentage of recovery versus pressure.

   (b) To demonstrate the effect of increased GOR on recovery, recalculate the recoveries, assuming that the cumulative produced GOR is 3300 SCF/STB Plot the percentage of recovery versus pressure on the same graph used for the previous problem.

(c) To a first approximation, what does tripling the produced GOR do to the percentage of recovery?

(d) Does this make it appear reasonable that to improve recovery high-ratio (GOR) wells should be worked over or shut in when feasible?

5.3 If 1 million STB of oil have been produced from the 3–A–2 reservoir at a cumulative produced GOR of 2700 SCF/STB, causing the reservoir pressure to drop from the initial reservoir pressure of 4400 psia to 2800 psia, what is the initial stock tank oil in place?

5.4 The following data are taken from an oil field that had no original gas cap and no water drive:

Due
3/2

        Oil pore volume of reservoir = 75 MM cu ft
        Solubility of gas in crude = 0.42 SCF/STB/psi
        Initial bottom-hole pressure = 3500 psia
        Bottom-hole temperature = 140°F
        Bubble-point pressure of the reservoir = 2400 psia
        Formation volume factor at 3500 psia = 1.333 bbl/STB
        Compressibility factor of the gas at 1500 psia and 140°F = 0.95
        Oil produced when pressure is 1500 psia = 1.0 MM STB
        Net cumulative produced GOR = 2800 SCF/STB

(a) Calculate the initial STB of oil in the reservoir.

(b) Calculate the initial SCF of gas in the reservoir.

(c) Calculate the initial dissolved GOR of the reservoir.

(d) Calculate the SCF of gas remaining in the reservoir at 1500 psia.

(e) Calculate the SCF of free gas in the reservoir at 1500 psia.

(f) Calculate the gas volume factor of the escaped gas at 1500 psia at standard conditions of 14.7 psia and 60°F.

(g) Calculate the reservoir volume of the free gas at 1500 psia.

(h) Calculate the total reservoir GOR at 1500 psia.

(i) Calculate the dissolved GOR at 1500 psia.

(j) Calculate the liquid volume factor of the oil at 1500 psia.

(k) Calculate the total, or two-phase, oil volume factor of the oil and its initial complement of dissolved gas at 1500 psia.

5.5 (a) Continuing the calculations of the Kelly-Snyder Field, calculate the fractional recovery and the gas saturation at 1400 psig.

(b) What is the deviation factor for the gas at 1600 psig and BHT 125°F?

5.6 The R Sand is a volumetric oil reservoir whose PVT properties are shown in Fig. 5.9. When the reservoir pressure dropped from an initial pressure of 2500 psia to an average pressure of 1600 psia, a total of 26.0 MM STB of oil had been produced. The cumulative GOR at 1600 psia was 954 SCF/STB, and the current GOR was 2250 SCF/STB. The average porosity for the field is 18% and average

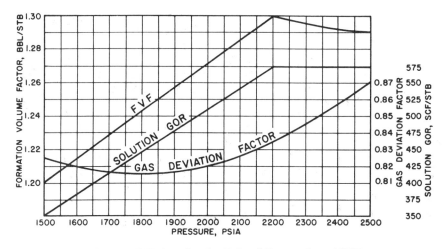

Fig. 5.9. PVT data for the R Sand Reservoir at 150°F.

connate water 18%. No appreciable amount of water was produced, and standard conditions were 14.7 psia and 60°F.

(a) Calculate the initial oil in place.

(b) Calculate the SCF of evolved gas remaining in the reservoir at 1600 psia.

(c) Calculate the average gas saturation in the reservoir at 1600 psia.

(d) Calculate the barrels of oil that would have been recovered at 1600 psia if all the produced gas had been returned to the reservoir.

(e) Calculate the two-phase volume factor at 1600 psia.

(f) Assuming no free gas flow, calculate the recovery expected by depletion drive performance down to 2000 psia.

(g) Calculate the initial SCF of free gas in the reservoir at 2500 psia.

5.7 If the reservoir of Prob. 5.6 had been a water-drive reservoir, in which 25×10^6 bbl of water had encroached into the reservoir when the pressure had fallen to 1600 psia, calculate the initial oil in place. Use the same current and cumulative GORs, the same PVT data, and assume no water production.

5.8 The following production and gas injection data pertain to a reservoir.

(a) Calculate the average producing GOR during the production interval from 6 MM to 8 MM STB.

(b) What is the cumulative produced GOR when 8 MM STB has been produced?

(c) Calculate the net average producing GOR during the production interval from 6 MM to 8 MM STB.

(d) Calculate the net cumulative produced GOR when 8 MM STB has been produced?

(e) Plot on the same graph the average daily GOR, the cumulative produced gas, the net cumulative produced gas, and the cumulative injected gas versus cumulative oil production.

Cumulative Oil Production, N_p MM STB	Average Daily Gas-Oil Ratio, R SCF/STB	Cumulative Gas Injected, G_I MM SCF
0	300	0
1	280	0
2	280	0
3	340	0
4	560	0
5	850	0
6	1120	520
7	1420	930
8	1640	1440
9	1700	2104
10	1640	2743

5.9 An undersaturated reservoir producing above the bubble point had an initial pressure of 5000 psia, at which pressure the oil volume factor was 1.510 bbl/STB. When the pressure dropped to 4600 psia, owing to the production of 100,000 STB of oil, the oil volume factor was 1.520 bbl/STB. The connate water saturation was 25%, water compressibility 3.2×10^{-6} psi^{-1}, and, based on an average porosity of 16%, the rock compressibility was 4.0×10^{-6} psi^{-1}. The average compressibility of the oil between 5000 and 4600 psia relative to the volume at 5000 psia was 17.00×10^{-6} psi^{-1}.

(a) Geologic evidence and the absence of water production indicated a volumetric reservoir. Assuming this was so, what was the calculated initial oil in place?

(b) It was desired to inventory the initial stock tank barrels in place at a second production interval. When the pressure had dropped to 4200 psia, formation volume factor 1.531 bbl/STB, 205 M STB had been produced. If the average oil compressibility was 17.65×10^{-6} psi^{-1}, what was the initial oil in place?

(c) When all cores and logs had been analyzed, the volumetric estimate of the initial oil in place was 7.5 MM STB. If this figure is correct, how much water entered the reservoir when the pressure declined to 4600 psia?

5.10 Estimate the fraction recovery from a sandstone reservoir by water drive if the permeability is 1500 md, the connate water is 20%, the reservoir oil viscosity is 1.5 cp, the porosity is 25%, and the average formation thickness is 50 ft.

5.11 The following PVT data are available for a reservoir, which from volumetric reserve estimation is considered to have 275 MM STB of oil initially in place. The original pressure was 3600 psia. The current pressure is 3400 psia, and 732,800 STB have been produced. How much oil will have been produced by the time the reservoir pressure is 2700 psia?

Pressure (psia)	Solution Gas Oil Ratio (SCF/STB)	Formation Volume Factor (bbl/STB)
3600	567	1.310
3200	567	1.317
2800	567	1.325
2500	567	1.333
2400	554	1.310
1800	434	1.263
1200	337	1.210
600	223	1.140
200	143	1.070

5.12 Production data, along with reservoir and fluid data, for an undersaturated reservoir follow. There was no measurable water produced, and it can be assumed that there was no free gas flow in the reservoir. Determine the following:

(a) Saturations of oil, gas, and water at a reservoir pressure of 2258.

(b) Has water encroachment occurred and, if so, what is the volume?

> Gas gravity = 0.78
> Reservoir temperature = 160°F
> Initial water saturation = 25%
> Original oil in place = 180 MM STB
> Bubble-point pressure = 2819 psia

The following expressions for B_o and R_{so} as functions of pressure were determined from laboratory data:

$$B_o = 1.00 + 0.00015p, \text{ bbl/STB}$$
$$R_{so} = 50 + 0.42p, \text{ SCF/STB}$$

Pressure (psia)	Cumulative Oil Produced (STB)	Cumulative Gas Produced (SCF)	Instantaneous GOR (SCF/STB)
2819	0	0	1000
2742	4.38 MM	4.38 MM	1280
2639	10.16 MM	10.36 MM	1480
2506	20.09 MM	21.295 MM	2000
2403	27.02 MM	30.26 MM	2500
2258	34.29 MM	41.15 MM	3300

5.13 The following table provides fluid property data for an initially undersaturated lense type of oil reservoir. The initial connate water saturation was 25%. Initial reservoir temperature and pressure were 97°F and 2110 psia, respectively. The bubble-point pressure was 1700 psia. Average compressibility factors between the initial and bubble-point pressures were $4.0 (10)^{-6} \text{ psia}^{-1}$ and $3.1 (10)^{-6} \text{ psia}^{-1}$ for the formation and water, respectively. The initial oil formation volume factor was 1.256 bbl/STB. The critical gas saturation is estimated to be 10%. Determine the recovery versus pressure curve for this reservoir.

Pressure (psia)	Oil Formation Volume Factor (bbl/STB)	Solution GOR (SCF/STB)	Gas Formation Volume Factor (ft³/SCF)
1700	1.265	540	0.007412
1500	1.241	490	0.008423
1300	1.214	440	0.009826
1100	1.191	387	0.011792
900	1.161	334	0.014711
700	1.147	278	0.019316
500	1.117	220	0.027794

5.14 The Wildcat reservoir was discovered in 1970. The reservoir had an initial pressure of 3000 psia, and laboratory data indicated a bubble point pressure of 2500 psia. The connate water saturation was 22%. Calculate the fractional recovery, N_p/N, from initial conditions down to a pressure of 2300 psia. State any assumptions which you make relative to the calculations.

Porosity = 0.165
Formation compressibility = 2.5 $(10)^{-6}$ psia^{-1}
Reservoir temperature = 150°F

Pressure (psia)	B_o (bbl/STB)	R_{so} (SCF/STB)	z	B_g (bbl/SCF)	Viscosity Ratio μ_o/μ_g
3000	1.315	650	0.745	0.000726	53.91
2500	1.325	650	0.680	0.000796	56.60
2300	1.311	618	0.663	0.000843	61.46

REFERENCES

[1] T. L. Kennerly, "Oil Reservoir Fluids (Sampling, Analysis, and Application of Data)," presented before the *Delta Section* of AIME, January, 1953. (Available from Core Laboratories, Inc., Dallas).

[2] Frank O. Reudelhuber, "Sampling Procedures for Oil Reservoir Fluids," *Jour. of Petroleum Technology* (December 1957) **9,** 15–18.

[3] Ralph H. Espach and Joseph Fry, "Variable Characteristics of the Oil in the Tensleep Sandstone Reservoir, Elk Basin Field, Wyoming and Montana," *Trans.* AIME (1951), **192,** 75.

[4] Cecil Q. Cupps, Philip H. Lipstate, Jr., and Joseph Fry, "Variance in Characteristics in the Oil in the Weber Sandstone Reservoir, Rangely Field, Colo.," *Bureau of Mines* R. I. 4761, U.S.D.I., April 1951, also *World Oil,* December, 1957, **133,** No. 7, 192.

[5] A. B. Cook, G. B. Spencer, F. P. Bobrowski, and Tim Chin, "A New Method of Determining Variations in Physical Properties of Oil in a Reservoir, with Application

to the Scurry Reef Field, Scurry County, Tex." U.S. Bureau of Mines R.I. 5106, U.S.D.I. February, 1955, pp. 12–23.

[6] D. R. McCord, "Performance Predictions Incorporating Gravity Drainage and Gas Cap Pressure Maintenance—LL-370 Area, Bolivar Coastal Field," *Trans.* AIME (1953), **198**, 232.

[7] J. J. Arps, "Estimation of Primary Oil Reserves," *Trans.* AIME (1956), **207**, 183–186.

[8] R. C. Craze and S. E. Buckley, "A Factual Analysis of the Effect of Well Spacing on Oil Recovery," *Drilling and Production Practice,* API (1945), pp. 144–155.

[9] J. J. Arps and T. G. Roberts, "The Effect of Relative Permeability Ratio, the Oil Gravity, and the Solution Gas-Oil Ratio on the Primary Recovery from a Depletion Type Reservoir," *Trans.* AIME (1955), **204**, 120–126.

[10] H. G. Botset and M. Muskat, "Effect of Pressure Reduction upon Core Saturation," *Trans.* AIME (1939), **132**, 172–183.

[11] R. K. Guthrie and Martin K. Greenberger, "The Use of Multiple Correlation Analyses for Interpreting Petroleum-Engineering Data," *Drilling and Production Practice,* API (1955), pp. 135–137.

[12] Stewart E. Buckley, "Petroleum Conservation. New York: American Institute of Mining and Metallurgical Engineers, 1951, p. 239.

[13] K. B. Barnes and R. F. Carlson, "Scurry Analysis," *Oil and Gas Journal* (1950), **48**, No. 51, 64.

[14] "Material-Balance Calculations, North Snyder Field Canyon Reef Reservoir," *Oil and Gas Journal* (1950), **49**, No. 1, 85.

[15] R. M. Dicharry, T. L. Perryman, and J. D. Ronquille, "Evaluation and Design of a CO_2 Miscible Flood Project—SACROC Unit, Kelly-Snyder Field," *Jour. of Petroleum Technology,* November 1973, 1309–1318.

[16] *Petroleum Reservoir Efficiency and Well Spacing.* (New Jersey: Standard Oil Development Company, 1943), p. 22.

[17] H. B. Hill and R. K. Guthrie, "Engineering Study of the Rodessa Oil Field in Louisiana, Texas, and Arkansas," U.S. Bureau of Mines R.I. 3715 (1943), p. 87.

[18] Howard N. Hall, "Compressibility of Reservoir Rocks," *Trans.* AIME (1953), **198**, 309.

Chapter 6

Saturated Oil Reservoirs

1. INTRODUCTION

The material balance equations discussed in Chapter 5 apply to volumetric and water-drive reservoirs in which there is no initial gas cap (i.e., they are initially undersaturated). However, the equations apply to reservoirs in which an artificial gas cap forms owing either to gravitational segregation of the oil and free gas phases below the bubble point, or to the injection of gas, usually in the higher structural portions of the reservoir. When there is an initial gas cap (i.e., the oil is initially saturated), there is negligible liquid expansion energy. However, the energy stored in the dissolved gas is supplemented by that in the cap, and it is not surprising that recoveries from gas cap reservoirs are generally higher than from those without caps, other things remaining equal. In gas cap drives, as production proceeds and reservoir pressure declines, the expansion of the gas displaces oil downward toward the wells. This phenomenon is observed in the increase of the gas-oil ratios in successively lower wells. At the same time, by virtue of its expansion, the gas cap retards pressure decline and therefore the liberation of solution gas within the oil zone, thus improving recovery by reducing the producing gas-oil ratios of the wells. This mechanism is most effective in those reservoirs of marked structural relief, which introduces a vertical component of fluid flow whereby gravitational segregation of the oil and free gas in the sand may occur.[*,1] The recoveries from volumetric gas cap reservoirs could range from the recoveries for undersaturated reservoirs up to 70 to 80% of the initial stock tank oil in place and will be higher for large gas caps, continuous uniform formations, and good gravitational segregation characteristics.

Large gas caps

The size of the gas cap is usually expressed relative to the size of the oil zone by the ratio *m*, as defined in Chapter 2.

Continuous uniform formations

Continuous uniform formations reduce the channeling of the expanding gas cap ahead of the oil and the bypassing of oil in the less permeable portions.

Good gravitational segregation characteristics

These characteristics include primarily (a) pronounced structure, (b) low oil viscosity, (c) high permeability, and (d) low oil velocities.

Water drive and *hydraulic control* are terms used in designating a mechanism that involves the movement of water into the reservoir as gas and oil are produced. Water influx into a reservoir may be edge water or bottom water, the latter indicating that the oil is underlain by a water zone of sufficient thickness so that the water movement is essentially vertical. The most common source of water drive is a result of expansion of the water and the compressibility of the rock in the aquifer; however, it may result from artesian flow. The important characteristics of a water-drive recovery process are the following:

1. The volume of the reservoir is constantly reduced by the water influx. This influx is a source of energy in addition to the energy of liquid expansion above the bubble point and the energy stored in the solution gas and in the free, or cap, gas.

2. The bottom-hole pressure is related to the ratio of water influx to voidage. When the voidage only slightly exceeds the influx, there is only a slight pressure decline. When the voidage considerably exceeds the influx, the pressure decline is pronounced and approaches that for gas cap or dissolved gas-drive reservoirs, as the case may be.

3. For edge-water drives, regional migration is pronounced in the direction of the higher structural areas.

4. As the water encroaches in both edge-water and bottom-water drives, there is an increasing volume of water produced, and eventually water is produced by all wells.

---

[*] References throughout the text are given at the end of each chapter.

5. Under favorable conditions, the oil recoveries are high and range from 60 to 80% of the oil in place.

2. MATERIAL BALANCE IN SATURATED RESERVOIRS

The general Schilthuis material balance equation was developed in Chapter 2 and is as follows:

$$N(B_t - B_{ti}) + \frac{NmB_{ti}}{B_{gi}} (B_g - B_{gi}) + (1 + m) \, N \, B_{ti} \left[\frac{c_w S_{wi} + c_f}{1 - S_{wi}} \right] \Delta \bar{p} + W_e$$

$$= N_p[B_t + (R_p - R_{soi})B_g] + B_w W_p \tag{2.7}$$

Equation (2.7) can be rearranged and solved for N, the initial oil in place:

$$N = \frac{N_p[B_t + (R_p - R_{soi})B_g] - W_e + B_w W_p}{B_t - B_{ti} + \dfrac{mB_{ti}}{B_{gi}}(B_g - B_{gi}) + (1 + m)B_{ti} \left[\dfrac{c_w S_{wi} + c_f}{1 - S_{wi}} \right] \Delta \bar{p}} \tag{6.1}$$

If the expansion term due to the compressibilities of the formation and connate water can be neglected, as they usually are in a saturated reservoir, then Eq. (6.1) becomes

$$N = \frac{N_p[B_t + (R_p - R_{soi})B_g] - W_e + B_w W_p}{B_t - B_{ti} + \dfrac{mB_{ti}}{B_{gi}}(B_g - B_{gi})} \tag{6.2}$$

Example 6.1 shows the application of Eq. (6.2) to the calculation of initial oil in place for a water-drive reservoir with an initial gas cap. The calculations are done once by converting all barrel units to cubic feet units and then converting all cubic feet units to barrel units. It does not matter which set of units is used, only that each term in the equation is consistent. Problems sometimes arise because gas formation volume factors are either reported in cu ft/SCF or in bbl/SCF. Usually when applying the material balance equation for a liquid reservoir, gas formation volume factors are reported in bbl/SCF. Use care in making sure that the units are correct.

Example 6.1 To calculate the stock tank barrels of oil initially in place in a combination drive reservoir.

Given:

Volume of bulk oil zone = 112,000 ac-ft
Volume of bulk gas zone = 19,600 ac-ft

Initial reservoir pressure 2710 psia
Initial FVF = 1.340 bbl/STB
Initial gas volume factor = 0.006266 cu ft/SCF
Initial dissolved GOR = 562 SCF/STB
Oil produced during the interval = 20 MM STB
Reservoir pressure at the end of the interval = 2000 psia
Average produced GOR = 700 SCF/STB
Two-phase FVF at 2000 psia = 1.4954 bbl/STB
Volume of water encroached = 11.58MM bbl
Volume of water produced = 1.05MM STB
FVF of the water = 1.028 bbl/STB
Gas volume factor at 2000 psia = 0.008479 cu ft/SCF

SOLUTION: In the use of Eq. (6.2):

B_{ti} = 1.3400 × 5.615 = 7.5241 cu ft/STB
B_t = 1.4954 × 5.615 = 8.3967 cu ft/STB
W_e = 11.58 × 5.615 = 65.02 MM cu ft
W_p = 1.05 × 1.028 × 5.615 = 6.06 MM res cu ft

Assuming the same porosity and connate water for the oil and gas zones:

$$m = \frac{19,600}{112,000} = 0.175$$

Substituting in Eq. (6.2):

$$N = \frac{20 \times 10^6\,[8.3967 + (700 - 562)0.008489] - (65.02 - 6.06) \times 10^6}{8.3967 - 7.5241 + 0.175\left(\dfrac{7.5241}{0.006266}\right)(0.008489 - 0.006266)}$$

= 98.97 MM STB

If B_t is in barrels per stock tank barrel, then B_g must be in barrels per standard cubic foot and W_e and W_p in barrles, and the substitution is as follows:

$$N = \frac{20 \times 10^6\,[1.4954 + (700 - 562)0.001510] - (11.58 - 1.05 \times 1.028) \times 10^6}{1.4954 - 1.3400 + 0.175\left(\dfrac{1.3400}{0.001116}\right)(0.001510 - 0.001116)}$$

= 98.97 MM STB

In Chapter 2, the concept of drive indexes, first introduced to the reservoir engineering literature by Pirson, was developed.[2] To illustrate the use of these drive indexes, calculations are performed on the Conroe Field, Texas.

Figure 6.1 shows the pressure and production history of the Conroe Field, and Fig. 6.2 gives the gas and two-phase oil formation volume factor for the reservoir fluids. Table 6.1 contains other reservoir and production data and summarizes the calculations in column form for three different periods.

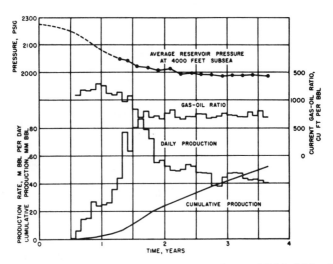

Fig. 6.1. Reservoir pressure and production data, Conroe Field. (*After* Schilthuis[3], *Trans.* AIME.)

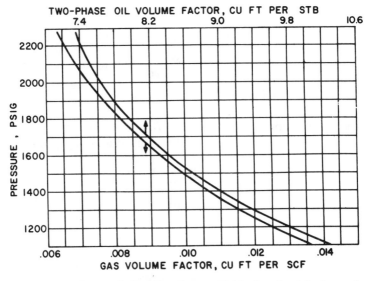

Fig. 6.2. Pressure volume relations for Conroe Field oil and original complement of dissolved gas. (*After* Schilthuis,[3] *Trans.* AIME.)

TABLE 6.1.
Material balance calculation of water influx or oil in place for oil reservoirs below the bubble-point pressure

For Conroe Field: $B_{ti} = 7.37$ cu ft/STB $\qquad$ $mB_{ti}/B_{gi} = 259$ SCF/STB
$B_{gi} = 0.00637$ cu ft/SCF $\qquad$ $R_{soi} = 600$ SCF/STB
(14.4 psia and 60°F)

$$m = \frac{181,225 \text{ ac-ft}}{810,000 \text{ ac-ft}} = 0.224$$

Line No.	Quantity	Units	Months after Start of Production				
			12	18	24	30	36
1	N_p	MM STB	9.070	22.34	32.03	40.18	48.24
2	R_p	SCF/STB	1630	1180	1070	1025	995
3	p	psig	2143	2108	2098	2087	2091
4	B_g	cu ft/SCF	0.00676	0.00687	0.00691	0.00694	0.00693
5	B_t	cu ft/STB	7.46	7.51	7.51	7.53	7.52
6	$N_p R_p$	MM SCF	14,800		34,400		48,100
7	$R_p - R_{soi}$	SCF/STB	1030		470		395
8	$(R_p - R_{soi}) B_g$	cu ft/STB	6.95		3.24		2.74
9	$(5) + (8)$	cu ft/STB	14.41		10.75		10.26
10	$B_g - B_{gi}$	cu ft/SCF	0.00039		0.00054		0.00056
11	$(10) \times (mB_{ti}/B_{gi})$	cu ft/STB	0.101		0.137		0.145
12	$B_t - B_{ti}$	cu ft/STB	0.09		0.14		0.15
13	$(11) + (12)$	cu ft/STB	0.191		0.277		0.295
14	$(1) \times (9)$	MM cu ft	131		345		495
15	$W_e - W_p$	MM cu ft	51.5		178		320
16	$(14) - (15)$	MM cu ft	79.5		167		175
17	$N = (16)/(13)$	MM STB	415		602		594
18	DDI	Fraction	0.285		0.244		0.180
19	SDI	Fraction	0.320		0.239		0.174
20	WDI	Fraction	0.395		0.516		0.646

The use of such tabular forms is common in many calculations of reservoir engineering in the interest of standardizing and summarizing calculations that may not be reviewed or repeated for intervals of months or sometimes longer. They also enable an engineer to take over the work of a predecessor with a minimum of briefing and study. Tabular forms also have the advantage of providing at a glance the component parts of a calculation, many of which have significance themselves. The more important factors can be readily distinguished from the less important ones, and trends in some of the component parts often provide insight into the reservoir behavior. For example, the values of line 11 in Table 6.1 show the expansion of the gas cap of the Conroe Field as the pressure declines. Line 17 shows the values of the initial oil in place calculated at three production intervals. These values and others calculated elsewhere are plotted versus cumulative production in Fig. 6.3, which also includes the recovery at each period, expressed as the percentage the cumulative oil is of the initial oil in place as calculated at that period. The increasing values of the initial oil during the early life of the field may be explained by some of the limitations of the material balance equation discussed in Chapter 2, particularly the average reservoir pressure. Lower values of the average reservoir pressure in the more permeable and in the developed portion of the reservoir cause the calculated values of the initial oil to be low through the effect on the oil and gas volume factors. The indications of Fig. 6.3 are that the reservoir contains approximately 600 MM STB of initial oil and that reliable values of the initial oil are not obtained until about 5% of the oil has been produced. This is not a universal figure but depends on a number of factors, particularly the amount of pressure decline. For Conroe, the drive indexes have been calculated at each of three periods, as given in lines 18, 19, and 20 of Table 6.1. For example, at the end of 12 months the calculated initial

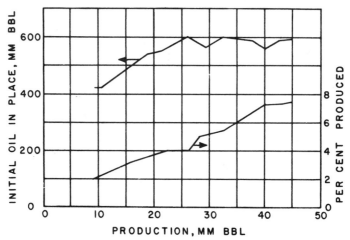

Fig. 6.3. Active oil, Conroe Field. (*After* Schilthuis[3], *Trans.* AIME.)

oil in place is 415 MM STB, and the value of $N_p[B_t + (R_p - R_{soi})B_g]$ given in line 14 is 131 MM cu ft. Then

$$DDI = \frac{415 \times 10^6 (7.46 - 7.37)}{131 \times ^6} = 0.285$$

$$SDI = \frac{\dfrac{415 \times 10^6 \times 0.224 \times 7.37}{0.00637}(0.00676 - 0.00637)}{131 \times 10^6} = 0.320$$

$$WDI = \frac{51.5 \times 10^6}{131 \times 10^6} = 0.395$$

These figures indicate that during the first 12 months 39.5% of the production was by water drive, 32.0% by gas cap expansion, and 28.5% by depletion drive. At the end of 36 months, as the pressure stabilized, the *current* mechanism was essentially 100% water drive and the *cumulative* mechanism increased to 64.6% by water drive. If figures for recovery by each of the three mechanisms could be obtained, the overall recovery could be estimated using the drive indexes. An increase in the depletion drive and gas-drive indexes would be reflected by declining pressures and increasing gas-oil ratios, and might indicate the need for water injection to supplement the natural water influx and to turn the recovery mechanism more toward water drive.

3. MATERIAL BALANCE AS A STRAIGHT LINE

In Chapter 2 Sect. 4, the method developed by Havlena and Odeh of applying the genereal material balance equation was presented.[4,5] This approach defines several new variables (see Chapter 2) and rewrites the material balance equation as Eq. (2.13):

$$F = NE_o + N(1 + m)B_{ti}E_{f,w} + \left[\frac{NmB_{ti}}{B_{gi}}\right] E_g + W_e \qquad (2.13)$$

This equation is then reduced for a particular application and arranged into a form of a straight line. When this is done, the slope and intercept often yield valuable assistance in determining such parameters as N and m. The usefulness of this approach is illustrated by applying the method to the data from the Conroe Field example discussed in the last section.

For the case of a saturated reservoir with an initial gas cap, such as the Conroe Field, and neglecting the compressibility term, $E_{f,w}$, Eq. (2.13) becomes

$$F = NE_o + \frac{NmB_{ti}}{B_{gi}}E_g + W_e \qquad (6.3)$$

If N is factored out of the first two terms on the right-hand side and both sides of the equation are divided by the expression remaining after factoring, we get

$$\frac{F}{E_o + \dfrac{mB_{ti}}{B_{gi}}E_g} = N + \frac{W_e}{E_o + \dfrac{mB_{ti}}{B_{gi}}E_g} \tag{6.4}$$

For the example of the Conroe Field in the previous section, the water production values were not known. For this reason, two dummy parameters are defined as $F' = F - W_p B_w$ and $W'_e = W_e - W_p B_w$. Equation (6.4) then becomes

$$\frac{F'}{E_o + \dfrac{mB_{ti}}{B_{gi}}E_g} = N + \frac{W'_e}{E_o + \dfrac{mB_{ti}}{B_{gi}}E_g} \tag{6.5}$$

Equation (6.5) is now in the desired form. If a plot of $F'/[E_o + mB_{ti}E_g/B_{gi}]$ as the ordinate and $W'_e/[E_o + mB_{ti}E_g/B_{gi}]$ as the abscissa is constructed, a straight line with slope equal to 1 and intercept equal to N is obtained. Table 6.2 contains the calculated values of the ordinate, line 5, and abscissa, line 7, using the Conroe Field data from Table 6.1. Figure 6.4 is a plot of these values.

If a least squares regression analysis is done on all three data points calculated in Table 6.2, the result is the solid line shown in Fig. 6.4. The line has a slope of 1.21 and an intercept, of N, of 396 MM STB. This slope is significantly larger than 1, which is what we should have obtained from the Havlena-Odeh method. If we now ignore the first data point, which represents the earliest production, and determine the slope and intercept of a line drawn through the remaining two points (the dashed line in Fig. 6.4), we get 1.00 for

TABLE 6.2.
Tabulated values from the Conroe Field for use in the Havlena-Odeh method

Line			Months after Start of Production		
No.	Quantity	Units	12	24	36
1	F'	MM cu ft	131	345	495
2	E_o	cu ft/STB	0.09	0.14	0.15
3	E_g	cu ft/SCF	0.00039	0.00054	0.00056
4	$E_o + m\dfrac{B_{ti}}{B_{gi}}E_g$	cu ft/STB	0.191	0.280	0.295
5	(1)/(4)	MM STB	686	1232	1678
6	W'_e	MM cu ft	51.5	178	320
7	(6)/(4)	MM STB	270	636	1085

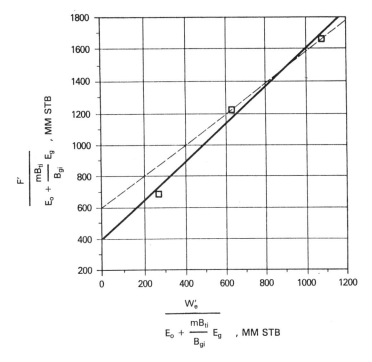

Fig. 6.4. Havlena-Odeh plot for the Conroe Field. Solid line represents line drawn through all the data points. Dashed line represents line drawn through data points from the later production periods.

a slope and 600 MM STB for N, the intercept. This value of the slope meets the requirement for the Havlena-Odeh method for this case. We should now raise the question: Can we justify ignoring the first point? If we realize that the production represents less than 5% of the initial oil in place and the fact that we have met the requirement for the slope of 1 for this case, then there is justification for not including the first point in our analysis. We conclude from our analysis that the initial oil in place is 600 MM STB for the Conroe Field.

You may take issue with the fact that an analsysis was done on only two points. Clearly, it would have been better to use more data points, but none were available in this particular example. As more production data are collected, then the plot in Fig. 6.4 can be updated and the calculation for N reviewed. The important point to remember is that if the Havlena-Odeh method is used, the condition of the slope and/or intercept must be met for the particular case you are working with. This imposes another restriction on the data and can be used to justify the exclusion of some data, as was done in the case of the Conroe Field example.

4. FLASH AND DIFFERENTIAL GAS LIBERATION

For heavy crudes whose dissolved gases are almost entirely methane and ethane, the manner of separation is relatively unimportant. For lighter crudes and heavier gases (i.e., for reservoir fluids with larger fractions of the *intermediate* hydrocarbons—mainly propane, butanes, and pentanes), the manner of separation raises some important questions. The nature of the difficulty lies mainly with these intermediate hydrocarbons which are, relatively speaking, intermediate between true gases and true liquids. They are therefore divided between the gas and liquid phases in proportions that are affected by the manner of separation. The situation may be explained with reference to two well-defined, isothermal, gas-liberation processes commonly used in laboratory PVT studies. In the *flash* liberation process, all the gas evolved during a reduction in pressure remains in contact and presumably in equilibrium with the liquid phase from which it is liberated. In the *differential* process, on the other hand, the gas evolved during a pressure reduction is removed from contact with the liquid phase as rapidly as it is liberated. Figure 6.5 shows the variation of solution gas with pressure for the differential process and the specific gravity of the gas that is being liberated at any pressure. Since the specific gravity of the gas is quite constant down to about 800 psia, it can be inferred that very close to the same quantity of gas would have been liberated by the flash process, down to 800 psia and *at the same temperature*. Below 800 psia, the vaporization of the intermediate hydrocarbons begins to be appreciable for the fluid of Fig. 6.5. In more volatile crudes, it begins at higher

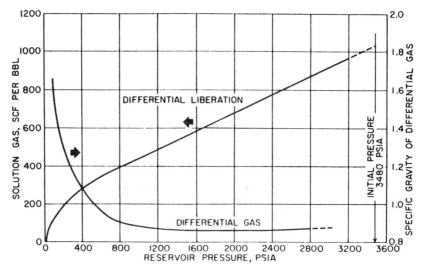

Fig. 6.5. Gas solubility and gas gravity by the differential liberation process on a subsurface sample from the Magnolia Field, Arkansas. (*After* Carpenter, Schroeder, and Cook,[6] *U.S. Bureau of Mines.*)

pressures, and vice versa. The vaporization is indicted by the rise in the gas gravity and by the increasing rate of gas liberation, indicated by the steepening of the slope dR_{so}/dp. If all the gas liberated down to, say, 400 psia remains in contact with the liquid phase, as in the flash process, more gas is liberated because the intermediate hydrocarbons in the liquid phase vaporize into the entire gas space in contact with the liquid until equilibrium is reached. Because gas is removed as rapidly as it is formed in the differential process, less vaporization of the intermediates occurs. The release of solution gas at lower pressures by the flash process, *at the same temperature,* is further accelerated because the loss of more of the intermediate hydrocarbons reduces the gas solubility. In some flash processes, the temperature is reduced at some pressure during the gas liberation process, whereas differential liberations are generally run at reservoir temperature. Because of the increased gas solubility and the lower volatility of the intermediates at lower temperatures, the quantity of gas released by the flash process is less at the lower temperatures and is commonly *less* than that by the differential process at reservoir temperature.

Table 6.3 gives the PVT data obtained from a laboratory study of a reservoir fluid sample at reservoir temperature of 220°F. The volumes given in Col. (2) are the result of the *flash* gas-liberation process and are shown plotted in Fig. 6.6. Below the bubble-point pressure at 2695 psig, the volumes include the volume of the liberated gas and are therefore two-phase volume factors. Since the stock tank oil remaining at atmospheric pressure depends on the pressure, temperature, and separation stages by which the gas is liberated at lower pressures, the volumes are reported relative to the volume at the bubble-point pressure, V_b. To relate the reservoir volumes to the stock tank oil volumes, additional tests are performed on other samples using small-scale separators, which are operated in the range of pressures and temperatures used in the field separation of the gas and oil. Table 6.4 shows the results of four laboratory tests at separator pressures of 0, 50, 100, and 200 psig, and separator temperatures from 74 to 77°F. The temperatures are lower for the lower separator pressures because of the greater cooling effect of the gas expansion and the greater vaporization of the intermediate hydrocarbon components at the lower pressures. At 100 psig and 76°F, the tests indicate that 505 SCF are liberated at the separator and 49 SCF in the stock tank, or a total of 554 SCF/STB by two-stage separation if the stock tank is called a stage. Otherwise, it is a single stage. Then the initial solution gas-oil ratio, R_{soi}, is 554 SCF/STB. The tests also shows that under these separation conditions 1.335 bbl of fluid at the bubble-point pressure yield 1.000 STB of oil. Hence, the formation volume factor at the bubble-point pressure is 1.335 bbl/STB and the two-phase flash FVF at 1773 psig is

$$B_{tf} = 1.335 \times 1.1814 = 1.577 \text{ bbl/STB}$$

The data of Table 6.4 indicate that both oil gravity and recovery can be

TABLE 6.3.
Reservoir fluid sample tabular data (*After* Kennerly, Core
Laboratories, Inc.)

| Pressure psig | Flash Liberation at 220°F, Relative Volume of Oil and Gas, V/V_b | Differential Liberation at 220°F | | |
		Liberated Gas-Oil Ratio, SCF/bbl of Residual Oil	Solution Gas-Oil Ratio, SCF/bbl of Residual Oil	Relative Oil Volume V/V_r
5000	0.9739			1.355
4700	0.9768			1.359
4400	0.9799			1.363
4100	0.9829			1.367
3800	0.9862			1.372
3600	0.9886			1.375
3400	0.9909			1.378
3200	0.9934			1.382
3000	0.9960			1.385
2900	0.9972			1.387
2800	0.9985			1.389
2695	1.0000	0	638	1.391
2663	1.0038			
2607	1.0101			
2512		42	596	1.373
2503	1.0233			
2358	1.0447			
2300		89	549	1.351
2197	1.0727			
2008		150	488	1.323
2000	1.1160	152	486	1.322
1773	1.1814			
1702		213	425	1.295
1550	1.2691			
1351	1.3792			
1315		290	348	1.260
1180	1.5117			
1010		351	287	1.232
992	1.7108			
711	2.2404			
705		412	226	1.205
540	2.8606			
410	3.7149			
405		474	164	1.175
289	5.1788			
150		539	99	1.141
0		638	0	1.066

Residual volume at 60°F		= 1.000
Residual oil gravity		= 28.8°API
Sp. gr. of liberated gas		= 1.0626

V = volume at given pressure.

V_b = volume at bubble-point pressure.

V_r = residual oil volume at 14.7 psia and 60°F by differential gas liberation at 220°F.

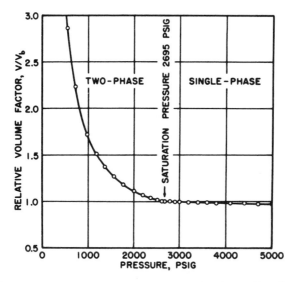

Fig. 6.6. Flash liberation PVT data for a reservoir fluid at 220°F. (*After* Kennerly, Core Laboratories, Inc.)

TABLE 6.4.
Separator tests of reservoir fluid sample (*After* Kennerly, Core Laboratories, Inc.)

Separator Pressure psig	Separator Temperature °F	Separator Gas-Oil Ratio[a] SCF/STB	Stock Tank Gas-Oil Ratio[a] SCF/STB	Stock Tank Gravity °API at 60°F	Formation Volume Factor[b] V_b/V_r	Specific Gravity of Flash Gas Air = 1.00
0	74	620	0	29.9	1.382	0.9725
50	75	539	23	31.5	1.340	
100	76	505	49	31.9	1.335	
200	77	459	98	31.8	1.337	

[a] Standard conditions 14.7 psia and 60°F.

[b] V_b/V_r = barrels of oil at the bubble-point pressure 2695 psig and 220°F per stock tank barrel at 14.7 psia and 60°F.

improved by using an optimum separator pressure of 100 psig, by reducing the loss of liquid components, particularly the intermediate hydrocarbons, to the separated gas. In reference to material balance calculations, they also indicate that the volume factors and solution gas-oil ratios depend on how the gas and oil are separated at the surface. When differing separation practices are used in the various wells owing to operator preference or to limitations of the flowing wellhead pressures, further complications are introduced. Figure 6.7 shows the variation in oil shrinkage with separator pressure for a West

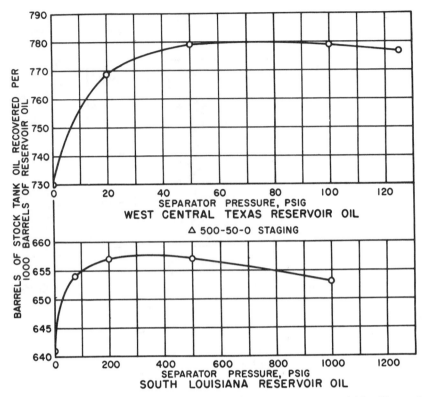

Fig. 6.7. Variation in stock tank recovery with separator pressure. (*After* Kennerly, Core Laboratories, Inc.)

Central Texas and a South Louisiana Field. Each crude oil has an optimum separator pressure at which the shrinkage is a minimum and stock tank oil gravity a maximum. For example, in the case of the West Central Texas reservoir oil, there is an increased recovery of 7% when the operating separator pressure is increased from atmospheric pressure to 70 psig. The effect of using two stages of separation with the South Louisiana reservoir oil is shown by the triangle.

The effect of changes in separator pressures and temperatures on gas-oil ratios, oil gravities, and shrinkage in reservoir oil was determined for the Scurry Reef Field by Cook, Spencer, Bobrowski, and Chin.[7,8] The data obtained from field and laboratory tests showed that the amount of gas liberated from the oil produced was affected materially by changes in both separator temperatures and pressures. For example, when the separator temperature was reduced to 62.5°F, the gas-oil ratio decreased from 1068 to 844 SCF/STB and the production increased from 125 to 135 STB/day. This was a decrease in gas-oil ratio of 21% and a production increase of 8%. Therefore, to yield the same volume of stock tank oil, the production of 8% more reservoir fluid was needed when the separator was operating at the higher temperature.

Table 6.3 also gives the solution gas and oil volume factors for the same reservoir fluid by *differential liberation* at 220°F all the way down to atmospheric pressure, whereas the *flash* tests were stopped at 289 psig, owing to limitations of the volume of the PVT cell. Figure 6.8 shows a plot of the oil (liquid) volume factor and the liberated gas-oil ratios relative to a barrel of *residual oil* (i.e., the oil remaining at 1 atm and 60°F after a differential liberation down to 1 atm at 220°F). The volume change from 1.066 at 220°F to 1.000 at 60°F is a measure of the coefficient of thermal expansion of the residual oil. In some cases, a barrel of *residual* oil by the *differential* process is close to a stock tank barrel by a particular flash process, and the two are taken as equivalent. In the present case, the volume factor at the bubble-point pressure is 1.335 bbl per stock tank barrel by the flash process, using separation at 100 psig and 76°F versus 1.391 bbl per residual barrel by the differential process. The initial solution gas-oil ratios are 554 SCF/STB versus 638 SCF/residual barrel, respectively.

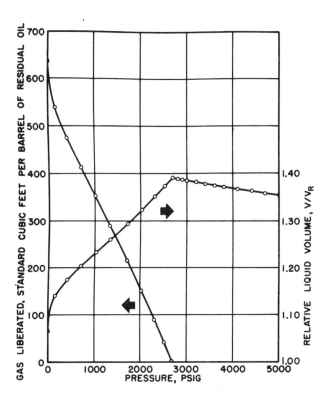

Fig. 6.8. PVT data for the differential gas liberation of a reservoir fluid at 220°F. (*After* Kennerly, Core Laboratories, Inc.)

In addition to the volumetric data of Table 6.3, PVT studies usually obtain values for (a) the specific volume of the bubble-point oil, (b) the thermal expansion of the saturated oil, and (c) the compressibility of the reservoir fluid at or above the bubble point. For the fluid of Table 6.3, the specific volume of the fluid at 220°F and 2695 psig is 0.02163 cu ft/lb, and the thermal expansion is 1.07741 volumes at 220°F and 5000 psia per volume at 74°F and 5000 psia, or a coefficient of 0.00053 per °F. The compressibility of the undersaturated liquid has been discussed and calculated from the data of Table 6.3 in Chapter 1, Sect. 6, as 10.27×10^{-6} psi^{-1} between 5000 psia and 4100 psia at 220°F.

The deviation factor of the gas released by the differential liberation process may be measured, or it may be estimated from the measured specific gravity. Alternatively, the gas composition may be calculated using a set of valid equilibrium constants and the composition of the reservoir fluid, and the gas deviation factor may be calculated from the gas composition.

5. THE CALCULATION OF FORMATION VOLUME FACTOR AND SOLUTION GAS-OIL RATIO FROM DIFFERENTIAL VAPORIZATION AND SEPARATOR TESTS

The data in Tables 6.3 and 6.4 can be combined to yield values for the oil formation volume factor and the solution gas-oil ratio. The formation volume factor is calculated from Eq. (6.6) or (6.7), depending on whether the pressure is above or below the bubble-point pressure:
For $p >$ bubble-point pressure

$$B_o = R.V. \, (B_{ofb}) \tag{6.6}$$

For $p <$ bubble-point pressure

$$B_o = B_{od} \left[\frac{B_{ofb}}{B_{odb}} \right] \tag{6.7}$$

where

$R.V.$—Relative volume from the flash liberation test listed in Table 6.3 as V/V_b

B_{ofb}—Formation volume factor from separator tests listed in Table 6.4 as V_b/V_r

B_{od}—Formation volume factor from differential liberation test listed in Table 6.3 as V/V_r

B_{odb}—Formation volume factor at the bubble point from differential liberation test

The solution gas-oil ratio can be calculated using Eq. (6.8).

$$R_{so} = R_{sofb} - \left[(R_{sodb} - R_{sod}) \frac{B_{ofb}}{B_{odb}} \right]$$ (6.8)

where

R_{sofb}—Sum of separator gas and the stock tank gas from separator tests listed in Table 6.4

R_{sod}—Solution gas-oil ratio from differential liberation test listed in Table 6.3

R_{sodb}—The value of R_{sod} at the bubble point

For example, at a pressure of 5000 psia and separator conditions of 200 psig and 77°F

$$B_o = 0.9739\,(1.337) = 1.302 \text{ bbl/STB}$$

and

$$R_{so} = 459 + 98 = 557 \text{ SCF/STB}$$

(Recall that $R_{so} = R_{sob}$ for pressures above the bubble point.)
At a pressure of 2512 psig, which is below the bubble-point pressure, B_o and R_{so} become

$$B_o = 1.373 \left[\frac{1.337}{1.391} \right] = 1.320 \text{ bbl/STB}$$

and

$$R_{so} = 557 - \left[(638 - 596) \left(\frac{1.337}{1.373} \right) \right] = 516 \text{ SCF/STB}$$

6. VOLATILE OIL RESERVOIRS

If all gas in reservoirs were methane and all oil were decane and heavier, the PVT properties of the reservoir fluids would be quite simple because the quantities of oil and gas obtained from a mixture of the two would be almost independent of the temperatures, pressures, and type of the gas liberation process by which the two are separated. Low volatility crudes approach this behavior, which is approximately indicated by reservoir temperatures below 150°F, solution gas-oil ratios below 500 SCF/bbl, and stock tank gravities below 35°API. Because the propane, butane, and pentane content of these fluids is low, the volatility is low.

For conditions above, but not too far above the approximate limits of low volatility fluids, satisfactory PVT data for material balance use are obtained by combining separator tests at appropriate temperatures and pressures with the flash and differential tests according to the procedure discussed in the previous section. Although this procedure is satisfactory for fluids of moderate volatility, it becomes less satisfactory as the volatility increases; and more complicated, extensive, and precise laboratory tests are necessary to provide PVT data that are realistic in the application, particularly to reservoirs of the depletion type.

With present-day deeper drilling, many reservoirs of higher volatility are being discovered that include the gas-condensate reservoirs discussed in Chapter 4. The volatility is higher because of the higher reservoir temperatures at depth, approaching 500°F in some cases, and also because of the composition of the fluids, which are high in propane through decane. The *volatile oil reservoir* is recognized as a type intermediate in volatility between the moderately volatile reservoir and the gas-condensate reservoir. Jacoby and Berry have approximately defined the volatile type of reservoir as one containing relatively large proportions of ethane through decane at a reservoir temperature near or above 250°F with a high formation volume factor and stock tank oil gravity above 45°API.[9] The fluid of the Elk City Field, Oklahoma, is an example. The reservoir fluid at the initial pressure of 4364 psia and reservoir temperature of 180°F had a formation volume factor of 2.624 bbl/STB and a solution gas-oil ratio of 2821 SCF/STB, both relative to production through a single separator operating at 50 psig and 60°F. The stock tank gravity was 51.4°API for these separator conditions. Cook, Spencer, and Bobrowski described the Elk City Field and a technique for predicting recovery by depletion drive-performance.[10] Reudelhuber and Hinds, and Jacoby and Berry also described somewhat similar laboratory techniques and prediction methods for the depletion drive performance of these volatile oil reservoirs.[11,9] The methods are similar to those used for gas-condensate reservoirs discussed in Chapter 4.

A typical laboratory method of estimating the recovery from volatile reservoirs is as follows. Samples of primary separator gas and liquid are obtained and analyzed for composition. With these compositions and a knowledge of separator gas and oil flow rates, the reservoir fluid composition can be calculated. Also, by recombining the separator fluids in the appropriate ratio, a reservoir fluid sample can be obtained. This reservoir fluid sample is placed in a PVT cell and brought to reservoir temperature and pressure. At this point, several tests are conducted. A constant composition expansion is performed to determine relative volume data. These data are the flash liberation volume data listed in Table 6.3. On a separate reservoir sample, a constant volume expansion is performed while the volumes and compositions of the produced phases are monitored. The produced phases are passed through a separator system that simulates the surface facilities. By expanding the original reservoir fluid from the initial reservoir pressure down to an abandonment pressure, the actual production process from the reservoir is simulated. Using the data from the laboratory expansion, the field production can be estimated with a procedure similar to the one used in Ex. 4.3 to predict performance from a gas-condensate reservoir.

7. MAXIMUM EFFICIENT RATE (MER)

Many studies indicate that the recovery from true solution gas-drive reservoirs by primary depletion is essentially independent of both individual well rates and total or reservoir production rates. Keller, Tracy, and Roe showed that this is true even for reservoirs with severe permeability stratification where the strata are separated by impermeable barriers and are hydraulically connected only at the wells.[12] The Gloyd-Mitchell zone of the Rodessa Field (see Chapter 5, Sect. 5) is an example of a solution gas-drive reservoir that is essentially not *rate sensitive* (i.e., the recovery is unrelated to the rate at which the reservoir is produced). The recovery from very permeable, uniform reservoirs under very active water drives may also be essentially independent of the rates at which they are produced.

Many reservoirs are clearly *rate-sensitive,* and there is a *maximum efficient rate* (MER) above which there will be a significant reduction in the practical ultimate oil recovery.[13, 14] Rate-sensitive reservoirs imply that there is some mechanism(s) at work in the reservoir, which, in a practical period, can substantially improve the recovery of the oil in place. These mechanisms include (a) partial water drive, (b) gravitational segregation, and (c) those effective in reservoirs of heterogeneous permeability.

When initially undersaturated reservoirs are produced under partial water drive at voidage rates (gas, oil, and water) considerably in excess of the natural influx rate, they are produced essentially as solution gas-drive reservoirs modified by a small water influx. Assuming that recovery by water displacement is considerably larger than by solution gas drive, there will be a considerable loss in recoverable oil by the high production rate, even when the oil zone is eventually entirely invaded by water. The loss is caused by the increase in the viscosity of the oil and the decrease in the volume factor of the oil at lower pressures, and also to the earlier abandonment of the wells that must be produced by artificial lift. Because of the higher oil viscosity at the lower pressure, producing water-oil ratios will be higher, and the economic limit of production rate will be reached at lower oil recoveries. Because of the lower oil volume factor at the lower pressure, at the same residual oil saturation in the invaded area, more stock tank oil will be left at low pressure. There are, of course, additional benefits to be realized by producing at such a rate so as to maintain high reservoir pressure. If there is no appreciable gravitational segregation and the effects of reservoir heterogeneity are small, the MER for a partial water-drive reservoir can be inferred from a study of the effect of the net reservoir voidage rate on reservoir pressure, and the consequent effect of pressure on the gas saturation relative to the critical gas saturation, (i.e., on gas-oil ratios). The MER may also be inferred from studies of the drive indexes, Eq. (6.3). The presence of a gas cap in a partial water-drive field introduces complications in determining the MER, which is affected by the relative size of the gas cap and the relative efficiencies of oil displacement by the expanding gas cap and by the encroaching water.

The Gloyd-Mitchell zone of the Rodessa Field was not rate sensitive because there was no water influx and because there was essentially no gravitational segregation of the free gas released from solution and the oil. If there had been substantial segregation, the well completion and well workover measures, which were taken in an effort to reduce gas-oil ratios, would have been effective as they are in many solution gas-drive reservoirs. In some cases a gas cap forms in the higher portions of the reservoir, and, when high gas-oil ratio wells are penalized or shut in, there may be a substantial improvement in recovery, as indicated by Eq. (5.11) for a reduction in the value of the produced gas-oil ratio R_p. Under these conditions, the MER is that rate at which gravitational segregation is substantial for practical producing rates.

Gravitational segregation is also important in many gas cap reservoirs. The effect of displacement rate on recovery by gas cap expansion when there is substantial segregation of the oil and gas is discussed in Chapter 9. The studies presented on the Mile Six Pool show that at the adopted displacement rate the recovery will be approximately 52.4%. If the displacement rate is doubled, the recovery will be reduced to about 36.0%, and at very high rates it will drop to 14.4% for negligible gravity segregation.

Gravitational segregation also occurs in the displacement of oil by water, and like the gas-oil segregation, it is also dependent on the time factor. Gravity segregation is generally of less relative importance in water drive than in gas cap drive because of the much higher recoveries usually obtained by water drive. The MER for water-drive reservoirs is that rate above which there will be insufficient time for effective segregation and, therefore, a substantial loss of recoverable oil. The rate may be inferred from calculations similar to those used for gas displacement in Chapter 9 or from laboratory studies. It is interesting that in the case of gravitational segregation, the reservoir pressure is not the index of the MER. In an active water-drive field, for example, there may be no appreciable difference in the reservoir pressure decline for a several-fold change in the production rate, and yet recovery at the lower rate may be substantially higher if gravity segregation is effective at the lower rate but not at the higher.

As water invades a reservoir of heterogeneous permeability, the displacement is more rapid in the more permeable portions, and considerable quantities of oil may be bypassed if the displacement rate is too high. At lower rates there is time for water to enter the less permeable portions of the rock and recover a larger portion of the oil. As the water level rises, water is sometimes imbibed or drawn into the less permeable portions by capillary action, and this may also help to recover oil from the less permeable areas. Because water imbibition and the consequent capillary expulsion of oil is far from instantaneous, if appreciable additional oil can be recovered by this mechanism, the displacement rate should be lowered when practicable to take advantage of it. Although the maximum efficient rate under these circumstances is more difficult to establish, it may be inferred from the degree of the reservoir heterogeneity and the capillary pressure characteristics of the reservoir rocks.

In the present discussion of MER, it is realized that the recovery of oil is also affected by the reservoir mechanisms, fluid injection, gas-oil and water-oil ratio control, and other factors, and that it is difficult to speak of rate-sensitive mechanisms entirely independently of these other factors, which in many cases are far more important.

PROBLEMS

6.1 Calculate the values for the second and fourth periods through the fourteenth step of Table 6.1 for Conroe Field.

6.2 Calculate the drive indexes at Conroe for the second and fourth periods.

6.3 If the recovery by water drive at Conroe is 70%, by segregation drive 50%, and by depletion drive 25%, using the drive indexes for the fifth period calculate the ultimate oil recovery expected at Conroe.

6.4 Explain why the first material balance calculation at Conroe gives a low value for the initial oil in place.

6.5 (a) Calculate the single-phase formation volume factor on a stock tank basis, from the PVT data given in Tables 6.3 and 6.4, at a reservoir pressure of 1702 psig, for separator conditions of 100 psig and 76°F.

(b) Calculate the solution GOR at 1702 psig on a stock tank basis for the same separator conditions.

(c) Calculate the two-phase formation volume factor by flash separation at 1550 psig for separator conditions of 100 psig and 76°F.

6.6 From the core data that follow, calculate the initial volume of oil and free gas in place by the volumetric method. Then, using the material balance equation, calculate the cubic feet of water that have encroached into the reservoir at the end of the four periods for which production data are given.

Pressure (psia)	B_t (bbl/STB)	B_g (ft^3/SCF)	N_p (STB)	R_p (SCF/STB)	W_p (STB)
3480	1.4765	0.0048844	0	0	0
3190	1.5092	0.0052380	11.17 MM	885	224.5 M
3139	1.5159	0.0053086	13.80 MM	884	534.2 M
3093	1.5223	0.0053747	16.41 MM	884	1005.0 M
3060	1.5270	0.0054237	18.59 MM	896	1554.0 M

Average porosity = 16.8%
Connate water saturation = 27%
Productive oil zone volume = 346,000 ac-ft
Productive gas zone volume = 73,700 ac-ft
B_w = 1.025 bbl/STB
Reservoir temperature = 207°F
Initial reservoir pressure = 3480 psia

6.7 The following PVT data are for the Aneth Field in Utah:

Pressure (psia)	B_o (bbl/STB)	R_{so} (SCF/STB)	B_g (bbl/SCF)	μ_o/μ_g
2200	1.383	727	—	—
1850	1.388	727	0.00130	35
1600	1.358	654	0.00150	39
1300	1.321	563	0.00182	47
1000	1.280	469	0.00250	56
700	1.241	374	0.00375	68
400	1.199	277	0.00691	85
100	1.139	143	0.02495	130
40	1.100	78	0.05430	420

The initial reservoir temperature was 133°F. The initial pressure was 2200 psia, and the bubble-point pressure was 1850 psia. There was no active water drive. From 1850 psia to 1300 psia a total of 720 MM STB of oil were produced and 590.6 MMM SCF of gas.

(a) How many reservoir barrels of oil were in place at 1850 psia?

(b) The average porosity was 10%, and connate water saturation was 28%. The field covered 50,000 acres. What is the average formation thickness in feet?

6.8 You have been asked to review the performance of a combination solution gas, gas-cap drive reservoir. Well test and log information show that the reservoir initially had a gas cap half the size of the initial oil volume. Initial reservoir pressure and solution gas-oil ratio were 2500 psia and 721 SCF/STB, respectively. Using the volumetric approach, initial oil in place was found to be 56 MM STB. As you procede with the analysis, you discover that your boss has not given you all the data you need to make the analysis. The missing information is that at some point in the life of the project a pressure maintenance program was initiated using gas injection. The time of the gas injection and the total amount of gas injected are not known. There was no active water drive or water production. PVT and production data are in the following table:

Pressure (psia)	B_g (bbl/SCF)	B_t (bbl/STB)	N_p (STB)	R_p (SCF/STB)
2500	0.001048	1.498	0	0
2300	0.001155	1.523	3.741MM	716
2100	0.001280	1.562	6.849MM	966
1900	0.001440	1.620	9.173MM	1297
1700	0.001634	1.701	10.99MM	1623
1500	0.001884	1.817	12.42MM	1953
1300	0.002206	1.967	14.39MM	2551
1100	0.002654	2.251	16.14MM	3214
900	0.003300	2.597	17.38MM	3765
700	0.004315	3.209	18.50MM	4317
500	0.006163	4.361	19.59MM	4839

(a) At what point (i.e., pressure) did the pressure maintenance program begin?

(b) How much gas in SCF had been injected when the reservoir pressure is 500 psia? Assume that the reservoir gas and the injected gas have the same compressibility factor.

6.9 An oil reservoir initially contains 4 MM STB of oil at its bubble-point pressure of 3150 psia with 600 SCF/STB of gas in solution. When the average reservoir pressure has dropped to 2900 psia, the gas in solution is 550 SCF/STB. B_{oi} was 1.34 bbl/STB and B_o at a pressure of 2900 psia is 1.32 bbl/STB.

Due 3/2

Other data:
R_p = 600 SCF/STB at 2900 psia
S_{wi} = 0.25
B_g = 0.0011 bbl/SCF at 2900 psia
  volumetric reservoir
  no original gas cap

(a) How many STB of oil will be produced when the pressure has decreased to 2900 psia?

(b) Calculate the free gas saturation that exists at 2900 psia.

6.10 Given the following data from laboratory core tests, production data, and logging information:

well spacing = 320 ac
net pay thickness = 50 ft with the gas/oil contact 10 ft from the top
porosity = 0.17
initial water saturation = 0.26
initial gas saturation = 0.15
bubble-point pressure = 3600 psia
initial reservoir pressure = 3000 psia
reservoir temperature = 120°F
B_{oi} = 1.26 bbl/STB
B_o = 1.37 bbl/STB at the bubble-point pressure
B_o = 1.19 bbl/STB at 2000 psia
N_p = 2.0 MM STB at 2000 psia
G_p = 2.4 MMM SCF at 2000 psia

gas compressibility factor, $z = 1.0 - 0.0001p$
solution gas-oil ratio, $R_{so} = 0.2p$

Calculate the amount of water that has influxed and the drive indexes at 2000 psia.

6.11 From the following information determine:

(a) Cumulative water influx at pressures 3625, 3530, and 3200 psia.

(b) Water-drive index for the pressures in (a).

Pressure (psia)	N_p (STB)	G_p (SCF)	W_p (STB)	B_g (bbl/SCF)	R_{so} (SCF/STB)	B_t (bbl/STB)
3640	0	0	0	0.000892	888	1.464
3625	0.06 MM	0.49 MM	0	0.000895	884	1.466
3610	0.36 MM	2.31 MM	0.001 MM	0.000899	880	1.468
3585	0.79 MM	4.12 MM	0.08 MM	0.000905	874	1.469
3530	1.21 MM	5.68 MM	0.26 MM	0.000918	860	1.476
3460	1.54 MM	7.00 MM	0.41 MM	0.000936	846	1.482
3385	2.08 MM	8.41 MM	0.60 MM	0.000957	825	1.491
3300	2.58 MM	9.71 MM	0.92 MM	0.000982	804	1.501
3200	3.40 MM	11.62 MM	1.38 MM	0.001014	779	1.519

pve 3/2

6.12 The cumulative oil production, N_p, and cumulative gas oil ratio, R_p, as functions of the average reservoir pressure over the first 10 years of production for a gas cap reservoir follow. Use the Havlena-Odeh approach to solve for the initial oil and gas (both free and solution) in place.

Pressure (psia)	N_p (STB)	R_p (SCF/STB)	B_o (bbl/STB)	R_{so} (SCF/STB)	B_g (bbl/SCF)
3330	0	0	1.2511	510	0.00087
3150	3.295 MM	1050	1.2353	477	0.00092
3000	5.903 MM	1060	1.2222	450	0.00096
2850	8.852 MM	1160	1.2122	425	0.00101
2700	11.503 MM	1235	1.2022	401	0.00107
2550	14.513 MM	1265	1.1922	375	0.00113
2400	17.730 MM	1300	1.1822	352	0.00120

6.13 Using the following data, determine the original oil in place by the Havlena-Odeh method. Assume there is no water influx and no initial gas cap. The bubble-point pressure is 1800 psia.

Pressure (psia)	N_p (STB)	R_p (SCF/STB)	B_t (bbl/STB)	R_{so} (SCF/STB)	B_g (bbl/SCF)
1800	0	0	1.268	577	0.00097
1482	2.223 MM	634	1.335	491	0.00119
1367	2.981 MM	707	1.372	460	0.00130
1053	5.787 MM	1034	1.540	375	0.00175

REFERENCES

[1] *Petroleum Reservoir Efficiency and Well Spacing* (New Jersey: Standard Oil Development Company, 1934), p. 24.

[2] Sylvain J. Pirson, *Elements of Oil Reservoir Engineering,* 2nd ed. New York: McGraw-Hill, 1958, pp. 635–693.

[3] Ralph J. Schilthuis, "Active Oil and Reservoir Energy," *Trans.* AIME (1936), **118,** 33.

[4] D. Havlena and A. S. Odeh, "The Material Balance as an Equation of a Straight Line," *Jour. of Petroleum Technology* (August 1963), 896–900.

[5] D. Havlena and A. S. Odeh, "The Material Balance as an Equation of a Straight Line. Part II—Field Cases," *Jour. of Petroleum Technology* (July 1964), 815–822.

[6] Charles B. Carpenter, H. J. Shroeder, and Alton B. Cook, "Magnolia Oil Field, Columbia County, Arkansas," U.S. Bureau of Mines, R. I. 3720 (1943), pp. 46, 47, 82.

[7] Alton B. Cook, G. B. Spencer, F. P. Bobrowski, and Tim Chin, "Changes in Gas-Oil Ratios with Variations in Separator Pressures and Temperatures," *Petroleum Engineer* (March 1954), **26,** B 77–B 82.

[8] Alton B. Cook, G. B. Spencer, F. P. Bobrowski, and Tim Chin, "A New Method of Determining Variations in Physical Properties in a Reservoir, with Application to the Scurry Reef Field, Scurry County, Texas," U.S. Bureau of Mines, R. I. 5106, February 1955, pp. 10–11.

[9] R. H. Jacoby and V. J. Berry, Jr., "A Method for Predicting Depletion Performance of a Reservoir Producing Volatile Crude Oil," *Trans.* AIME (1957), **201,** 27.

[10] Alton B. Cook, G. B. Spencer, and F. P. Bobrowski, "Special Considerations in Predicting Reservoir Performance of Highly Volatile Type Oil Reservoirs," *Trans.* AIME (1951), **192,** 37–46.

[11] F. O. Reudelhuber and Richard F. Hinds, "A Compositional Material Balance Method for Prediction of Recovery from Volatile Oil Depletion Drive Reservoirs," *Trans.* AIME (1957), **201,** 19–26.

[12] W. O. Keller, G. W. Tracy, and R. P. Roe, "Effects of Permeability on Recovery Efficiency by Gas Displacement," *Drilling and Production Practice,* API (1949), p. 218.

[13] Edgar Kraus, "MER—A History," *Drilling and Production Practice,* API (1947), pp. 108–110.

[14] Stewart E. Buckley, *Petroleum Conservation.* New York: American Institute of Mining and Metallurgical Engineers, 1951, pp. 151–163.

Chapter 7

Single-Phase Fluid Flow in Reservoirs

1. INTRODUCTION

In the previous four chapters, the material balance equations for each of the four reservoir types defined in Chapter 1 were developed. These material balance equations may be used to calculate the production of oil and/or gas as a function of reservoir pressure. The reservoir engineer, however, would like to know the production as a function of time. To learn this, it is necessary to develop a model containing time or some related property such as flow rate. This chapter discusses the flow of fluids in reservoirs and the models that are used to relate reservoir pressure to flow rate. The discussion in this chapter is limited to single-phase flow. Some multiphase flow considerations are presented in Chapters 9 and 10.

2. DARCY'S LAW AND PERMEABILITY

In 1856, as a result of experimental studies on the flow of water through unconsolidated sand filter beds, Henry Darcy formulated the law that bears his name. This law has been extended to describe, with some limitations, the movement of other fluids, including two or more immiscible fluids, in consolidated rocks and other porous media. Darcy's law states that the velocity of a homogeneous fluid in a porous medium is proportional to the driving force and inversely proportional to the fluid viscosity, or

$$v = -0.001127 \frac{k}{\mu} \left[\frac{dp}{ds} - 0.433 \; \gamma' \cos \alpha \right] \qquad (7.1)$$

where,

v = the apparent velocity, bbls/day-ft^2

k = permeability, millidarcies (md)

μ = fluid viscosity, cp

p = pressure, psia

s = distance along flow path in ft

γ' = fluid specific gravity (always relative to water)

α = the angle measured counterclockwise from the downward vertical to the positive s direction

and the term

$$\left[\frac{dp}{ds} - 0.433 \; \gamma' \cos \alpha \right]$$

represents the driving force. The driving force may be caused by fluid pressure gradients (dp/ds) and/or hydraulic (gravitational) gradients (0.433 $\gamma' \cos \alpha$). In many cases of practical interest, the hydraulic gradients, although always present, are small compared with the fluid pressure gradients, and are frequently neglected. In other cases, notably production by pumping from reservoirs whose pressures have been depleted, and gas cap expansion reservoirs with good gravity drainage characteristics, the hydraulic gradients are important, and must be considered.

The apparent velocity, v, is equal to qB/A, where q is the volumetric flow rate in STB/day, B is the formation volume factor, and A is the apparent or total cross-sectional area of the rock in square feet. In other words, A includes the area of the rock material as well as the area of the pore channels. The fluid pressure gradient, dp/ds, is taken in the same direction as v and q. The negative sign in front of the constant 0.001127 indicates that if the flow is taken as positive in the positive s-direction, then the pressure decreases in that direction so that the slope dp/ds is negative.

Darcy's law applies only in the region of laminar flow; in turbulent flow, which occurs at higher velocities, the pressure gradient increases at a greater rate than does the flow rate. Fortunately, except for some instances of quite large production or injection rates in the vicinity of the well bore, flow in the reservoir and in most laboratory tests is by design, streamlined and Darcy's law is valid. Darcy's law does not apply to flow within individual pore channels but to portions of a rock the dimensions of which are reasonably large compared with the size of the pore channels—that is, it is a statistical law that averages the behavior of many pore channels. For this reason, although samples with dimensions of a centimeter or two are satisfactory for permeability measurements on uniform sandstones, much larger samples are required for reliable measurements on rocks of the fracture and vugular types.

Owing to the porosity of the rock, the tortuosity of the flow paths, and the absence of flow in some of the (dead) pore spaces, the actual fluid velocity varies from point to point within the rock and maintains an average that is many times the apparent velocity. Because actual velocities are in general not measurable, and to keep porosity and permeability separated, *apparent* velocities form the basis of Darcy's law. This means that the actual average forward velocity of a fluid is the apparent velocity divided by the porosity where the fluid completely saturates the rock.

A basic unit of permeability is the *darcy* (d). A rock of 1 darcy permeability is one in which a fluid of 1 cp viscosity will move at a velocity of 1 cm/sec under a pressure gradient of 1 atm/cm. Since this is a fairly large unit for most producing rocks, permeabilities are commonly expressed in units one thousandth as large, the *millidarcy,* or 0.001 darcy. Throughout this text, the unit of permeability used is the millidarcy (md). Commercial oil and gas sands have permeabilities varying from a few millidarcies to several thousands. Intergranular limestone permeabilities may be only a fraction of a millidarcy and yet be commercial if the rock contains additional natural or artificial fractures or other kinds of openings. Fractured and vugular rocks may have enormous permeabilities, and some cavernous limestones approach the equivalent of underground tanks.

The permeability of a sample as measured in the laboratory may vary considerably from the average of the reservoir as a whole or a portion thereof, for there are often wide variations both laterally and vertically, the permeability sometimes changing several-fold within an inch in rock that appears quite uniform. Generally, the permeability measured parallel to the bedding planes of stratified rocks is larger than the vertical permeability. Also, in some cases the permeability along the bedding plane varies considerably and consistently with core orientation, owing presumably to the oriented deposition of more or less elongated particles and/or the subsequent leaching or cementing action of migrating waters. Some reservoirs show general permeability trends from one portion to another, and many reservoirs are closed on all or part of their boundaries by rock of very low permeability, certainly by the overlying caprock. The occurrence of one or more strata of consistent permeability over a portion or all of a reservoir is common. In the proper development of reservoirs, it is customary to core selected wells throughout the productive area, measuring the permeability and porosity on each foot of core recovered. The results are frequently handled statistically.[*,1,2]

---

[*] References throughout the text are given at the end of each chapter.

Hydraulic gradients in reservoirs vary from a maximum near 0.500 psi/ft for brines to 0.433 psi/ft for fresh water at 60°F, depending on the pressure, temperature, and salinity of the water. Reservoir oils and high-pressure gas and gas-condensate gradients lie in the range of 0.10 to 0.30 psi/ft, depending on the temperature, pressure, and composition of the fluid. Gases at low pressure will have very low gradients (e.g., about 0.002 psi/ft for natural gas at 100 psia). The figures given are the vertical gradients. The effective gradient is reduced by the factor cos α. Thus a reservoir oil with a reservoir specific gravity of 0.60 will have a vertical gradient of 0.260 psi/ft; however, if the fluid is constrained to flow along the bedding plane of its stratum, which dips at 15° (α = 75°), then the effective hydraulic gradient is only 0.26 cos 75°, or 0.067 psi/ft. Although these hydraulic gradients are small compared with usual reservoir pressures, the fluid pressure gradients, except in the vicinity of well bores, are also quite small and in the same range. Fluid pressure gradients within a few feet of well bores may be as high as tens of psi per foot but will fall off rapidly away from the well, inversely with the radius.

Measured static well pressures are usually corrected to the top of the production (perforated) interval using gradients measured within the well bore, and thence up or down to a datum level using the gradient of the reservoir fluid. The datum level is selected near the center of gravity of the initial hydrocarbon accumulation. Isobaric maps prepared from pressures corrected to datum may be used to find the average pressure of the hydrocarbons and also to determine the direction and magnitude of fluid movement from one portion of a reservoir to another.

Equation (7.1) suggests that the velocity and pressure gradient are related by the *mobility*. The mobility, given the symbol λ, is the ratio of permeability to viscosity, k/μ. The mobility appears in all equations describing the flow of single-phase fluids in reservoir rocks. When two fluids are flowing simultaneously—for example, gas and oil to a well bore, it is the ratio of the mobility of the gas, λ_g, to that of the oil, λ_o, which determines their individual flow rates. The mobility ratio M (see Chapter 9) is an important factor affecting the displacement efficiency of oil by water. When one fluid displaces another, the standard notation for the mobility ratio is the mobility of the displacing fluid to that of the displaced fluid. For water displacing oil, it is λ_w/λ_o.

3. THE CLASSIFICATION OF RESERVOIR FLOW SYSTEMS

Reservoir flow systems are usually classed according to (a) the type of fluid, (b) the geometry of the reservoir or portion thereof, and (c) the relative rate at which the flow approaches a steady-state condition following a disturbance.

For most engineering purposes, the reservoir fluid may be classed either as (a) incompressible, (b) slightly compressible, or (c) compressible. The concept of the incompressible fluid, volume of which does not change with pressure, simplifies the derivation and the final form of many equations, which are sufficiently accurate for many purposes. However, the engineer should realize that there are no true incompressible fluids.

A slightly compressible fluid, which is the description of nearly all liquids, is sometimes defined as one whose volume change with pressure is quite small and expressible by the equation:

$$V = V_R e^{c(p_R - p)} \tag{7.2}$$

where

R = reference conditions.

Equation (7.2) may be derived by integrating Eq. (1.1), which defines compressibility between limits, assuming an average compressibility c, as

$$\int_{p_R}^{p} -c\,dp = \int_{V_R}^{V} \frac{dV}{V}$$

$$c(p_R - p) = \ln\left(\frac{V}{V_R}\right)$$

$$e^{c(p_R - p)} = \frac{V}{V_R}$$

$$V = V_R\left(e^{c(p_R - p)}\right)$$

But e^x may be represented by a series expansion as

$$e^x = 1 + x + \frac{x^2}{2!} + \frac{x^3}{3!} + \cdots \frac{x^n}{n!}$$

Where x is small, the first two terms, $1 + x$, suffice, and where the exponent x is $c(p_R - p)$. the equation may be written as

$$V = V_R[1 + c(p_R - p)] \tag{7.3}$$

A compressible fluid is one in which the volume has a strong dependence on pressure. All gases are in this category. In Chapter 1, the real gas law was used to describe how gas volumes vary with pressure:

$$V = \frac{znR'T}{p} \tag{1.7}$$

Unlike the case of the slightly compressible fluids, the gas isothermal compressibility, c_g, cannot be treated as a constant with varying pressure. In fact, the following expression for c_g was developed:

$$c_g = \frac{1}{p} - \frac{1}{z}\frac{dz}{dp}$$ (1.19)

Although fluids are typed mainly by their compressibilities, in addition there may be single phase or multiphase flow. Many systems are either only gas, oil, or water, and most of the remainder are either gas-oil or oil-water systems. For the purposes of this chapter, discussion is restricted to cases where there is only a single phase flowing.

The two geometries of greatest practical interest are those that give rise to *linear* and to *radial* flow. In linear flow, as shown in Fig. 7.1, the flow lines are parallel and the cross section exposed to flow is constant. In radial flow, the flow lines are straight and converge in two dimensions toward a common center (i.e., a well). The cross section exposed to flow decreases as the center is approached. Occasionally, *spherical* flow is of interest, in which the flow lines are straight and converge toward a common center in three dimensions. Although the actual paths of the fluid particles in rocks are irregular due to the shape of the pore spaces, the overall or average paths may be represented as straight lines in linear, radial, or spherical flow.

Actually, none of these geometries is found precisely in petroleum reservoirs, but for many engineering purposes the actual may often be closely represented by one of the these idealizations. In some types of reservoir studies (i.e., water flooding and gas cycling), these idealizations are inadequate, and more sophisticated models are commonly used in their stead.

Flow systems in reservoir rocks are classified, according to their time dependence, as steady-state, transient, late transient, or pseudosteady-state. During the life of a well or reservoir, the type of system can change several times, which suggests that it is critical to know as much about the flow system as possible in order to use the appropriate model to describe the relationship between the pressure and the flow rate. In steady-state systems, the pressure and fluid saturations at every point throughout the system adjust instantaneously to a change in pressure or flow rate in any part of the system. Of course, no real system can respond instantaneously, but some reservoir flow

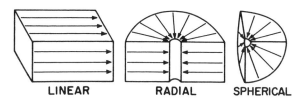

LINEAR   RADIAL   SPHERICAL

Fig. 7.1. Common flow geometries.

systems do resemble a steady-state process. This resemblance occurs when any production from a reservoir is replaced with an equal mass of fluid from some external source. In Chapter 8, we consider a case of water influx that comes close to meeting this requirement, but in general there are very few systems that can be assumed to be at steady-state conditions.

To consider the remaining three classifications of time dependence, we discuss the movement of pressure that occurs when a change in the flow rate of a well located in the center of a reservoir, as illustrated in Figure 7.2, causes a pressure disturbance in the reservoir. The discussion assumes the following: (1) the flow system is made up of a reservoir of constant thickness and rock properties; (2) the radius of the circular reservoir is r_e; and (3) the flow rate, after the disturbance has occurred, is constant. As the flow rate is changed at the well, the movement of pressure begins to occur away from the well. The movement of pressure is a diffusion phenomenon and is modeled by the diffusivity equation (see Sect. 5). The pressure moves at a rate proportional to the formation diffusivity, η,

$$\eta = \frac{k}{\phi \mu c_t} \tag{7.4}$$

where k is the effective permeability of the flowing phase, ϕ is the total effective porosity, μ is the fluid viscosity of the flowing phase, and c_t is the total compressibility. The total compressibility is obtained by weighting the compressibility of each phase by its saturation and adding the formation compressibility, or:

$$c_t = c_g S_g + c_o S_o + c_w S_w + c_f \tag{7.5}$$

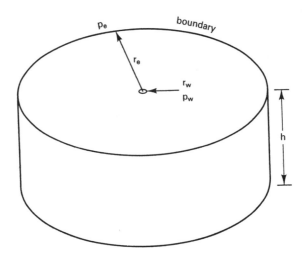

Fig. 7.2. Schematic of a single well in a circular reservoir.

The formation compressibility, c_f, should be expressed in change in pore volume per unit *pore* volume per psi. During the time the pressure is traveling at this rate, the flow is said to be transient. While the pressure is in this transient region, the outer boundary of the reservoir has no influence on the pressure movement, and the reservoir acts as if it were infinite in size.

The late transient region is the period after the pressure has reached the outer boundary of the reservoir and before the pressure behavior has had time to stabilize in the reservoir. In this region, the pressure no longer travels at a rate proportional to η. It is very difficult to describe the pressure behavior during this period.

The fourth period, the pseudosteady-state, is the period after the pressure behavior has stabilized in the reservoir. During this period, the pressure at every point throughout the reservoir is changing at a constant rate and as a linear function of time. This period has been incorrectly referred to as the steady-state period.

An estimation for the time when a flow system of the type shown in Figure 7.2 reaches pseudosteady-state can be made from the following equation

$$t_{pss} = \frac{1200r_e^2}{\eta} = \frac{1200\phi\mu c_t r_e^2}{k} \tag{7.6}$$

where t_{pss} is the time to reach pseudosteady-state expressed in hours.[3] For a well producing an oil with a reservoir viscosity of 1.5 cp and a total compressibility of 15×10^{-6} psi^{-1}, from a circular reservoir of 1000-ft radius with a permeability of 100 md and a total effective porosity of 20%:

$$t_{pss} = \frac{1200(0.2)(1.5)(15(10)^{-6})(1000^2)}{100} = 54 \text{ hr}$$

This means that approximately 54 hours, or 2.25 days, is required for the flow in this reservoir to reach pseudosteady-state conditions after a well located in its center is opened to flow, or following a change in the well flow rate. It also means that if the well is shut in, it will take approximately this time for the pressure to equalize throughout the drainage area of the well, so that the measured subsurface pressure equals the average drainage area pressure of the well.

This same criterion may be applied approximately to gas reservoirs but with less certainty because the gas is more compressible. For a gas viscosity of 0.015 cp and a compressibility of 400×10^{-6} psi^{-1},

$$t_{pss} = \frac{1200(0.2)(0.015)(400(10)^{-6})(1000^2)}{100} = 14.4 \text{ hr}$$

Thus, under somewhat comparable conditions (i.e., the same r_e and k), gas reservoirs reach pseudosteady-state conditions more rapidly than do oil reservoirs. This is due to the much lower viscosity of gases, which more than offsets the increase in fluid compressibility. On the other hand, gas wells are usually drilled on wider spacings so that the value of r_e generally is larger for gas wells than for oil wells, thus increasing the time required to reach pseudosteady-state. Many gas reservoirs, such as those found in the overthrust belt, are associated with sands of low permeability. If we consider an r_e value of 2500 ft and a permeability of 1 md, which would represent a tight gas sand, then we calculate the following value for t_{pss}:

$$t_{pss} = \frac{1200(0.2)(0.015)(400(10)^{-6})(2500^2)}{1} = 9000 \text{ hr}$$

The calculations suggest that reaching pseudosteady-state conditions in a typical tight gas reservoir takes a very long time compared to a typical oil reservoir. In general, pseudosteady-state mechanics suffice when the time required to reach pseudosteady-state is small compared with the time between substantial changes in the flow rate, or, in the case of reservoirs, small compared with the total producing life (time) of the reservoir.

4. STEADY-STATE FLOW SYSTEMS

Now that Darcy's law has been reviewed and the classification of flow systems has been discussed, the actual models that relate flow rate to reservoir pressure can be developed. The next several sections discuss the steady-state models. In this discussion, both linear and radial flow geometries are discussed since there are many applications for these types of systems. For both the linear and radial geometries, equations are developed for all three general types of fluids (i.e., incompressible, slightly compressible, and compressible).

4.1. Linear Flow of Incompressible Fluids, Steady State

Figure 7.3 represents linear flow through a body of constant cross section, where both ends are entirely open to flow, and where no flow crosses the sides, top, or bottom. If the fluid is incompressible, or essentially so for all en-

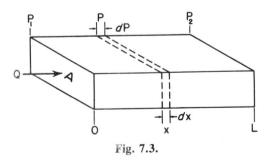

Fig. 7.3.

gineering purposes, then the velocity is the same at all points, as is the total flow rate across any cross section, so that,

$$v = \frac{qB}{A_c} = -0.001127\frac{k}{\mu}\frac{dp}{dx}$$

Separating variables and integrating over the length of the porous body;

$$\frac{qB}{A_c}\int_0^L dx = -0.001127\frac{k}{\mu}\int_{p_1}^{p_2} dp$$

$$q = 0.001127\frac{kA_c(p_1-p_2)}{B\mu L} \qquad (7.7)$$

For example, under a pressure differential of 100 psi for a permeability of 250 md, a fluid viscosity of 2.5 cp, a formation volume factor of 1.127 bbl/STB, a length of 450 ft, and a cross-sectional area of 45 sq ft, the flow rate is

$$q = 0.001127\frac{(250)(45)(100)}{(1.127)(2.5)(450)} = 1.0 \text{ STB/day}$$

In this integration B, q, μ, and k have been removed from the integral sign, assuming they are invariant with pressure. Actually, for flow above the bubble point, the volume, and hence the rate of flow, varies with the pressure as expressed by Eq. (7.2). The formation volume factor and viscosity also vary with pressure, as explained in Chapter 1. Fatt and Davis have shown a variation in permeability with net overburden pressure for several sandstones.[4] The net overburden pressure is the gross less the internal fluid pressure; therefore, a variation of permeability with pressure is indicated, particularly in the shallower reservoirs. Because none of these effects is serious for a few hundred psi difference, values at the average pressure may be used for most purposes.

4.2. Linear Flow of Slightly Compressible Fluids, Steady State

The equation for flow of slightly compressible fluids is modified from what was just derived in the previous section since the volume of slightly compressible fluids increases as pressure decreases. Earlier in this chapter, Eq. (7.3) was derived which describes the relationship between pressure and volume for a slightly compressible fluid. The product of the flow rate, defined in STB units, and the formation volume factor has a similar dependence on pressure and is given by

$$qB = q_R[1 + c(p_R - p)] \qquad (7.8)$$

where q_R is the flow rate at some reference pressure, p_R. If Darcy's law is written for this case, variables separated, and the resulting equation integrated over the length of the porous body, the following is obtained:

$$\frac{q_R}{A_c} \int_0^L dx = -0.001127 \frac{k}{\mu} \int_{p_1}^{p_2} \frac{dp}{1 + c(p_R - p)}$$

$$q_R = \frac{0.001127 k A_c}{\mu L c} \ln \left[\frac{1 + c(p_R - p_2)}{1 + c(p_R - p_1)} \right] \tag{7.9}$$

This integration assumes a constant compressibility over the entire pressure drop. For example, under a pressure differential of 100 psi for a permeability of 250 md, a fluid viscosity of 2.5 cp, a length of 450 ft, a cross-sectional area of 45 sq ft, a constant compressibility of $65(10^{-6})$ psi^{-1}, and choosing p_1 as the reference pressure, the flow rate is

$$q_1 = \frac{(0.001127)(250)(45)}{(2.5)(450)(65 \times 10^{-6})} \ln \left[\frac{1 + 65 \times 10^{-6}(100)}{1} \right] = 1.123 \text{ bbl/day}$$

When compared with the flow rate calculation in the preceding section, q_1 is found to be different due to the assumption of a slightly compressible fluid in the calculation rather than an incompressible fluid. Note also, that the flow rate is not in STB units because the calculation is being done at a reference pressure that is not the standard pressure. If p_2 is chosen to be the reference pressure, then the result of the calculation will be q_2, and the value of the calculated flow rate will be different still because of the volume dependence on the reference pressure:

$$q_2 = \frac{(0.001127)(250)(45)}{(2.5)(450)(65 \times 10^{-6})} \ln \left[\frac{1}{1 + 65 \times 10^{-6}(-100)} \right] = 1.131 \text{ bbl/day}$$

The calculations show that q_1 and q_2 are not largely different, which confirms what was discussed earlier: the fact that volume is not a strong function of pressure for slightly compressible fluids.

4.3. Linear Flow of Compressible Fluids, Steady State

The rate of flow of gas expressed in standard cubic feet per day is the same at all cross sections in a steady-state, linear system. However, because the gas expands as the pressure drops, the velocity is greater at the downstream end than at the upstream end, and consequently the pressure gradient increases toward the downstream end. The flow at any cross section x of Fig. 7.3 where the pressure is p may be expressed in terms of the flow in standard cubic feet per day by substituting the definition of the gas formation volume factor

$$qB_g = \frac{q p_{sc} T z}{5.615 T_{sc} p}$$

Substituting in Darcy's law:

$$\frac{qp_{sc}Tz}{5.615T_{sc}pA_c} = -0.001127\frac{k}{\mu}\frac{dp}{dx}$$

Separating variables and integrating;

$$\frac{qp_{sc}Tz\mu}{(5.615)(0.001127)kT_{sc}A_c}\int_0^L dx = -\int_{p_1}^{p_2} pdp = \frac{1}{2}(p_1^2 - p_2^2)$$

Finally

*Compressible flow
steady-state*

$$q = \frac{0.003164T_{sc}A_ck\,(p_1^2 - p_2^2)}{p_{sc}TzL\mu} \tag{7.10}$$

For example, where $T_{sc} = 60°F$, $A_c = 45$ sq ft, $k = 125$ md, $p_1 = 1000$ psia, $p_2 = 500$ psia, $p_{sc} = 14.7$ psia, $T = 140°F$, $z = 0.92$, $L = 450$ ft, and $\mu = 0.015$ cp

$$q = \frac{0.003164(520)(45)(125)\,(1000^2 - 500^2)}{14.7(600)(0.92)(450)(0.015)} = 126.7 \text{ M SCF/day}$$

Here again, T, k, and the product μz have been withdrawn from the integrals as if they were invariant with pressure, and as before, average values may be used in this case. At this point, it is instructive to examine an observation that Wattenbarger and Ramey made about the behavior of the gas deviation factor—viscosity product as a function of pressure.[5] Figure 7.4 is a typical plot of μz versus pressure for a real gas. Note that the product, μz, is nearly constant for pressures less than about 2000 psia. Above 2000 psia, the product $\mu z/p$ is nearly constant. Although the shape of the curve varies slightly for different gases at different temperatures, the pressure dependence is representative of most natural gases of interest. The pressure at which the curve bends varies from about 1500 to 2000 psia for various gases. This variation suggests that Eq. (7.10) is valid only for pressures less than about 1500 to 2000 psia, depending on the properties of the flowing gas. Above this pressure range, it would be more accurate to assume that the product $\mu z/p$ is constant. For the case of $\mu z/p$ constant, the following is obtained:

$$\frac{qp_{sc}T(z\mu/p)}{(5.615)(0.001127)kT_{sc}A_c}\int_0^L dx = -\int_{p_1}^{p_2} dp = p_1 - p_2$$

$$q = \frac{0.006328kT_{sc}A_c(p_1 - p_2)}{p_{sc}T(z\mu/p)} \tag{7.11}$$

In applying Eq. (7.11), the product $\mu z/p$ should be evaluated at the average pressure between p_1 and p_2.

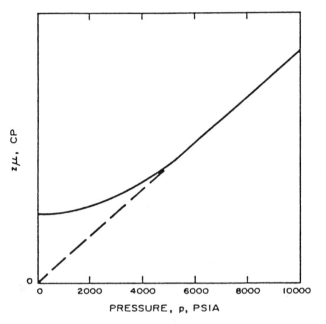

Fig. 7.4. Isothermal variation of μz with pressure.

4.4. Permeability Variations in Linear Systems

Consider two or more beds of equal cross section but of unequal lengths and permeabilities (Fig. 7.5) in which the same linear flow rate q exists, assuming an incompressible fluid. Obviously the pressure drops are additive, and

$$(p_1 - p_4) = (p_1 - p_2) + (p_2 - p_3) + (p_3 - p_4)$$

Substituting the equivalents of these pressure drops from Equation (7.7),

$$\frac{q_t B \mu L_t}{0.001127 k_{\text{avg}} A_c} = \frac{q_1 B \mu L_1}{0.001127 k_1 A_{c1}} + \frac{q_2 B \mu L_2}{0.001127 k_2 A_{c2}} + \frac{q_3 B \mu L_3}{0.001127 k_3 A_{c3}}$$

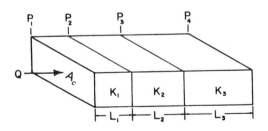

Fig. 7.5. Series flow in linear beds.

But since the flow rates, the cross sections, the viscosities, and the formation volume factors (neglecting the change with pressure) are equal in all beds

$$\frac{L_t}{k_{avg}} = \frac{L_1}{k_1} + \frac{L_2}{k_2} + \frac{L_3}{k_3}$$

Or

$$k_{avg} = \frac{L_t}{\dfrac{L_1}{k_1} + \dfrac{L_2}{k_2} + \dfrac{L_3}{k_3}} = \frac{\Sigma \, L_i}{\Sigma \, L_i/k_i} \qquad (7.12)$$

The average permeability as defined by Eq. (7.12) is that permeability to which a number of beds of various geometries and permeabilities could be changed, and yet the same total flow rate under the same applied pressure drop could be obtained.

Equation (7.12) was derived using the incompressible fluid equation. Because the permeability is a property of the rock and not of the fluids flowing through it, except for gases at low pressure, the average permeability must be equally applicable to gases. This requirement may be demonstrated by observing that for pressures below 1500 to 2000 psia:

$$(p_1^2 - p_4^2) = (p_1^2 - p_2^2) + (p_2^2 - p_3^2) + (p_3^2 - p_4^2)$$

Substituting the equivalents from Eq. (7.10), the same Eq. (7.12) is obtained.

The average permeability of 10 md, 50 md, and 1000 md beds, which are 6 ft, 18 ft, and 40 ft in length, respectively, but of equal cross section, when placed in series is

$$k_{avg} = \frac{\Sigma \, L_i}{\Sigma \, L_i/k_i} = \frac{6 + 18 + 40}{6/10 + 18/50 + 40/1000} = 64 \text{ md}$$

Consider two or more beds of equal length but unequal cross sections and permeabilities flowing the same fluid in linear flow under the same pressure drop (p_1 to p_2) as shown in Fig. 7.6.
Obviously the total flow is the sum of the individual flows, or

$$q_t = q_1 + q_2 + q_3$$

And

$$\frac{k_{avg} A_{ct}(p_1 - p_2)}{B\mu L} = \frac{k_1 A_{c1}(p_1 - p_2)}{B\mu L} + \frac{k_2 A_{c2}(p_1 - p_2)}{B\mu L} + \frac{k_3 A_{c3}(p_1 - p_2)}{B\mu L}$$

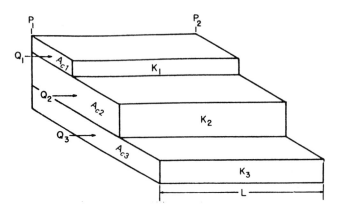

Fig. 7.6. Parallel flow in linear beds.

Cancelling

$$k_{avg} A_{ct} = k_1 A_{c1} + k_2 A_{c2} + k_3 A_{c3}$$

$$k_{avg} = \frac{\Sigma\, k_i A_{ci}}{\Sigma\, A_{ci}} \qquad (7.13)$$

And where all beds are of the same width, so that their areas are proportional to their thicknesses,

$$k_{avg} = \frac{\Sigma\, k_i h_i}{\Sigma\, h_i} \qquad (7.14)$$

Where the parallel beds are homogeneous in permeability and fluid content, the pressure and the pressure gradient are the same in all beds at equal distances. Thus there will be no cross-flow between beds, owing to fluid pressure differences. However, when water displaces oil—for example from a set of parallel beds—the rates of advance of the flood fronts will be greater in the more permeable beds. Because the mobility of the oil (k_o/μ_o) ahead of the flood front is different from the mobility of water (k_w/μ_w) behind the flood front, the pressure gradients will be different. In this instance, there will be pressure differences between two points at the same distance through the rock, and cross flow will take place between the beds if they are not separated by impermeable barriers. Under these circumstances, Eqs. (7.13) and (7.14) are not strictly applicable, and the average permeability changes with the stage of displacement. Water may also move from the more permeable to the less permeable beds by capillary action, which further complicates the study of parallel flow.

The average permeability of three beds of 10 md, 50 md, and 1000 md, and 6 ft, 18 ft, and 36 ft respectively in thickness but of equal width, when placed in parallel is

$$k_{avg} = \frac{\Sigma k_i h_i}{\Sigma h_i} = \frac{10 \times 6 + 18 \times 50 + 36 \times 1000}{6 + 18 + 36} = 616 \text{ md}$$

4.5. Flow Through Capillaries and Fractures

Although the pore spaces within rocks seldom resemble straight, smooth-walled capillary tubes of constant diameter, it is often convenient and instructive to treat these pore spaces as if they were composed of bundles of parallel capillary tubes of various diameters. Consider a capillary tube of length L and inside radius r_o which is flowing an incompressible fluid of μ viscosity in laminar or viscous flow under a pressure difference of $(p_1 - p_2)$. From fluid dynamics, Poiseuille's law, which describes the total flow rate through the capillary, can be written as:

$$q = 1.30(10)^{10} \frac{\pi r_o^4 (p_1 - p_2)}{B \mu L} \tag{7.15}$$

Darcy's law for the linear flow of incompressible fluids in permeable beds, Eq. (7.7), and Poiseuille's law for incompressible fluid capillary flow, Eq. (7.15), are quite similar:

$$q = 0.001127 \frac{k A_c (p_1 - p_2)}{B \mu L} \tag{7.7}$$

Writing $A_c = \pi r_0^2$ for area in Eq. (7.7), and equating it to Eq. (7.15);

$$k = 1.15(10)^{13} r_0^2 \tag{7.16}$$

Thus the permeability of a rock composed of closely packed capillaries, each having a radius of $4.17(10)^{-6}$ foot (0.00005 in.), is about 200 md. And if only 25% of the rock consists of pore channels (i.e., it has 25% porosity), the permeability is about one-quarter as large, or about 50 md.

An equation for the viscous flow of incompressible wetting fluids through smooth fractures of constant width may be obtained as:

$$q = 8.7(10)^9 \frac{W^2 A_c (p_1 - p_2)}{B \mu L} \tag{7.17}$$

In Eq. (7.17), W is the width of the fracture, A_c is the cross-sectional area of the fracture, which equals the product of the width W and lateral extent of the fracture, and the pressure difference is that which exists between the ends of the fracture of length L. Equation (7.17) may be combined with Eq. (7.7) to obtain an expression for the permeability of a fracture as:

$$k = 7.7(10)^{12}W^2 \tag{7.18}$$

The permeability of a fracture only $8.33(10)^{-5}$ ft wide $(0.001$ in.$)$ is 53,500 md.

    Fractures and solution channels account for economic production rates in many dolomite, limestone, and sandstone rocks, which could not be produced economically if such openings did not exist. Consider, for example, a rock of very low primary or matrix permeability, say 0.01 md, but which contains on the average a fracture $4.17(10)^{-4}$ ft wide and 1 ft in lateral extent per square foot of rock. Assuming the fracture is in the direction in which flow is desired, the law of parallel flow, Eq. (7.13) will apply, and

$$k_{avg} = \frac{0.01[1 - (1)(4.17(10)^{-4})] + (7.7(10)^{12}(4.17(10)^{-4})^2[1(4.17(10)^{-4})]}{1}$$

$$k_{avg} = 558 \text{ md}$$

4.6. Radial Flow of Incompressible Fluid, Steady State

Consider radial flow toward a vertical wellbore of radius r_w in a horizontal stratum of uniform thickness and permeability, as shown in Fig. 7.7. If the fluid is incompressible, the flow across any circumference is a constant. Let p_w be the pressure maintained in the wellbore when the well is flowing q STB/day and a pressure p_e is maintained at the external radius r_e. Let the pressure at any radius r be p. Then at this radius r

$$v = \frac{qB}{A_c} = \frac{qB}{2\pi r h} = -0.001127 \frac{k\,dp}{\mu\,dr}$$

where positive q is in the positive r direction. Separating variables and integrating between any two radii, r_1 and r_2, where the pressures are p_1 and p_2, respectively

$$\int_{r_1}^{r_2} \frac{qB\,dr}{2\pi r h} = -0.001127 \int_{p_1}^{p_2} \frac{k}{\mu}\,dp$$

$$q = -\frac{0.00708\ kh\ (p_2 - p_1)}{\mu B\ \ln\ (r_2/r_1)}$$

The minus sign is usually dispensed with, for where p_2 is greater than p_1, the flow is known to be negative—that is, in the negative r direction, or toward the wellbore:

$$q = \frac{0.00708\ kh\ (p_2 - p_1)}{\mu B\ \ln\ (r_2/r_1)}$$

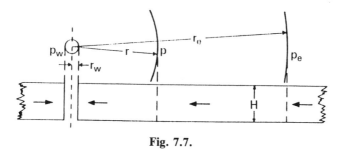

Fig. 7.7.

Frequently the two radii of interest are the wellbore radius r_w and the external or drainage radius r_e. Then

$$q = \frac{0.00708 \, kh \, (p_e - p_w)}{\mu B \, \ln \, (r_e/r_w)} \tag{7.19}$$

The external radius is usually inferred from the well spacing. For example, a circle of 660 ft radius can be inscribed within a square 40 ac unit; so 660 ft is commonly used for r_e with 40 ac spacing. Sometimes a radius of 745 ft is used, this being the radius of a circle 40 ac in area. The wellbore radius is usually assigned from the bit diameter, the casing diameter, or a caliper survey. In practice, neither the external radius nor the wellbore radius is generally known with precision. Fortunately, they enter the equation as a logarithm, so that the error in the equation will be much less than the errors in the radii. Since wellbore radii are about 1/3 ft and 40 ac spacing ($r_e = 660$ ft) is quite common, a ratio 2000 is quite commonly used for r_e/r_w. Since ln 2000 is 7.60 and ln 3000 is 8.00, a 50% increase in the value of r_e/r_w gives only a 5.3% increase in the value of the logarithm.

The external pressure p_e used in Eq. (7.19) is generally taken as the static well pressure corrected to the middle of the producing interval, and the flowing well pressure p_w is the flowing well pressure also corrected to the middle of the producing interval during a period of stabilized flow at rate q. When reservoir pressure stabilizes as under natural water drive or pressure maintenance, Eq. (7.19) is quite applicable because the pressure is maintained at the external boundary, and the fluid produced at the well is replaced by fluid crossing the external boundary. The flow, however, may not be strictly radial.

4.7. Radial Flow of Slightly Compressible Fluids, Steady State

Equation (7.3) is again used to express the volume dependence on pressure for slightly compressible fluids. If this equation is substituted into the radial form of Darcy's law, the following is obtained:

$$qB = \frac{q_R[1 + c(p_R - p)]}{2\pi rh} = -0.001127\frac{k}{\mu}\frac{dp}{dr}$$

Separating the variables, assuming a constant compressibility over the entire pressure drop, and integrating over the length of the porous medium,

$$q_R = \frac{0.00708\ kh}{\mu c\ \ln\ (r_2/r_1)}\ln\left[\frac{1 + c(p_R - p_2)}{1 + c(p_R - p_1)}\right] \tag{7.20}$$

4.8. Radial Flow of Compressible Fluids, Steady State

The flow of a gas at any radius r of Fig. 7.7, where the pressure is p, may be expressed in terms of the flow in standard cubic feet per day by:

$$qB_g = \frac{qp_{sc}Tz}{5.615T_{sc}p}$$

Substituting in the radial form of Darcy's law

$$\frac{qp_{sc}Tz}{5.615T_{sc}p(2\pi rh)} = -0.001127\frac{k}{\mu}\frac{dp}{dr}$$

Separating variables and integrating

$$\frac{qp_{sc}Tz\mu}{5.615(0.001127)(2\pi)T_{sc}kh}\int_{r_1}^{r_2}\frac{dr}{r} = -\int_{p_1}^{p_2}pdp = \frac{1}{2}(p_1^2 - p_2^2)$$

or

$$\frac{qp_{sc}Tz\mu}{0.01988T_{sc}kh}\ln\ (r_2/r_1) = p_1^2 - p_2^2$$

Finally

$$q = \frac{0.01988T_{sc}kh\ (p_1^2 - p_2^2)}{p_{sc}T(z\mu)\ln\ (r_2/r_1)} \tag{7.21}$$

The product μz has been assumed to be constant for the derivation of Eq. (7.21). It was pointed out in Sect. 4.3, that this is usually true only for pressures less than about 1500 to 2000 psia. For greater pressures, it was stated that a better assumption was that the product $\mu z/p$ was constant. For this case, the following is obtained:

$$q = \frac{0.03976T_{sc}kh\ (p_1 - p_2)}{p_{sc}T(z\mu/p)\ln\ (r_2/r_1)} \tag{7.22}$$

In applying Eqs. (7.21) and (7.22), the products μz and $\mu z/p$ should be calculated at the average pressure between p_1 and p_2.

4.9. Permeability Variations in Radial Flow

Many producing formations are composed of strata or stringers that may vary widely in permeability and thickness, as illustrated in Fig. 7.8. If these strata are producing fluid to a common wellbore under the same drawdown and from the same drainage radius, then

$$q_t = q_1 + q_2 + q_3 \cdots + q_n$$

$$\frac{0.00708\, k_{avg} h_t (p_e - p_w)}{\mu B \ \ln\,(r_e/r_w)} = \frac{0.00708\, k_1 h_1 (p_e - p_w)}{\mu B \ \ln\,(r_e/r_w)} + \frac{0.00708\, k_2 h_2 (p_e - p_w)}{\mu B \ \ln\,(r_e/r_w)} + \text{etc.}$$

Then cancelling

$$k_{avg} h_t = k_1 h_1 + k_2 h_2 + \cdots + k_n h_n$$

$$k_{avg} = \frac{\Sigma k_i h_i}{\Sigma h_i} \tag{7.23}$$

This equation is the same as for parallel flow in linear beds with the same bed width. Here, again, average permeability refers to that permeability by which all beds could be replaced and still obtain the same production rate under the same drawdown. The product kh is called the *capacity* of a bed or stratum, and the *total capacity* of the producing formation, $\Sigma k_i h_i$, is usually expressed in *millidarcy-feet*. Because the rate of flow is directly proportional to the capacity, Eq. (7.19), a 10 ft bed of 100 md will have the same production rate as a 100 ft bed of 10 md permeability, other things being equal. There are limits of formation capacity below which production rates are not economic, just as there are limits of net productive formation thicknesses below which wells will never pay out. Of two formations with the same capacity, the one with the

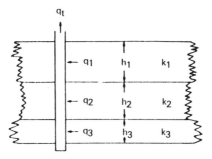

Fig. 7.8. Radial flow in parallel beds.

lower oil viscosity may be economic but the other not; and the available pressure drawdown enters in similarly. Net sand thicknesses of the order of 5 ft and capacities of the order of a few hundred millidarcy-feet are likely to be uneconomic, depending on other factors such as available drawdown, viscosity, porosity, connate water, depth, and the like. The capacity of the formation together with the viscosity also determines to a large extent whether a well will flow or whether artifical lift must be used. The amount of solution gas is an important factor. With the discovery of formation fracturing, to be discussed later, the productivity of low capacity wells in many cases has been greatly improved.

We now consider a radial flow system of constant thickness with a permeability of k_e between the drainage radius r_e and some lesser radius r_a, and an *altered* permeability k_a between the radius r_a and the wellbore radius r_w as shown in Fig. 7.9. The pressure drops are additive, and

$$(p_e - p_w) = (p_e - p_a) + (p_a - p_w)$$

Then from Eq. (7.19)

$$\frac{q \mu B \ln (r_e/r_w)}{0.00708 \, k_{avg} h} = \frac{q \mu B \ln (r_e/r_a)}{0.00708 \, k_e h} + \frac{q \mu B \ln (r_a/r_w)}{0.00708 \, k_a h}$$

Cancelling and solving for k_{avg}

$$k_{avg} = \frac{k_a k_e \ln (r_e/r_w)}{k_a \ln (r_e/r_a) + k_e \ln (r_a/r_w)} \tag{7.24}$$

Equation (7.24) may be extended to include three or more zones in series. This equation is important in studying the effect of a decrease or increase of permeability in the zone about the wellbore on the well productivity.

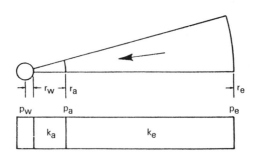

Fig. 7.9. Radial flow in beds in series.

5. DEVELOPMENT OF THE RADIAL DIFFERENTIAL EQUATION

The radial differential equation, which is the general differential equation used to model time-dependent flow systems, is now developed. Consider the volume element shown in Fig. 7.10. The element has a thickness Δr and is located r distance from the center of the well. Mass is allowed to flow into and out of the volume element during a period Δt. The volume element is in a reservoir of constant thickness and constant properties. Flow is allowed in only the radial direction. The following nomenclature, which is the same nomenclature defined previously, is used:

q = volume flow rate, STB/day for incompressible and slightly compressible fluids and SCF/day for compressible fluids

ρ = density of flowing fluid at reservoir conditions, lb/ft^3

r = distance from wellbore, ft

h = formation thickness, ft

υ = velocity of flowing fluid, $bbl/day\text{-}ft^2$

t = hours

ϕ = porosity, fraction

k = permeability, md

μ = flowing fluid viscosity, cp

With these assumptions and definitions, a mass balance can be written around the volume element over the time interval Δt. In word form, the mass balance is written as:

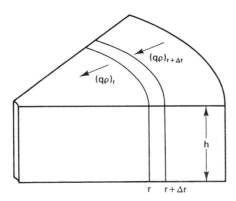

Fig. 7.10. Volume element used in the development of the radial differential equation.

Mass entering   Mass leaving   Rate at which
volume element   −   volume element   =   mass accumulates
during interval Δt during interval Δt during interval Δt

The mass entering the volume element during Δt is given by:

$$(qB\rho)_{r+\Delta r} = 2\pi(r + \Delta)rh\,(\rho\upsilon(5.615/24))_{r+\Delta r} \qquad (7.25)$$

The mass leaving the volume element during Δt is given by:

$$(qB\rho)_r = 2\pi rh\,(\rho\upsilon(5.615/24))_r \qquad (7.26)$$

The rate at which mass accumulates during the interval Δt is given by:

$$\frac{2\pi r\,\Delta rh\,[(\phi\rho)_{t+\Delta t} - (\phi\rho)_t]}{\Delta t} \qquad (7.27)$$

Combining Eqs. (7.25), (7.26), and (7.27), as suggested by the word "equation" written above,

$$2\pi(r + \Delta r)h\,(\rho\upsilon(5.615/24))_{r+\Delta r} - 2\pi rh\,(\rho\upsilon(5.615/24))_r = \frac{2\pi r\,\Delta rh\,[(\phi\rho)_{t+\Delta t} - (\phi\rho)_t]}{\Delta t}$$

If both sides of this equation are divided by $2\pi r\,\Delta rh$ and the limit is taken in each term as Δr and Δt approach zero, the following is obtained:

$$\frac{\partial}{\partial r}(0.234\,\rho\upsilon) + \frac{1}{r}(0.234\,\rho\upsilon) = \frac{\partial}{\partial t}(\phi\rho)$$

or

$$\frac{0.234}{r}\frac{\partial}{\partial r}(r\rho\upsilon) = \frac{\partial}{\partial t}(\phi\rho) \qquad (7.28)$$

Equation (7.28) is the continuity equation and is valid for any flow system of radial geometry. To obtain the radial differential equation that will be the basis for time-dependent models, pressure must be introduced and ϕ eliminated from the partial derivative term on the right-hand side of Eq. (7.28). To do this, Darcy's equation must be introduced to relate the fluid flow rate to reservoir pressure:

$$\upsilon = -0.001127\,\frac{k}{\mu}\frac{\partial p}{\partial r}$$

Realizing that the minus sign can be dropped from Darcy's equation because of the sign convention for fluid flow in porous media and substituting Darcy's equation into Eq. (7.28):

$$\frac{0.234}{r}\frac{\partial}{\partial r}\left(0.001127\frac{k}{\mu}\rho r\frac{\partial p}{\partial r}\right)=\frac{\partial}{\partial t}(\phi\rho) \qquad (7.29)$$

The porosity from the partial derivative term on the right-hand side is eliminated by expanding the right-hand side by taking the indicated derivatives:

$$\frac{\partial}{\partial t}(\phi\rho)=\phi\frac{\partial\rho}{\partial t}+\rho\frac{\partial\phi}{\partial t} \qquad (7.30)$$

It can be shown that porosity is related to the formation compressibility by the following:

$$c_f=\frac{1}{\phi}\frac{\partial\phi}{\partial p} \qquad (7.31)$$

Applying the chain rule of differentiation to $\partial\phi/\partial t$,

$$\frac{\partial\phi}{\partial t}=\frac{\partial\phi}{\partial p}\frac{\partial p}{\partial t}$$

Substituting Eq. (7.31) into this equation,

$$\frac{\partial\phi}{\partial t}=\phi c_f\frac{\partial p}{\partial t}$$

Finally, substituting this equation into Eq. (7.30) and the result into Eq. (7.29),

$$\frac{0.234}{r}\frac{\partial}{\partial r}\left(0.001127\frac{k}{\mu}\rho r\frac{\partial p}{\partial r}\right)=\rho\phi c_f\frac{\partial p}{\partial t}+\phi\frac{\partial p}{\partial t} \qquad (7.32)$$

Equation 7.32 is the general partial differential equation used to describe the flow of any fluid flowing in a radial direction in porous media. In addition to the initial assumptions, Darcy's equation has been added, which implies that the flow is laminar. Otherwise, the equation is not restricted to any type of fluid or any particular time region.

6. TRANSIENT FLOW SYSTEMS

By applying appropriate boundary and initial conditions, particular solutions to the differential equation derived in the preceding section can be discussed. The solutions obtained pertain to the transient and pseudosteady-state flow periods for both slightly compressible and compressible fluids. Since the incompressible fluid does not exist, solutions involving this type of fluid are not discussed. Only the radial flow geometry is considered because it is the most useful and applicable geometry. If you are interested in linear flow, Matthews and Russell present the necessary equations.[6] Also, due to the complex nature of the pressure behavior during the late-transient period, solutions of the differential equation for this time region are not considered. To further justify not considering flow models from this period, it is true that the most practical applications involve the transient and pseudosteady-state periods.

6.1. Radial Flow of Slightly Compressible Fluids, Transient Flow

If Eq. (7.2) is expressed in terms of density, ρ, which is the inverse of specific volume, then the following is obtained:

$$\rho = \rho_R e^{c(p-p_R)} \tag{7.33}$$

where p_R is some reference pressure and ρ_R is the density at that reference pressure. Inherent in this equation is the assumption that the compressibility of the fluid is constant. This is nearly always a good assumption over the pressure range of a given application. Substituting Eq. (7.33) into Eq. (7.32),

$$\frac{0.234}{r} \frac{\partial}{\partial r}\left(0.001127 \frac{k}{\mu}[\rho_R e^{c(p-p_R)}] r \frac{\partial p}{\partial r}\right) = [\rho_R e^{c(p-p_R)}]\phi c_f \frac{\partial p}{\partial t} + \phi \frac{\partial}{\partial t}[\rho_R e^{c(p-p_R)}]$$

To simplify this equation, one must make the assumption that k and μ are constant over the pressure, time, and distance ranges in our applications. This is rarely true about k. However, if k is assumed to be a volumetric average permeability over these ranges, then the assumption is good. In addition, it has been found that viscosities of liquids do not change significantly over typical pressure ranges of interest. Making this assumption allows k/μ to be brought outside the derivative. Taking the necessary derivatives and simplifying,

$$\frac{\partial^2 p}{\partial r^2} + \frac{1}{r} \frac{\partial p}{\partial r} + c\left[\frac{\partial p}{\partial r}\right]^2 = \frac{\phi\mu}{0.0002637k}(c_f + c)\frac{\partial p}{\partial t}$$

or

$$\frac{\partial^2 p}{\partial r^2} + \frac{1}{r} \frac{\partial p}{\partial r} + c\left[\frac{\partial p}{\partial r}\right]^2 = \frac{\phi\mu c_t}{0.0002637k}\frac{\partial p}{\partial t} \tag{7.34}$$

The last term on the left-hand side of Eq. (7.34) causes this equation to be nonlinear and very difficult to solve. However, it has been found that the term is very small for most applications of fluid flow involving liquids. When this term becomes negligible for the case of liquid flow, Eq. (7.34) reduces to:

$$\frac{\partial^2 p}{\partial r^2} + \frac{1}{r}\frac{\partial p}{\partial r} = \frac{\phi\mu c_t}{0.0002637k}\frac{\partial p}{\partial t} \tag{7.35}$$

This equation is the diffusivity equation in radial form. The name comes from its application to the radial flow of the diffusion of heat. Basically, the flow of heat, the flow of electricity, and the flow of fluids in permeable rocks can be described by the same mathematical forms. The group of terms $\phi\mu c_t/k$ was previously defined to be equal to $1/\eta$, where η is called the diffusivity constant (see Sect. 3). This same constant was encountered in Eq. (7.6) for the readjustment time.

To obtain a solution to Eq. (7.35), it is necessary first to specify one initial and two boundary conditions. The initial condition is simply that at time $t = 0$, the reservoir pressure is equal to the initial reservoir pressure, p_i. The first boundary condition is given by Darcy's equation if it is required that there be a constant rate at the wellbore:

$$q = -0.001127\frac{kh}{B\mu}(2\pi r)\left(\frac{\partial p}{\partial r}\right)_{r=r_w}$$

The second boundary condition is given by the fact that the desired solution is for the transient period. For this period, the reservoir behaves as if it were infinite in size. This suggests that at $r = \infty$, the reservoir pressure will remain equal to the initial reservoir pressure, p_i. With these conditions, Matthews and Russel gave the following solution:

$$p(r,t) = p_i - \frac{70.6q\mu B}{kh}\left[-E_i\left(-\frac{\phi\mu c_t r^2}{0.00105\,kt}\right)\right] \tag{7.36}$$

where all variables are consistent with units that have been defined previously—that is, $p(r,t)$ and p_i are in psia, q is in STB/day, μ is in cp, B (formation volume factor) is in bbl/STB, k is in md, h is in ft, c_t is in psi^{-1}, r is in ft, and t is in hr.[6] Equation (7.36) is called the *line source solution* to the diffusivity equation and is used to predict the reservoir pressure as a function of time and position. The mathematical function, E_i, is the exponential integral and is defined by:

$$E_i(-x) = -\int_x^\infty \frac{e^{-u}\,du}{u} = \left[\ln x - \frac{x}{1!} + \frac{x^2}{2(2!)} - \frac{x^3}{3(3!)} + \text{etc.}\right]$$

This integral has been calculated as a function of x and is presented in Table 7.1, from which Fig. 7.11 was developed.

TABLE 7.1.
Values of $-E_i(-x)$ as a function of x

x	$-E_i(-x)$	x	$-E_i(-x)$	x	$-E_i(-x)$
0.1	1.82292	4.3	0.00263	8.5	0.00002
0.2	1.22265	4.4	0.00234	8.6	0.00002
0.3	0.90568	4.5	0.00207	8.7	0.00002
0.4	0.70238	4.6	0.00184	8.8	0.00002
0.5	0.55977	4.7	0.00164	8.9	0.00001
0.6	0.45438	4.8	0.00145	9.0	0.00001
0.7	0.37377	4.9	0.00129	9.1	0.00001
0.8	0.31060	5.0	0.00115	9.2	0.00001
0.9	0.26018	5.1	0.00102	9.3	0.00001
1.0	0.21938	5.2	0.00091	9.4	0.00001
1.1	0.18599	5.3	0.00081	9.5	0.00001
1.2	0.15841	5.4	0.00072	9.6	0.00001
1.3	0.13545	5.5	0.00064	9.7	0.00001
1.4	0.11622	5.6	0.00057	9.8	0.00001
1.5	0.10002	5.7	0.00051	9.9	0.00000
1.6	0.08631	5.8	0.00045	10.0	0.00000
1.7	0.07465	5.9	0.00040		
1.8	0.06471	6.0	0.00036		
1.9	0.05620	6.1	0.00032		
2.0	0.04890	6.2	0.00029		
2.1	0.04261	6.3	0.00026		
2.2	0.03719	6.4	0.00023		
2.3	0.03250	6.5	0.00020		
2.4	0.02844	6.6	0.00018		
2.5	0.02491	6.7	0.00016		
2.6	0.02185	6.8	0.00014		
2.7	0.01918	6.9	0.00013		
2.8	0.01686	7.0	0.00012		
2.9	0.01482	7.1	0.00010		
3.0	0.01305	7.2	0.00009		
3.1	0.01149	7.3	0.00008		
3.2	0.01013	7.4	0.00007		
3.3	0.00894	7.5	0.00007		
3.4	0.00789	7.6	0.00006		
3.5	0.00697	7.7	0.00005		
3.6	0.00616	7.8	0.00005		
3.7	0.00545	7.9	0.00004		
3.8	0.00482	8.0	0.00004		
3.9	0.00427	8.1	0.00003		
4.0	0.00378	8.2	0.00003		
4.1	0.00335	8.3	0.00003		
4.2	0.00297	8.4	0.00002		

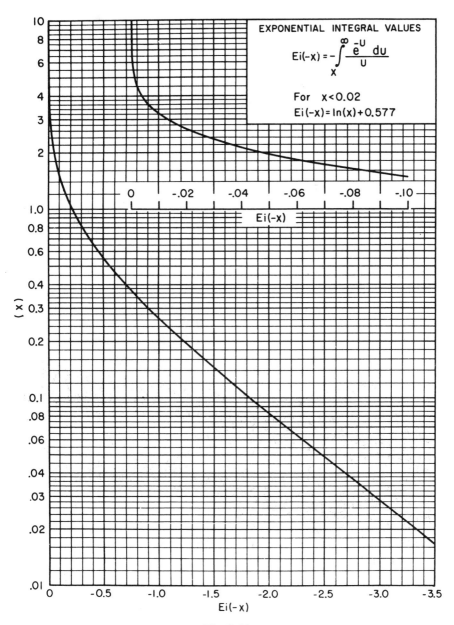

Fig. 7.11.

Equation (7.36) can be used to find the pressure drop $(p_i - p)$ that will have occurred at any radius about a flowing well after the well has flowed at a rate, q, for some time, t. For example, consider a reservoir where oil is flowing and $\mu_o = 0.72$ cp; $B_o = 1.475$ bbl/STB; $k = 100$ md; $h = 15$ ft; $c_t = 15 \times 10^{-6}$ psi^{-1}; $\phi = 23.4\%$; $p_i = 3000$ psia. After a well is produced at 200 STB/day for 10 days, the pressure at a radius of 1000 ft will be:

$$p = 3000 - \frac{70.6(200)(0.72)(1.475)}{100(15)}\left[-E_i\left(-\frac{0.234(0.72)15(10)^{-6}(1000)^2}{0.00105(100)(10)(24)}\right)\right]$$

Then

$$p = 3000 + 10.0\ E_i(-0.10)$$

From Fig. 7.11, $E_i(-0.10) = -1.82$. Therefore

$$p = 3000 + 10.0\ (-1.82) = 2981.8 \text{ psia}$$

Figure 7.12 shows this pressure plotted on the 10-day curve and shows the pressure distributions at 0.1, 1.0, and 100 days for the same flow conditions.

It has been shown that for values of the E_i function argument less than 0.01 the following approximation can be made:

$$-E_i(-x) = -\ln(x) - 0.5772$$

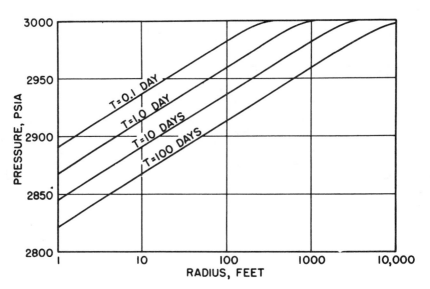

Fig. 7.12. Pressure distribution about a well at four time periods after start of production.

This suggests that

$$\frac{\phi\mu c_t r^2}{0.00105kt} < 0.01$$

By rearranging the equation and solving for t, the time required to make this approximation valid for the pressure determination 1000 ft from the producing well can be found:

$$t > \frac{0.234(0.72)15(10)^{-6}(1000)^2}{0.00105(100)(0.01)} \approx 2400 \text{ hr} = 100 \text{ days}$$

To determine if the approximation to the E_i function is valid when calculating the pressure at the sandface of a producing well, it is necessary to assume a wellbore radius, r_w, (0.25 ft) and to calculate the time that would make the approximation valid. The following is obtained:

$$t > \frac{0.234(0.72)15(10)^{-6}(0.25)^2}{0.00105(100)(0.01)} \approx 0.0002 \text{ hours}$$

It is apparent from these calculations that whether the approximation can be used is a strong function of the distance from the pressure disturbance to the point at which the pressure determination is desired or, in this case, from the producing well. For all practical purposes, the assumption is valid when considering pressures at the point of the disturbance. Therefore, at the wellbore and wherever the assumption is valid, Eq. (7.36) can be rewritten as:

$$p(r,\ t) = p_i - \frac{70.6q\mu B}{kh}\left[-\ln\left(-\frac{\phi\mu c_t r^2}{0.00105kt}\right) - 0.5772\right]$$

Substituting the log base 10 into this equation for the ln term, rearranging and simplifying, one gets:

$$p(r,\ t) = p_i - \frac{162.6\ q\mu B}{kh}\left[\log\left(\frac{kt}{\phi\mu c_t r^2}\right) - 3.23\right] \tag{7.37}$$

Equation (7.37) serves as the basis for a well testing procedure called *transient well testing*, a very useful technique that is discussed later in this chapter.

6.2. Radial Flow of Compressible Fluids, Transient Flow

In Sect. 5, Eq. (7.32)

$$\frac{0.234}{r}\frac{\partial}{\partial r}\left(0.001127\frac{k}{\mu}\rho r\frac{\partial p}{\partial r}\right) = \rho\phi c_f\frac{\partial p}{\partial t} + \phi\frac{\partial p}{\partial t} \tag{7.32}$$

was developed to describe the flow of any fluid flowing in a radial geometry in porous media. To develop a solution to Eq. (7.32) for the compressible fluid, or gas, case, two additional equations are required: (1) an equation of state, usually the real gas law, which is Eq. (1.7); and (2) Eq. (1.19), which describes how the gas isothermal compressibility varies with pressure:

$$pV = znR'T \tag{1.7}$$

$$c_g = \frac{1}{p} - \frac{1}{z}\frac{dz}{dp} \tag{1.19}$$

These three equations can be combined to yield

$$\frac{1}{r}\frac{\partial}{\partial r}\left(r\frac{p}{\mu z}\frac{\partial p}{\partial r}\right) = \frac{\phi c_t p}{0.0002637 kz}\frac{\partial p}{\partial t} \tag{7.38}$$

Al-Hussainy, Ramey, and Crawford and Russel, Goodrich, Perry, and Bruskotter introduced a transformation of variables to obtain a solution to Eq. (7.38).[7,8] The transformation involves the real gas pseudopressure, $m(p)$, which has units of psia²/cp in standard field units and is defined as:

$$m(p) = 2\int_{p_R}^{p}\frac{p}{\mu z}dp \tag{7.39}$$

where p_R is a reference pressure, usually chosen to be 14.7 psia, from which the function is evaluated. Since μ and z are only functions of pressure for a given reservoir system, which we have assumed to be isothermal, Eq. (7.39) can be differentiated and the chain rule of differentiation applied to obtain the following relationships:

$$\frac{\partial m(p)}{\partial p} = \frac{2p}{\mu z} \tag{7.40}$$

$$\frac{\partial m(p)}{\partial r} = \frac{\partial m(p)}{\partial p}\frac{\partial p}{\partial r} \tag{7.41}$$

$$\frac{\partial m(p)}{\partial t} = \frac{\partial m(p)}{\partial p}\frac{\partial p}{\partial t} \tag{7.42}$$

Substituting Eq. (7.40) into Eq. (7.41) and (7.42) yields

$$\frac{\partial p}{\partial r} = \frac{\mu z}{2p}\frac{\partial m(p)}{\partial r} \tag{7.43}$$

$$\frac{\partial p}{\partial t} = \frac{\mu z}{2p}\frac{\partial m(p)}{\partial t} \tag{7.44}$$

Combining Eq. (7.43), (7.44), and (7.38) yields

$$\frac{\partial^2 m(p)}{\partial r^2} + \frac{1}{r}\frac{\partial m(p)}{\partial r} = \frac{\phi\mu c_t}{0.0002637k}\frac{\partial m(p)}{\partial t} \tag{7.45}$$

Equation (7.45) is the diffusivity equation for compressible fluids, and it has a very similar form to Eq. (7.35), which is the diffusivity equation for slightly compressible fluids. The only difference in the appearance of the two equations is that Eq. (7.45) has the real gas pseudopressure, $m(p)$, substituted for p. There is another significant difference that is not apparent from looking at the two equations. This difference is in the assumption concerning the magnitude of the $c(\partial p/\partial r)^2$ term in Eq. (7.34) which represents the pressure gradient. To linearize Eq. (7.34), it is necessary to limit the term to a small value so that it results in a negligible quantity, which is normally the case for liquid flow applications. This limitation is not necessary for the gas equation. Since pressure gradients around the gas wells can be very large, the transformation of variables has led to a much more practical and useful equation for gases.

Equation (7.45) is still a nonlinear differential equation because of the dependence of μ and c_t on pressure or the real gas pseudopressure. Thus, there is no analytical solution for Eq. (7.45). Al-Hussainy and Ramey, however, used finite difference techniques to obtain an approximate solution to Eq. (7.45).[9] The result of their studies for pressures at the wellbore (i.e., where the logarithm approximation to the E_i function can be made) is the following equation:

$$m(p_{wf}) = m(p_i) - \frac{1637(10)^3 qT}{kh}\left[\log\left(\frac{kt}{\phi\mu_i c_{ti} r_w^2}\right) - 3.23\right] \tag{7.46}$$

where p_{wf} is the flowing pressure at the wellbore, p_i is the initial reservoir pressure, q is the flow rate in SCF/day at standard conditions of $60°F$ and 14.7 psia, T is the reservoir temperature in $°R$, k is in md, h is in ft, t is in hr, μ_i is in cp and is evaluated at the initial pressure, p_i, c_{ti} is in psi^{-1} and is also evaluated at p_i, and r_w is the wellbore radius in feet. Equation (7.46) can be used to calculate the flowing pressure at the sandface of a gas well.

The use of Eq. (7.46) requires values of the real gas pseudopressure. The procedure used to find values of $m(p)$ has been discussed in the literature.[10, 11] The procedure involves determining μ and z for several pressures over the pressure range of interest, using the methods of Chapter 1. Values of $p/\mu z$ are then calculated, and a plot of $p/\mu z$ versus p is made, as illustrated in Fig. 7.13. A numerical integration scheme such as Simpson's rule is then used to determine the value of the area from the reference pressure up to a pressure of interest, p_1. The value of $m(p_1)$ that corresponds with pressure, p_1, is given by:

$$m(p_1) = 2 \, (\text{area}_1)$$

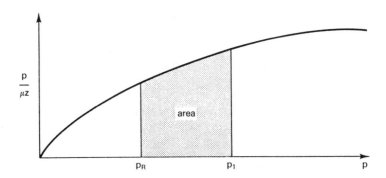

Fig. 7.13. Graphical determination of $m(p)$.

where

$$\text{area}_1 = \int_{p_R}^{p_1} \frac{p}{\mu z} \, dp$$

The real gas pseudopressure method can be applied at any pressure of interest if the data are available.

7. PSEUDOSTEADY-STATE FLOW SYSTEMS

For the transient flow cases that were considered in the previous section, the well was assumed to be located in a very large reservoir. This assumption was made so that the flow from or to the well would not be affected by boundaries that would inhibit the flow. Obviously, the time that this assumption can be made is a finite amount and often is very short in length. As soon as the flow begins to feel the affect of a boundary, it is no longer in the transient regime. At this point, it becomes necessary to make a new assumption that will lead to a different solution to the radial diffusivity equation. The following sections discuss solutions to the radial diffusivity equation that allow calculations during the pseudosteady-state flow regime.

7.1. Radial Flow of Slightly Compressible Fluids, Pseudosteady-State Flow

Once the pressure disturbance has been felt throughout the reservoir including at the boundary, then the reservoir can no longer be considered as being infinite in size and the flow is not in the transient regime. This situation necessitates another solution to Eq. (7.35), using a different boundary condition at the outer boundary. The initial condition remains the same as before (i.e., the reservoir pressure is p_i throughout the reservoir at time $t = 0$). The flow rate is again treated as constant at the wellbore. The new boundary condition used to find a solution to the radial diffusivity equation is that the outer boundary of the reservoir is a no-flow boundary. In mathematical terms,

$$\frac{\partial p}{\partial r} = 0 \quad \text{at } r = r_e$$

Applying these conditions to Eq. (7.35), the solution for the pressure at the wellbore becomes

$$p_{wf} = p_i - \frac{162.6q\mu B}{kh} \log\left[\frac{4A}{1.781C_A r_w^2}\right] - \frac{0.2339qBt}{Ah\phi c_t} \tag{7.47}$$

where A is the drainage area of the well in square feet and C_A is a reservoir shape factor. Values of the shape factor are given in Table 7.2 for several reservoir types. Eq. (7.47) is valid only for sufficiently long enough times for the flow to have reached the pseudosteady-state time period.

After reaching pseudosteady-state flow, the pressure at every point in the reservoir is changing at the same rate, which suggests that the average reservoir pressure is also changing at the same rate. The volumetric average reservoir pressure, which is usually designated as $\bar{p}$ and is the pressure used to calculate fluid properties in material balance equations, is defined as:

$$\bar{p} = \frac{\sum\limits_{j=1}^{n} \bar{p}_j V_j}{\sum\limits_{j=1}^{n} V_j} \tag{7.48}$$

where p_j is the average pressure in the jth drainage volume and V_j is the volume of the jth drainage volume. It is useful to rewrite Eq. (7.47) in terms of the average reservoir pressure, $\bar{p}$:

$$p_{wf} = \bar{p} - \frac{162.6q\mu B}{kh} \log\left[\frac{4A}{1.781C_A r_w^2}\right] \tag{7.49}$$

For a well in the center of a circular reservoir with a distance to the outer boundary of r_e, Eq. (7.49) reduces to:

$$p_{wf} = \bar{p} - \frac{70.6q\mu B}{kh}\left[\ln\left(\frac{r_e^2}{r_w^2}\right) - 1.5\right]$$

If this equation is rearranged and solved for q,

$$q = \frac{0.00708kh}{\mu B}\left[\frac{\bar{p} - p_{wf}}{\ln(r_e/r_w) - 0.75}\right] \tag{7.50}$$

TABLE 7.2.
Shape factors for various single-well drainage areas (*after* Earlougher.[3])

In Bounded Reservoirs	C_A	$\ln C_A$	$\frac{1}{2} \ln\left(\frac{2.2458}{C_A}\right)$	Exact for $t_{DA} >$	Less than 1% Error for $t_{DA} >$	Use Infinite System Solution with Less than 1% Error for $t_{DA} <$
	31.62	3.4538	−1.3224	0.1	0.06	0.10
	31.6	3.4532	−1.3220	0.1	0.06	0.10
	27.6	3.3178	−1.2544	0.2	0.07	0.09
	27.1	3.2995	−1.2452	0.2	0.07	0.09
	21.9	3.0865	−1.1387	0.4	0.12	0.08
	0.098	−2.3227	1.5659	0.9	0.60	0.015

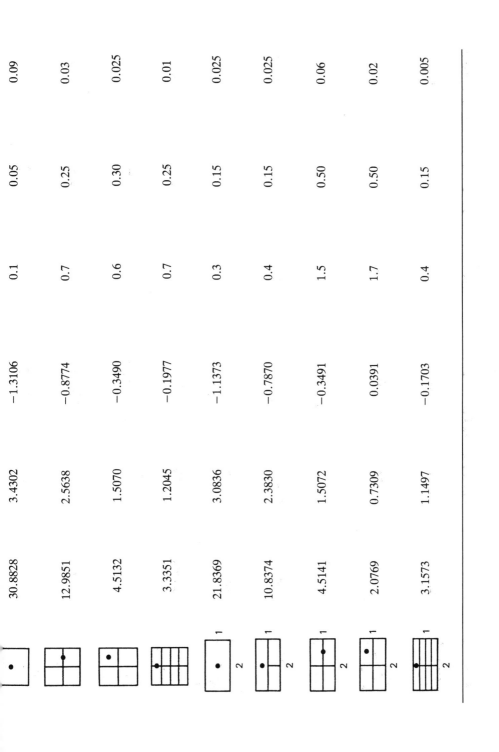

30.8828	3.4302	−1.3106	0.1	0.05	0.09
12.9851	2.5638	−0.8774	0.7	0.25	0.03
4.5132	1.5070	−0.3490	0.6	0.30	0.025
3.3351	1.2045	−0.1977	0.7	0.25	0.01
21.8369	3.0836	−1.1373	0.3	0.15	0.025
10.8374	2.3830	−0.7870	0.4	0.15	0.025
4.5141	1.5072	−0.3491	1.5	0.50	0.06
2.0769	0.7309	0.0391	1.7	0.50	0.02
3.1573	1.1497	−0.1703	0.4	0.15	0.005

7.2. Radial Flow of Compressible Fluids, Pseudosteady-State Flow

The differential equation for the flow of compressible fluids in terms of the real gas pseudopressure was derived in Eq. (7.45). When the appropriate boundary conditions are applied to Eq. (7.45), the pseudosteady-state solution rearranged and solved for q yields Eq. (7.51):

$$q = \frac{19.88(10)^{-6}khT_{sc}}{Tp_{sc}} \left[\frac{m(\bar{p}) - m(p_{wf})}{\ln(r_e/r_w) - 0.75} \right] \tag{7.51}$$

8. PRODUCTIVITY INDEX (PI)

The ratio of the rate of production, expressed in STB/day for liquid flow, to the pressure drawdown at the midpoint of the producing interval, is called the *productivity index*, symbol J.

$$J = \frac{q}{\bar{p} - p_{wf}} \tag{7.52}$$

The PI is a measure of the well potential, or the ability of the well to produce, and is a commonly measured well property. To calculate J from a production test, it is necessary to flow the well a sufficiently long time to reach pseudosteady-state flow. Only during this flow regime will the difference between $\bar{p}$ and p_{wf} be constant. It was pointed out in Section 3 that once the pseudosteady-state period had been reached, then the pressure changes at every point in the reservoir at the same rate. This is not true for the other periods, and a calculation of productivity index during other periods would not be accurate.

In some wells, the PI remains constant over a wide variation in flow rate such that the flow rate is directly proportional to the bottom-hole pressure drawdown. In other wells, at higher flow rates the linearity fails, and the PI index declines, as shown in Fig. 7.14. The cause of this decline may be (a) turbulence at increased rates of flow, (b) decrease in the permeability to oil due to presence of free gas caused by the drop in pressure at the well bore, (c) the increase in oil viscosity with pressure drop below bubble point, and/or (d) reduction in permeability due to formation compressibility.

In depletion reservoirs, the productivity indexes of the wells decline as depletion proceeds, owing to the increase in oil viscosity as gas is released from solution and to the decrease in the permeability of the rock to oil as the oil saturation decreases. Since each of these factors may change from a few to several-fold during depletion, the PI may decline to a small fraction of the initial value. Also, as the permeability to oil decreases, there is a corresponding increase in the permeability to gas, which results in rising gas-oil ratios. The maximum rate at which a well can produce depends on the productivity

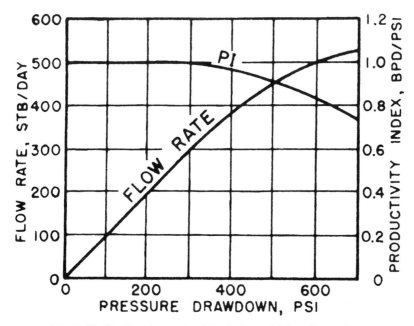

Fig. 7.14. Decline in productivity index at higher flow rates.

index at prevailing reservoir conditions and on the available pressure draw-down. If the producing bottom-hole pressure is maintained near zero by keeping the well "pumped off," then the available drawdown is the prevailing reservoir pressure, and the maximum rate is $\bar{p} \times J$.

In wells producing water, the PI, which is based on dry oil production, declines as the water cut increases because of the decrease in oil permeability, even though there is no substantial drop in reservoir pressure. In the study of these "water wells," it is sometimes useful to place the PI on the basis of total flow, including both oil and water, where in some cases the water cut may rise to 99% or more.

The *injectivity index* is used with salt water disposal wells and with *injection wells* for secondary recovery or pressure maintenance. It is the ratio of the injection rate in STB per day to the excess pressure above reservoir pressure that causes that injection rate, or

$$\text{Injectivity index} = I = \frac{q}{p_{wf} - \bar{p}} \text{ STB/day/psi} \qquad (7.53)$$

With both productivity index and injectivity index, the pressures referred to are sandface pressures, so that frictional pressure drops in the tubing or casing are not included. In the case of injecting or producing at high rates, these pressure losses may be appreciable.

In comparing one well with another in a given field, particularly when there is a variation in net productive thickness but when the other factors affecting the productivity index are essentially the same, the specific productivity index J_s is sometimes used, which is the productivity index divided by the net feet of pay, or,

$$\text{Specific productivity index} = J_s = \frac{J}{h} = \frac{q}{h(\bar{p} - p_{wf})} \text{ STB/day/psi/ft} \qquad (7.54)$$

8.1. Productivity Ratio (PR)

In evaluating well performance, the standard usually referred to is the productivity index of an open hole that completely penetrates a circular formation normal to the strata, and in which no alteration in permeability has occurred in the vicinity of the wellbore. Substituting Eq. (7.50) into Eq. (7.52) we get

$$J = 0.00708 \frac{kh}{\mu B (\ln(r_e/r_w) - 0.75)} \qquad (7.55)$$

The PR then is the ratio of the PI of a well in any condition to the PI of this *standard* well:

$$PR = \frac{J}{J_{sw}} \qquad (7.56)$$

Thus, the productivity ratio may be less than one, greater than one, or equal to one. Although the productivity index of the standard well is generally unknown, the relative effect of certain changes in the well system may be evaluated from theoretical considerations, laboratory models, or well tests. For example, the theoretical productivity ratio of a well reamed from an 8-in. borehole diameter to 16 in. is derived by Eq. (7.55):

$$PR = \frac{J_{16}}{J_8} = \frac{\ln (r_e/0.333) - 0.75}{\ln (r_e/0.667) - 0.75}$$

Assuming $r_e = 660$ ft,

$$PR = \frac{\ln (660/0.333) - 0.75}{\ln (660/0.667) - 0.75} = 1.11$$

Thus, doubling the borehole diameter should increase the PI approximately 11%. An inspection of Eq. (7.55) indicates that the PI can be improved by increasing the average permeability k, decreasing the viscosity μ, or increasing the wellbore radius r_w.

9. SUPERPOSITION

Earlougher and others have discussed the application of the principle of super-position to fluid flow in reservoirs.[3, 12, 13, 14] This principle allows the use of the constant rate, single well equations that have been developed earlier in this chapter and applies them to a variety of other cases. We illustrate the application by examining the solution to Eq. (7.35), which is a linear, second-order differential equation. The principle of superposition can be stated as: The addition of solutions to a linear differential equation results in a new solution to the originial differential equation. For example, consider the reservoir system depicted in Fig. 7.15. In the example shown in Fig. 7.15 wells 1 and 2 are opened up to their respective flow rates, q_1 and q_2, and the pressure drop that occurs in the observation well is monitored. The principle of super-position states that the total pressure drop will be the sum of the pressure drop caused by the flow from well 1 and the pressure drop caused by the flow from well 2:

$$\Delta p_t = \Delta p_1 + \Delta p_2$$

Each of the individual Δp terms is given by Eq. (7.36), or:

$$\Delta p = p_i - p(r, t) = \frac{70.6q\mu B}{kh} \left[- E_i \left(-\frac{\phi \mu c_t r^2}{0.00105kt} \right) \right]$$

To apply the method of superposition, pressure drops or changes are added. It is not correct simply to add or subtract individual pressure terms. It is obvious that if there are more than two flowing wells in the reservoir system, the procedure is the same, and the total pressure drop is given by the following,

$$\Delta p_j = \sum_{j=1}^{N} \Delta p_j \qquad (7.57)$$

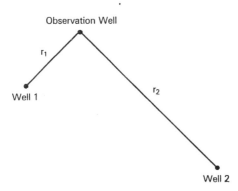

Fig. 7.15. Two flowing-well reservoir system to illustrate the principle of super-position.

where N equals the number of flowing wells in the system. Example 7.1 illustrates the calculations involved when more than one well affects the pressure of a point in a reservoir.

Example 7.1. For the well layout shown in Fig. 7.16, calculate the total pressure drop as measured in the observation well (well 3) caused by the four flowing wells (wells 1, 2, 4, and 5) after 10 days. The wells were shut in for a long time before opening them to flow.

Given: The following data apply to the reservoir system:

oil viscosity = 0.40 cp B_o = 1.50 bbl/STB
k = 47 md   formation thickness = 50 ft
porosity = 11.2% $c_t = 15 \times 10^{-6}$ psi^{-1}

Well	Flow Rate (STB/day)	Distance to Observation well (feet)
1	265	1700
2	270	1920
4	287	1870
5	260	1690

SOLUTION: The individual pressure drops can be calculated with Eq. (7.36), and the total pressure drop is given by Eq. (7.57). For well 1

$$\Delta p_1 = \frac{70.6(265)(0.40)(1.5)}{(47)(50)} \left[-E_i \left(-\frac{(.112)(0.40)(15 \times 10^{-6})(1700)^2}{0.00105(47)(240)} \right) \right]$$

$$\Delta p_1 = 4.78 \left[-E_i(-0.164) \right]$$

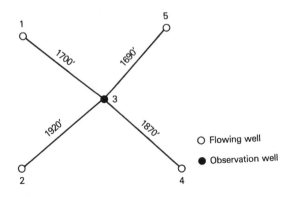

Fig. 7.16. Well layout for Ex. Prob. 7.1.

From Fig. 7.11

$$-E_i(-0.164) = 1.39$$

Therefore

$$\Delta p_1 = 4.78(1.39) = 6.6 \text{ psi}$$

Similarly, for wells 2, 4, and 5

$$\Delta p_2 = 4.87[-E_i(-0.209)] = 5.7 \text{ psi}$$
$$\Delta p_4 = 5.14[-E_i(-0.198)] = 6.4 \text{ psi}$$
$$\Delta p_5 = 4.69[-E_i(-0.162)] = 6.6 \text{ psi}$$

Using Eq. (7.57) to find the total pressure drop at the observation well, Well 3, the individual pressure drops are added together to give the total:

$$\Delta p_t = \Delta p_1 + \Delta p_2 + \Delta p_4 + \Delta p_5$$

or

$$\Delta p_t = 6.6 + 5.7 + 6.4 + 6.6 = 25.3 \text{ psi}$$

The superposition principle can also be applied in the time dimension, as is illustrated in Fig. 7.17. In this case, one well (which means the position where the pressure disturbances occur remains constant) has been produced at two flow rates. The change in the flow rate from q_1 to q_2 occurred at time t_1. Figure 7.17 shows that the total pressure drop is given by the sum of the pressure drop caused by the flow rate q_1 and the pressure drop caused by the change in flow rate $q_2 - q_1$. This new flow rate, $q_2 - q_1$, has flowed for time $t - t_1$.

The pressure drop for this flow rate, $q_2 - q_1$, is given by

$$\Delta p = p_i - p(r,\ t) = \frac{70.6(q_2 - q_1)\mu B}{kh}\left[-E_i\left(\frac{\phi\mu c_t r^2}{0.00105k(t-t_1)}\right)\right]$$

As in the case of the multiwell system just described, superposition can also be applied to multirate systems as well as the two rate examples depicted in Fig. 7.17.

9.1. Superposition in Bounded or Partially Bounded Reservoirs

Although Eq. (7.36) applies to infinite reservoirs, it may be used in conjunction with the superposition principle to simulate boundaries of closed or partially closed reservoirs. The effect of boundaries is always to cause greater

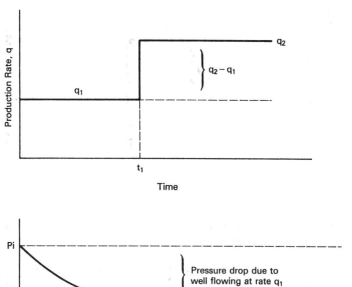

Fig. 7.17. Production rate and pressure history for a well with two flow rates.

pressure drops than those calculated for the infinite reservoirs. The method of *images* is useful in handling the effect of boundaries. For example, the pressure drop at point X (Fig. 7.18), owing to production in a well located a distance d from a sealing fault, will be the sum of the effects of the producing well and an image well that is superimposed at a distance d behind the fault. In this case the total pressure drop is given by Eq. (7.57), where the individual pressure drops are again given by Eq. (7.36), or for the case shown in Fig. 7.18:

$$\Delta p = \Delta p_1 + \Delta p_{\text{image}}$$

$$\Delta p_1 = \frac{70.6 q \mu B}{kh}\left[-E_i\left(\frac{\phi \mu c_t r_1^2}{-0.00105 kt}\right)\right]$$

$$\Delta p_{\text{image}} = \frac{70.6 q \mu B}{kh}\left[-E_i\left(\frac{\phi \mu c_t r_2^2}{-0.00105 kt}\right)\right]$$

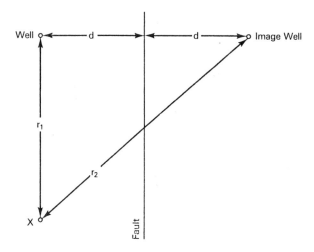

Fig. 7.18. Method of images used in the solution of boundary problems.

10. INTRODUCTION TO PRESSURE TRANSIENT TESTING

Pressure transient testing is an important diagnostic tool that can provide valuable information for the reservoir engineer. A transient test is initiated by creating a disturbance at a wellbore, (i.e., a change in the flow rate) and then monitoring the pressure as a function of time. An efficiently conducted test that yields good data can provide information such as average permeability, drainage volume, wellbore damage or stimulation, and reservoir pressure.

A pressure transient test does not always yield a unique solution. There are often anomalies associated with the reservoir system that yield pressure data that could lead to multiple conclusions. In these cases, the strength of transient testing is realized only when the procedure is used in conjunction with other diagnostic tools or other information.

In the next two subsections, the two most popular tests (i.e., the drawdown and buildup tests) are introduced. However, notice that the material is intended to be only an introduction. You must be aware of many other considerations in order to conduct a proper transient test. We refer you to some excellent books in this area by Earlougher, Matthews and Russell, and Lee.[3,6,15]

10.1. Introduction to Drawdown Testing

The drawdown test consists of flowing a well at a constant rate following a shut-in period. The shut-in period should be sufficiently long for the reservoir pressure to have stabilized. The basis for the drawdown test is found in Eq. (7.37),

$$p(r, t) = p_i - \frac{162.6\,q\mu B}{kh} \left[\log \frac{kt}{\phi\mu c_t r^2} - 3.23 \right] \qquad (7.37)$$

which predicts the pressure at any radius, *r*, as a function of time for a given reservoir flow system during the transient period. If $r = r_w$, then $p(r, t)$ will be the pressure at the wellbore. For a given reservoir system, p_i, q, μ, B, k, h, ϕ, c_t, and r_w are constant, and Eq. (7.37) can be written as

$$p_{wf} = b + m\log(t) \qquad (7.58)$$

where,

p_{wf} = flowing well pressure in psia

  b = constant

  t = time in hrs

$$m = \text{constant} = -\frac{162.6\,q\mu B}{kh} \qquad (7.59)$$

Equation (7.59) suggests that a plot of p_{wf} versus *t* on semilog graph paper would yield a straight line with slope *m* through the early time data that correspond with the transient time period. This is providing that the assumptions inherent in the derivation of Eq. (7.37) are met. These assumptions are the following:

1. Laminar, horizontal flow in a homogeneous reservoir.
2. Reservoir and fluid properties, k, ϕ, h, c_t, μ, and B are independent of pressure.
3. Single-phase liquid flow in the transient time region.
4. Negligible pressure gradients.

The expression for the slope, Eq. (7.59), can be rearranged to solve for the capacity, *kh*, of the drainage area of the flowing well. If the thickness is known, then the average permeability can be obtained by Eq. (7.60):

$$k = -\frac{162.6q\mu B}{mh} \qquad (7.60)$$

If the drawdown test is conducted long enough for the pressure transients to reach the pseudosteady-state period, then Eq. (7.47) is used to describe the pressure behavior:

$$p_{wf} = p_i - \frac{162.6q\mu B}{kh} \log\left[\frac{4A}{1.781C_A r_w^2} \right] - \frac{0.2339qBt}{Ah\phi c_t} \qquad (7.47)$$

Again grouping together the terms that are constant for a given reservoir system, Eq. (7.47) becomes

$$p_{wf} = b' + m't \qquad (7.61)$$

where

$b' = $ constant

$$m' = \text{constant} = -\frac{0.2339qB}{Ah\phi c_t} \qquad (7.62)$$

Now a plot of pressure versus time on regular cartesian graph paper yields a straight line with slope equal to m' through the late time data that correspond to the pseudosteady-state period. If Eq. (7.62) is rearranged, an expression for the drainage volume of the test well can be obtained:

$$Ah\phi = -\frac{0.2339qB}{m'c_t} \qquad (7.63)$$

The drawdown test can also yield information about damage that may have occurred around the wellbore during the initial drilling or during subsequent production. We now develop an equation that allows the calculation of a damage factor, using information from the transient flow region.

A damage zone yields an additional pressure drop because of the reduced permeability in that zone. Van Everdingen and Hurst developed an expression for this pressure drop and defined a dimensionless damage factor, S, called the skin factor[16, 17]:

$$\Delta p_{skin} = \frac{141.2q\mu B}{kh}S \qquad (7.64)$$

or

$$\Delta p_{skin} = -0.87mS \qquad (7.65)$$

From Eq. (7.65), a positive value of S causes a positive pressure drop and therefore represents a damage situation. A negative value of S causes a negative pressure drop that represents a stimulated condition like a fracture. Notice that these pressure drops caused by the skin factor are compared to the pressure drop that would normally occur through this affected zone as predicted by Eq. (7.37). Combining Eq. (7.37) and (7.65), the following expression is obtained for the pressure at the wellbore:

$$p_{wf} = p_i - \frac{162.6q\mu B}{kh}\left[log\frac{kt}{\phi\mu c_t r_w^2} - 3.23 + 0.87S \right] \qquad (7.66)$$

This equation can be rearranged and solved for the skin factor, S:

$$S = 1.151 \left[\frac{p_i - p_{wf}}{\frac{162.6 q \mu B}{kh}} - \log \frac{kt}{\phi \mu c_t r_w^2} + 3.23 \right]$$

The value of p_{wf} must be obtained from the straight line in the transient flow region. Usually a time corresponding to 1 hr is used, and the corresponding pressure is given by the designation, p_{1hr}. Substituting m into this equation and recognizing that the denominator of the first term within the brackets is actually $-m$,

$$S = 1.151 \left[\frac{p_{1hr} - p_i}{m} - \log \frac{k}{\phi \mu c_t r_w^2} + 3.23 \right] \tag{7.67}$$

Eq. (7.67) can be used to calculate a value for S from the slope of the transient flow region and the value of p_{1hr} also taken from the straight line in the transient period.

A drawdown test is often conducted during the initial production from a well since the reservoir has obviously been stabilized at the initial pressure, p_i. The difficult aspect of the test is maintaining a constant flow rate, q. If the flow rate is not kept constant during the length of the test, then the pressure behavior will reflect this varying flow rate and the correct straight line regions on the semilog and regular cartesian plots may not be identified. Other factors such as wellbore storage (shut-in well) or unloading (flowing well) can interfere with the analysis. Wellbore storage and unloading are phenomena that occur in every well to a certain degree and cause anomalies in the pressure behavior. Wellbore storage is caused by fluid flowing into the wellbore after a well has been shut in on the surface and by the pressure in the wellbore changing as the height of the fluid in the wellbore changes. Wellbore unloading in a flowing well will lead to more production at the surface than what actually occurs down hole. The effects of wellbore storage and unloading can be so dominating that they completely mask the transient time data. If this happens and if the engineer does not know how to analyze for these effects, the pressure data may be misinterpreted and errors in calculated values of permeability, skin, and the like may occur. Wellbore storage and unloading effects are discussed in detail by Earlougher.[3] These effects should always be taken into consideration when you evaluate pressure transient data.

The following example problem illustrates the analysis of drawdown test data.

Example 7.2. A drawdown test was conducted on a new oil well in a large reservoir. At the time of the test, the well was the only well that had been developed in the reservoir. Analysis of the data indicates that wellbore storage does not affect the pressure measurements. Use the data to calculate the average permeability of the area around the well the skin factor, and the drainage area of the well.

Given:

$p_i = 4000$ *psia*	formation thickness $= 20$ ft
$q = 500$ STB/day	$c_t = 30 \times 10^{-6}$ psia^{-1}
$\mu_o = 1.5$ cp	porosity $= 25\%$
$B_o = 1.2$ bbl/STB	$r_w = 0.333$ ft

Flowing pressure, p_{wf} psia	time, t hrs
3503	2
3469	5
3443	10
3417	20
3383	50
3368	75
3350	100
3306	150
3282	200
3250	300

SOLUTION: Figure 7.19 contains a semilog plot of the pressure data. The slope of the early time data, which are in the transient time region, was found to be -86 psi/cycle and the value of P_{1hr} was found to be 3526 psia. Equation (7.60) can now be used to calculate the permeability:

$$k = -\frac{162.6(500)(1.5)(1.2)}{-86(20)} = 85.1 \text{ md}$$

The skin factor is found from Eq. (7.67):

$$S = 1.151 \left[\frac{3526 - 4000}{-86} - \log \left(\frac{85.1}{(0.25)(1.5)(30 \times 10^{-6})(0.333)^2} \right) + 3.23 \right]$$

$$S = 1.04$$

This positive value for the skin factor suggests the well is slightly damaged. From the slope of a plot of P versus time on regular cartesian graph paper, shown in Fig. 7.20, and using Eq. (7.63), an estimate for the drainage area of the well can be obtained. From the semilog plot of pressure versus time, the first six data points fell on the straight line region indicating they

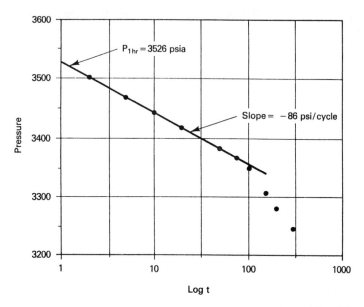

Fig. 7.19. Plot of pressure versus log time for the data of Ex. 7.2.

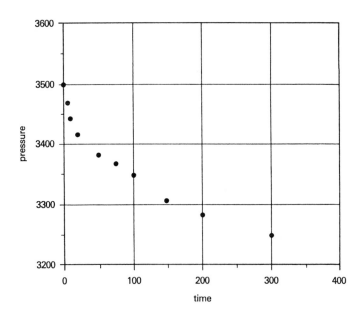

Fig. 7.20. Plot of pressure versus time for the data of Ex. 7.2.

were in the transient time period. Therefore, the last two to three points of the data are in the pseudosteady-state period and can be used to calculate the drainage area. The slope of a line drawn through the last three points is -0.373. Therefore,

$$A = -\frac{0.2339(500)(1.2)}{(-.373)(30 \times 10)^{-6}(20)(0.25)} = 2,508,000 \ ft^2 = 57.6 \ ac$$

10.2. Introduction to Buildup Testing

The buildup test is the most popular transient test used in the industry. It is conducted by shutting in a producing well that has obtained a stabilized pressure in the pseudosteady-state time region by flowing the well at a constant rate for a sufficiently long period. The pressure is then monitored during the length of the shut-in period. The primary reason for the popularity of the buildup test is the fact that it is easy to maintain the flow rate constant at zero during the length of the test. The main disadvantage of the buildup test over that of the drawdown test is that there is no production during the test and therefore no subsequent income.

A pressure buildup test is simulated mathematically by using the principle of superposition. Before the shut-in period, a well is flowed at a constant flow rate, q. At the time corresponding to the point of shut-in, t_p, a second well, superimposed over the location of the first well, is opened to flow at a rate equal to $-q$, while the first well is allowed to continue to flow at rate q. The time that the second well is flowed is given the symbol of Δt. When the effects of the two wells are added, the result is that a well has been allowed to flow at rate q for time t_p and then shut in for time Δt. This simulates the actual test procedure, which is shown schematically in Fig. 7.21. The time corresponding to the point of shut-in, t_p, can be estimated from the following equation,

$$t_p = \frac{N_p}{q} \tag{7.68}$$

where, N_p = cumulative production that has occurred during the time before shut-in that the well was flowed at the constant flow rate q.

Equations (7.37) and (7.57) can be used to describe the pressure behavior of the shut-in well:

$$p_{ws} = p_i - \frac{162.6q\mu B}{kh}\left[\log\frac{k(t_p + \Delta t)}{\phi\mu c_t r_w^2} - 3.23\right] - \frac{162.6(-q)\mu B}{kh}\left[\log\frac{k\Delta t}{\phi\mu c_t r_w^2} - 3.23\right]$$

Expanding this equation and cancelling terms,

$$p_{ws} = p_i - \frac{162.6q\mu B}{kh}\left[\log\frac{(t_p + \Delta t)}{\Delta t}\right] \tag{7.69}$$

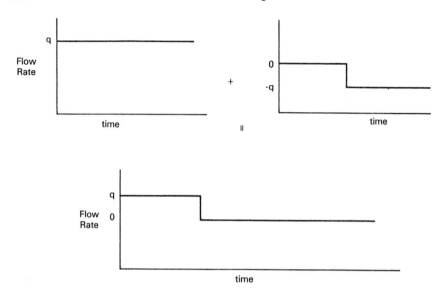

Fig. 7.21. Graphical simulation of pressure buildup test using superposition.

where,

p_{ws} = bottom-hole shut-in pressure

Equation (7.69) is used to calculate the shut-in pressure as a function of the shut-in time and suggests that a plot of this pressure versus the ratio of $(t_p + \Delta t)/\Delta t$ on semilog graph paper will yield a straight line. This plot is referred to as a Horner plot after the man who introduced it into the petroleum literature.[18] The slope of the Horner plot is equal to m, or

$$m = -\frac{162.6 q \mu B}{kh}$$

This equation can be rearranged to solve for the permeability:

$$k = -\frac{162.6 q \mu B}{mh} \qquad (7.70)$$

The skin factor equation for buildup is found by combining Eq. (7.67), written for $t = t_p$ ($\Delta t = 0$), and Eq. (7.69):

$$S = 1.151 \left[\frac{p_{wf}(\Delta t = 0) - p_{ws}}{m} - \log \frac{k t_p \Delta t}{\phi \mu c_t r_w^2 (t_p + \Delta t)} + 3.23 \right]$$

The shut-in pressure, p_{ws}, can be taken at any Δt on the straight line of the transient flow period. For convenience, Δt is set equal to 1 hr, and p_{ws} is taken it that point. At a time of $\Delta t = 1$ hr, t_p is much larger than Δt for most tests, and $t_p + \Delta t \approx t_p$. With these considerations, the skin factor equation becomes

$$S = 1.151 \left[\frac{p_{wf}(\Delta t = 0) - p_{1hr}}{m} - \log \frac{k}{\phi \mu c_t r_w^2} + 3.23 \right] \qquad (7.71)$$

We conclude this section with an example problem illustrating the analysis of a buildup test. Notice again that there is much more to this overall area of pressure transient testing. Pressure transient testing is a very useful diagnostic tool for the reservoir engineer if used correctly. The intent of this section was simply to introduce these important concepts. You should pursue the indicated references if you need a more thorough coverage of the material.

Example 7.3. Calculation of permeability and skin from a pressure buildup test.

Given:

flow rate before shut in period = 280 STB/day

N_p during constant rate period before shut in = 2682 STB

p_{wf} at the time of shut-in = 1123 psia

From the foregoing data and Eq. (7.68), t_p can be calculated

$$t_p = \frac{N_p}{q} = \left(\frac{2682}{280} \right) 24 = 230 \text{ hours}$$

Other given data are

$$B_o = 1.31 \text{ bbl/STB} \qquad \mu_o = 2.0 \text{ cp}$$
$$h = 40 \text{ ft} \qquad c_t = 15 \times 10^{-6} \text{ psi}^{-1}$$
$$\phi = 0.10 \qquad r_w = 0.333 \text{ ft}$$

time after shut in, Δt hours	Pressure, p_{ws} psia	$\dfrac{t_p + \Delta t}{\Delta t}$
2	2290	116.0
4	2514	58.5
8	2584	29.8
12	2612	20.2
16	2632	15.4
20	2643	12.5
24	2650	10.6
30	2658	8.7

SOLUTION: The slope of the straight line region (notice the difficulty in identifying the straight line region) of the Horner plot in Fig. 7.22 is -170 psi/cycle. From Eq. (7.70)

$$k = -\frac{162.6\,(280)(2.0)(1.31)}{(-170)(40)} = 17.5 \text{ md}$$

Again from the Horner plot, p_{1hr} is 2435 psia and from Eq. (7.71)

$$S = 1.151\left[\frac{1123 - 2435}{-170} - \log\left(\frac{17.5}{(0.1)(2.0)(15 \times 10^{-6})(0.333)^2}\right) + 3.23\right]$$

$$S = 3.71$$

The difficulty in identifying the straight line of the transient time region results from the two earliest data points being dominated by wellbore storage effects. The time ratio $(t_p + \Delta t)/\Delta t$ decreases as Δt increases. Therefore, the early time data are on the right of Fig. 7.22. This difficulty in identifying the proper straight line region points out the importance of a thorough understanding of wellbore storage and other anomalies that could affect the pressure transient data.

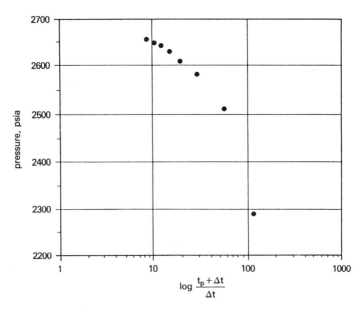

Fig. 7.22. Plot of pressure versus time ratio for Ex. 7.3.

PROBLEMS

7.1 Two wells are located 2500 ft apart. The static well pressure at the top of
perforations (9332 ft subsea) in well A is 4365 psia and at the top of perforations
(9672 ft subsea) in well B is 4372 psia. The reservoir fluid gradient is 0.25 psi/ft,
reservoir permeability is 245 md, and reservoir fluid viscosity is 0.63 cp.

  (a) Correct the two static pressures to a datum level of 9100 ft subsea

  (b) In what direction is the fluid flowing between the wells?

  (c) What is the average effective pressure gradient between the wells?

  (d) What is the fluid velocity?

  (e) Is this the total velocity or only the component of the velocity in the direction
  between the two wells?

  (f) Show that the same fluid velocity is obtained using Eq. (7.1).

7.2 A sand body is 1500 ft long, 300 ft wide, and 12 ft thick. It has a uniform
permeability of 345 md to oil at 17% connate water saturation. The porosity is
32%. The oil has a reservoir viscosity of 3.2 cp and B_o of 1.25 bbl/STB at the
bubble point.

  (a) If flow takes place above the bubble-point pressure, what pressure drop will
  cause 100 reservoir bbl/day to flow through the sand body, assuming the
  fluid behaves essentially as an incompressible fluid? What for 200 reservoir
  bbl/day?

  (b) What is the apparent velocity of the oil in feet per day at the 100 bbl/day flow
  rate?

  (c) What is the actual average velocity?

  (d) What time will be required for complete displacement of the oil from the
  sand?

  (e) What pressure gradient exists in the sand?

  (f) What will be the effect of raising both the upstream and downstream pressures by, say, 1000 psi?

  (g) Considering the oil as a fluid with a very high compressibility of $65(10)^{-6}$
  psi^{-1}, how much greater is the flow rate at the downstream end than the
  upstream end at 100 bbl/day?

  (h) What pressure drop will be required to flow 100 bbl/day, measured at the
  upstream pressure, through the sand if the compressibility of the oil is
  $65(10)^{-6}$ psi^{-1}? Consider the oil to be a slightly compressible fluid.

  (i) What will be the downstream flow rate?

  (j) What conclusion can be drawn from these calculations concerning the use of
  the incompressible flow equation for the flow of slightly compressible liquids, even with high compressibilities?

7.3 If the sand body of Prob. 7.2 had been a gas reservoir with a bottom-hole
temperature of 140°F but with the same connate water and permeability to gas,
calculate the following:

  (a) With an upstream pressure of 2500 psia, what downstream pressure will cause 5.00 MM SCF/day to flow through the sand? Assume an average gas viscosity of 0.023 cp and an average gas deviation factor of 0.88.

(b) What downstream pressure will cause 25 MM SCF/day to flow if the gas viscosity and deviation factors remain the same?

(c) Explain why it takes more than five times the pressure drop to cause five times the gas flow.

(d) What is the pressure at the midpoint of the sand when 25 MM SCF/day is flowing?

(e) What is the mean pressure at 25 MM SCF/day?

(f) Why is there a greater pressure drop in the downstream half of the sand body than in the upstream half?

(g) From the gas law calculate the rate of flow at the mean pressure p_m, and show that the equation in terms of q_m is valid by numerical substitution.

7.4 **(a)** Plot pressure versus distance through the sand of the previous problem at the 25 MM SCF/day flow rate.

(b) Plot the pressure gradient versus distance through the sand body.

7.5 A rectangular sand body is flowing gas at 10 MM SCF/day under a downstream pressure of 1000 psia. Standard conditions are 14.4 psia and 80°F. The average deviation factor is 0.80. The sand body is 1000 ft long, 100 ft wide, and 10 ft thick. Porosity is 22%, and average permeability to gas at 17% connate water is 125 md. Bottom-hole temperature is 160°F, and gas viscosity is 0.029 cp.

(a) What is the upstream pressure?

(b) What is the pressure gradient at the midpoint of the sand?

(c) What is the average pressure gradient throughout the sand?

(d) Where does the mean pressure occur?

7.6 A horizontal pipe 10 cm in diameter (I.D.) and 3000 cm long is filled with a sand of 20% porosity. It has a connate water saturation of 30% and, at that water saturation, a permeability of oil of 200 md. The viscosity of the oil is 0.65 cp and the water is immobile.

(a) What is the apparent velocity of the oil under a 100 psi pressure differential?

(b) What is the flow rate?

(c) Calculate the oil contained in the pipe and the time needed to displace it at the rate of 0.055 cu cm/sec.

(d) From this actual time and the length of the pipe, calculate the actual average velocity.

(e) Calculate the actual average velocity from the apparent velocity, porosity, and connate water.

(f) Which velocity is used to calculate flow rates, and which is used to calculate displacement times?

(g) If the oil is displaced with water so that 20% unrecoverable (or residual) oil saturation is left behind the water flood front, what are the apparent and actual average velocities in the watered zone behind the flood front if the oil production rate is maintained at 0.055 cu cm/sec? Assume piston-like displacement of the oil by the water.

(h) What is the rate of advance of the flood front?

(i) How long will it take to obtain all the recoverable oil, and how much will be recovered?

(j) How much pressure drop wil be required to produce oil at the rate of 0.055 cu cm/sec when the water flood front is at the midpoint of the pipe?

7.7 **(a)** Three beds of equal cross section have permeabilities of 50, 200, and 500 md and lengths of 40, 10, and 75 ft, respectively. What is the average permeability of the beds placed in series?

(b) What are the ratios of the pressure drops across the individual beds for liquid flow?

(c) For gas flow will the overall pressure drop through beds in series be the same for flow in either direction? Will the individual pressure drops be the same?

(d) The gas flow constant for a given linear system is 900, so that $p_1^2 - p_2^2 = 900$ L/k. If the upstream pressure is 500 psia, calculate the pressure drops in each of two beds for series flow in both directions. The one bed is 10 ft long and 100 md; the second is 70 ft and 900 md.

(e) A producing formation from top to bottom consists of 10 ft of 350 md sand, 4 in. of 0.5 md shale, 4 ft of 1230 md sand, 2 in. of 2.4 md shale, and 8 ft of 520 md sand. What is the average vertical permeability?

(f) If the 8 ft of 520 md sand is in the lower part of the formation and carries water, what well completion technique will you use to keep the water-oil ratio low for the well? Discuss the effect of the magnitude of the lateral extent of the shale breaks on the well production.

7.8 **(a)** Three beds of 40, 100, and 800 md, and 4, 6, and 10 ft thick, respectively, are conducting fluid in parallel flow. If all are of equal length and width, what is the average permeability?

(b) In what ratio are the separate flows in the three beds?

7.9 As project supervisor for an in situ uranium leaching project, you have observed that to maintain a constant injection rate in well A, the pump pressure has had to be increased so that $p_e - p_w$ has increased by a factor of 20 from the value at startup. An average permeability of 100 md was measured from plugs cored before the injection of leachant. You suspect buildup of a calcium carbonate precipitate has damaged the formation near the injection well. If the permeability of the damaged section can be assumed to be 1 md, find the extent of the damage. The wellbore radius is 0.5 ft, and the distance to the outer boundary of the uranium deposit is estimated to be 1000 ft.

7.10 A well was given a large fracture treatment, creating a fracture that extends to a radius of about 150 ft. The effective permeability of the fracture area was estimated to be 200 md. The permeability of the area beyond the fracture is 15 md. Assume that the flow is steady-state, single-phase, incompressible flow. The outer boundary at $r = r_e = 1500$ ft has a pressure of 2200 psia and the wellbore pressure is 100 psia ($r_w = 0.5$ ft). The reservoir thickness is 20 ft and the porosity is 18%. The flowing fluid has a formation volume factor of 1.12 bbl/STB and a viscosity of 1.5 cp.

(a) Calculate the flow rate in STB/day.

(b) Calculate the pressure in the reservoir at a distance of 300 ft from the center of the wellbore.

7.11 (a) A limestone formation has a matrix (primary or intergranular) permeability of less than 1 md. However, it contains 10 solution channels per square foot, each 0.02 in. in diameter. If the channels lie in the direction of fluid flow, what is the permeability of the rock?

(b) If the porosity of the matrix rock is 10%, what percentage of the fluid is stored in the primary pores, and what in the secondary pores (vugs, fractures, etc.)?

(c) If the secondary pore system is well connected throughout a reservoir, what conclusions must be drawn concerning the probable result of gas or water drive on the recovery of either oil, gas, or gas-condensate? What then are the means of recovering the hydrocarbons from the primary pores?

7.12 During a gravel rock operation the 6 in. I.D. liner became filled with gravel, and a layer of mill scale and dirt accumulated to a thickness of 1 in. on top of the gravel within the pipe. If the permeability of the accumulation is 1000 md, what additional pressure drop is placed on the system when pumping a 1 cp fluid at the rate of 100 bbl/hr?

7.13 One hundred capillary tubes of 0.02 in. ID and 50 capillary tubes of 0.04 in. ID, all of equal length, are placed inside a pipe of 2 in. inside diameter. The space between the tubes is filled with wax so that flow is only through the capillary tubes. What is the permeability of this 'rock'?

7.14 Suppose, after cementing, an opening 0.01 in. wide is left between the cement and an 8 in. diameter hole. If this circular fracture extends from the producing formation through an impermeable shale 20 ft thick to an underlying water sand, at what rate will water enter the producing formation (well) under a 100 psi pressure drawdown? The water contains 60,000 ppm salt and the bottom-hole temperature is 150°F.

7.15 A high water-oil ratio is being produced from a well. It is thought that the water is coming from an underlying aquifer 20 ft from the oil producing zone. In between the aquifer and the producing zone is an impermeable shale zone. Assume that the water is coming up through an incomplete cementing job that left an opening 0.01 in. wide between the cement and the 8 in. hole. The water has a viscosity of 0.5 cp. Determine the rate at which water is entering the well at the producing formation level if the pressure in the aquifer is 150 psi greater than the pressure in the well at the producing formation level.

7.16 Derive the equation for the steady-state, semispherical flow of an incompressible fluid.

7.17 A well has a shut-in bottom-hole pressure of 2300 psia and flows 215 bbl/day of oil under a drawdown of 500 psi. The well produces from a formation of 36 ft net productive thickness. Use $r_w = 6$ in.; $r_e = 660$ feet; $\mu = 0.88$ cp; $B_o = 1.32$ bbl/STB.

(a) What is the productivity index of the well?

(b) What is the average permeability of the formation?

(c) What is the capacity of the formation?

7.18 A producing formation consists of two strata: one 15 ft thick and 150 md in permeability; the other 10 ft thick and 400 md in permeability.

(a) What is the average permeability?

(b) What is the capacity of the formation?

(c) If during a well workover the 150 md stratum permeability is reduced to 25 md out to a radius of 4 ft, and the 400 md stratum is reduced to 40 md out to an 8 ft radius, what is the average permeability after the workover, assuming no cross-flow between beds? Use $r_e = 500$ ft and $r_w = 0.5$ ft.

(d) To what percentage of the original productivity index will the well be reduced?

(e) What is the capacity of the damaged formation?

7.19 **(a)** Plot pressure versus radius on both linear and semilog paper at 0.1, 1.0, 10, and 100 days for $p_e = 2500$ psia, $q = 300$ STB/day; $B_o = 1.32$ bbl/STB; $\mu = 0.44$ cp; $k = 25$ md; $h = 43$ ft; $c_t = 18 \times 10^{-6}$ psi^{-1}; $\phi = 0.16$.

(b) Assuming that a pressure drop of 5 psi can be easily detected with a pressure gauge, how long must the well be flowed to produce this drop in a well located 1200 ft away?

(c) Suppose the flowing well is located 200 ft due east of a north-south fault. What pressure drop will occur after 10 days of flow, in a shut-in well located 600 ft due north of the flowing well?

(d) What will the pressure drop be in a shut-in well 500 ft from the flowing well when the flowing well has been shut in for one day following a flow period of 5 days at 300 STB/day?

7.20 A shut-in well is located 500 ft from one well and 1000 ft from a second well. The first well flows for 3 days at 250 STB/day, at which time the second well begins to flow at 400 STB/day. What is the pressure drop in the shut-in well when the second well has been flowing for 5 days (i.e., the first has been flowing a total of 8 days)? Use the reservoir constants of Prob. 7.19.

7.21 A well is opened to flow at 200 STB/day for 1 day. The second day its flow is increased to 400 STB/day and the third to 600 STB/day. What is the pressure drop caused in a shut-in well 500 ft away after the third day? Use the reservoir constants of Prob. 7.19.

7.22 The following data pertain to a volumetric gas reservoir:

Net formation thickness = 15 ft

Hydrocarbon porosity = 20%

Initial reservoir pressure = 6000 psia

Reservoir temperature = 190°F

Gas viscosity = 0.020 cp

Casing diameter = 6 in.

Average formation permeability = 6 md

(a) Assuming ideal gas behavior and uniform permeability, calculate the percentage of recovery from a 640 ac unit for a producing rate of 4.00 MM SCF/day when the flowing well pressure reaches 500 psia.

(b) If the average reservoir permeability had been 60 md instead of 6 md, what recovery would be obtained at 4.00 MM SCF/day and a flowing well pressure of 500 psia?

(c) Recalculate part (a) for a production rate of 2.00 MM SCF/day.

(d) Suppose four wells are drilled on the 640 ac unit, and each is produced at 4.00 MM SCF/day. For 6 md and 500 psia minimum flowing well pressure, calculate the recovery.

7.23 A sandstone reservoir, producing well above its bubble-point pressure, contains only one producing well, which is flowing only oil at a constant rate of 175 STB/day. Ten weeks after this well began producing, another well was completed 660 ft away in the same formation. On the basis of the reservoir properties that follow, estimate the initial formation pressure that should be encountered by the second well at the time of completion.

$$\phi = 15\% \qquad\qquad h = 30 \text{ ft}$$

$$c_o = 18(10)^{-6} \text{ psi}^{-1} \qquad \mu = 2.9 \text{ cp}$$

$$c_w = 3(10)^{-6} \text{ psi}^{-1} \qquad k = 35 \text{ md}$$

$$c_f = 4.3(10)^{-6} \text{ psi}^{-1} \qquad r_w = 0.33 \text{ ft}$$

$$S_w = 33\% \qquad\qquad p_i = 4300 \text{ psia}$$

$$B_o = 1.25 \text{ bbl/STB}$$

7.24 Develop an equation to calculate and then calculate the pressure at well 1, illustrated in Fig. 7.23, if the well has flowed for 5 days at a flow rate of 200 STB/day.

$$\phi = 25\% \qquad\qquad h = 20 \text{ ft}$$

$$c_t = 30(10)^{-6} \text{ psi}^{-1} \qquad \mu = 0.5 \text{ cp}$$

$$k = 50 \text{ md} \qquad\qquad B_o = 1.32 \text{ bbl/STB}$$

$$r_w = 0.33 \text{ ft} \qquad\qquad p_i = 4000 \text{ psia}$$

7.25 A pressure drawdown test was conducted on the discovery well in a new reservoir to estimate the drainage volume of the reservoir. The well was flowed at a constant rate of 125 STB/day. The bottom-hole pressure data, as well as other rock and fluid property data, follow. What are the drainage volume of the well and the average permeability of the drainage volume? The initial reservoir pressure was 3900 psia.

$$B_o = 1.1 \text{ bbl/STB} \qquad \mu_o = 0.80 \text{ cp}$$

$$\phi = 20\% \qquad\qquad h = 22 \text{ ft}$$

$$S_o = 80\% \qquad\qquad S_w = 20\%$$

$$c_o = 10(10)^{-6} \text{ psi}^{-1} \qquad c_w = 3(10)^{-6} \text{ psi}^{-1}$$

$$c_f = 4(10)^{-6} \text{ psi}^{-1} \qquad r_w = 0.33 \text{ ft}$$

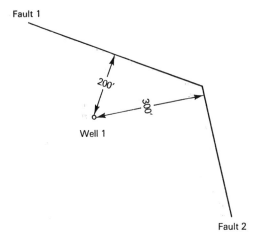

Fig. 7.23. Well layout for Prob. 7.24.

Time in Hours	p_{wf}, psi
0.5	3657
1.0	3639
1.5	3629
2.0	3620
3.0	3612
5.0	3598
7.0	3591
10.0	3583
20.0	3565
30.0	3551
40.0	3548
50.0	3544
60.0	3541
70.0	3537
80.0	3533
90.0	3529
100.0	3525
120.0	3518
150.0	3505

7.26 The initial average reservoir pressure in the vicinity of a new well was 4150 psia. A pressure drawdown test was conducted while the well was flowed at a constant oil flow rate of 550 STB/day. The oil had a viscosity of 3.3 cp and a formation volume factor of 1.55 bbl/STB. Other data, along with the bottom-hole pressure data recorded during the drawdown test, follow. Assume that wellbore storage considerations may be neglected, and determine the following:

(a) The permeability of the formation around the well.

(b) Any damage to the well.

(c) The drainage volume of the reservoir communicating to the well.

$\phi = 34.3\%$ $h = 93$ ft

$c_t = 1(10)^{-5}$ psi^{-1} $r_w = 0.5$ ft

Time in Hours	p_{wf}, psi
1	4025
2	4006
3	3999
4	3996
6	3993
8	3990
10	3989
20	3982
30	3979
40	3979
50	3978
60	3977
70	3976
80	3975

7.27 The first oil well in a new reservoir was flowed at a constant flow rate of 195 STB/day until a cumulative volume of 361 STB had been produced. After this production period, the well was shut in and the bottom-hole pressure monitored for several hours. The flowing pressure just as the well was being shut in was 1790 psia. For the data that follow, calculate the formation permeability and the initial reservoir pressure.

$B_o = 2.15$ bbl/STB $\mu_o = 0.85$ cp

$\phi = 11.5\%$ $h = 23$ ft

$c_t = 1(10)^{-5}$ psi^{-1} $r_w = 0.33$ ft

Δt in Hours	p_{ws}, psi
0.5	2425
1.0	2880
2.0	3300
3.0	3315
4.0	3320
5.0	3324
6.0	3330
8.0	3337
10.0	3343
12.0	3347
14.0	3352
16.0	3353
18.0	3356

7.28 A well located in the center of several other wells in a consolidated sandstone reservoir was chosen for a pressure buildup test. The well had been put on production at the same time as the other wells and had been produced for 80 hr at a constant oil flow rate of 375 STB/day. The wells were drilled on 80 ac spacing. For the pressure buildup data and other rock and fluid property data that follow, estimate a value for the formation permeability and determine if the well is damaged. The flowing pressure at shut-in was 3470 psia.

$B_o = 1.31$ bbl/STB $\mu_o = 0.87$ cp

$\phi = 25.3\%$ $h = 22$ ft

$S_o = 80\%$ $S_w = 20\%$

$c_o = 17(10)^{-6}$ psi^{-1} $c_w = 3(10)^{-6}$ psi^{-1}

$c_f = 4(10)^{-6}$ psi^{-1} $r_w = 0.33$ ft

Δt in Hours	p_{ws}, psi
0.114	3701
0.201	3705
0.432	3711
0.808	3715
2.051	3722
4.000	3726
8.000	3728
17.780	3730

REFERENCES

[1] W. T. Cardwell, Jr., and R. L. Parsons, "Average Permeabilities of Heterogeneous Oil Sands," *Trans.* AIME (1945), **160,** 34.

[2] J. Law, "A Statistical Approach to the Interstitial Heterogeneity of Sand Reservoirs," *Trans.* AIME (1948), **174,** 165.

[3] Robert C. Earlougher, Jr., "Advances in Well Test Analysis," Monograph Vol. 5, Society of Petroleum Engineers of AIME. Dallas, TX: Millet the Printer, 1977.

[4] I. Fatt and D. H. Davis, "Reduction in Permeability with Overburden Pressure," *Trans.* AIME (1952), **195,** 329.

[5] Robert A. Wattenbarger and H. J. Ramey, Jr., "Gas Well Testing With Turbulence. Damage and Wellbore Storage," *Jour. of Petroleum Technology* (Aug. 1968), 877–887; *Trans.* AIME, **243.**

[6] C. S. Matthews and D. G. Russell, "Pressure Buildup and Flow Tests in Wells," Monograph Vol. 1, Society of Petroleum Engineers of AIME. Dallas, TX: Millet the Printer, 1967.

[7] R. Al-Hussainy, H. J. Ramey, Jr., and P. B. Crawford, "The Flow of Real Gases Through Porous Media," *Trans.* AIME (1966), **237,** 624.

[8] D. G. Russell, J. H. Goodrich, G. E. Perry, and J. F. Bruskotter, "Methods for Predicting Gas Well Performance," *Jour. of Petroleum Technology* (Jan. 1966), 99–108; *Trans.* AIME, **237.**

[9] R. Al-Hussainy and H. J. Ramey, Jr., "Application of Real Gas Flow Theory to Well Testing and Deliverability Forecasting," *Jour. of Petroleum Technology* (May 1966), 637–642; *Trans.* AIME, **237.** Also *Reprint Series. No. 9—Pressure Analysis Methods,* Society of Petroleum Engineers of AIME, Dallas (1967) 245–250.

[10] *Theory and Practice of the Testing of Gas Wells,* 3rd ed. Calgary, Canada: Energy Resources Conservation Board, 1975.

[11] L. P. Dake, *Fundamentals of Reservoir Engineering.* Amsterdam: Elsevier, 1978.

[12] A. F. van Everdingen and W. Hurst, "The Application of the Laplace Transformation to Flow Problems in Reservoirs," *Trans.* AIME (1949), **186,** 305–324.

[13] D. R. Horner, "Pressure Build-Up in Wells," *Proc.,* Third World Pet. Cong., The Hague (1951), Sec. II, 503–523. Also *Reprint Series, No. 9—Pressure Analysis Methods,* Society of Petroleum Engineers of AIME, Dallas, TX (1967) 25–43.

[14] Royal Eugene Collins, *Flow of Fluids Through Porous Materials.* New York: Reinhold, 1961, 108–123.

[15] W. John Lee, *Well Testing,* Society of Petroleum Engineers, Dallas, TX (1982).

[16] A. F. van Everdingen, "The Skin Effect and Its Influence on the Productive Capacity of a Well," *Trans.,* AIME (1953), **198,** 171.

[17] W. Hurst, "Establishment of the Skin Effect and Its Impediment to Fluid Flow into a Wellbore," *Petroleum Engineering* (Oct. 1953), **25,** B–6.

[18] D. R. Horner, "Pressure Build-Up in Wells," *Proc.,* Third World Petroleum Congress, The Hague (1951), Sec. II, 503. Also *Reprint Series, No. 9—Pressure Analysis Methods,* Society of Petroleum Engineers of AIME, Dallas, TX (1967), 25.

Chapter **8**

Water Influx

1. INTRODUCTION

Many reservoirs are bounded on a portion or all of their peripheries by water-bearing rocks called *aquifers* (Latin:aqua—water, ferre—to bear). The aquifers may be so large compared with the reservoirs they adjoin as to appear infinite for all practical purposes, and they may range down to those so small as to be negligible in their effect on reservoir performance. The aquifer itself may be entirely bounded by impermeable rock so that the reservoir and aquifer together form a closed, or volumetric, unit (Fig. 8.1). On the other hand, the reservoir may outcrop at one or more places where it may be replenished by surface waters (Fig. 8.2). Finally, an aquifer may be essentially horizontal with the reservoir it adjoins, or it may rise, as at the edge of structural basins, considerably above the reservoir to provide some artesian kind of flow of water to the reservoir.*

In response to a pressure drop in the reservoir, the aquifer reacts to offset, or retard, pressure decline by providing a source of water influx or encroachment by (a) expansion of the water; (b) expansion of other known or unknown hydrocarbon accumulations in the aquifer rock; (c) compressibility

---

* References throughout the text are given at end of each chapter.

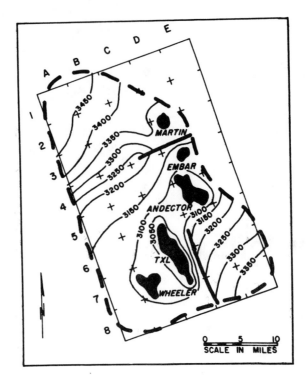

Fig. 8.1. A reservoir analyzer study of five fields completed in a closed aquifer in the Ellenburger formation in West Texas. (*After* Moore and Truby.[1])

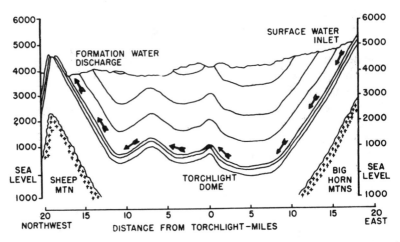

Fig. 8.2. Geologic cross section through the Torchlight Tensleep Reservoir, Wyoming. (*After* Stewart, Callaway, and Gladfelter.[2])

of the aquifer rock; and/or (d) artesian flow, which occurs when the aquifer rises to a level above the reservoir, whether it outcrops or not, and whether or not the outcrop is replenished by surface water.

To determine the effect that an aquifer has on the production from a hydrocarbon reservoir, it is necessary to be able to calculate the amount of water that has influxed into the reservoir from the aquifer. This calculation can be made using the material balance equation when the initial hydrocarbon amount and the production are known. The Havlena and Odeh approach to material balance calculations, presented in Chapter 2, can sometimes be used to obtain an estimate for both water influx and initial hydrocarbon amount.[3,4] For the case of a water-drive reservoir, no original gas cap, and negligible compressibilities, Eq. (2.13) reduces to the following:

$$F = NE_o + W_e$$

or

$$\frac{F}{E_o} = N + \frac{W_e}{E_o}$$

If correct values of W_e are placed in this equation as a function of reservoir pressure, then the equation should plot as a straight line with intercept, N, and slope equal to unity. The procedure to solve for both W_e and N in this case involves assuming a model for W_e as a function of pressure, calculating W_e, making the plot of F/E_o versus W_e/E_o, and observing if a straight line is obtained. If a straight line is not obtained, then a new model for W_e is assumed and the procedure repeated.

Choosing an appropriate model for water influx involves many uncertainties. Some of these include the size and shape of the aquifer and aquifer properties such as porosity and permeability. Normally, little is known about these parameters largely because the cost to drill into the aquifer to obtain the necessary data is not often justified.

In this chapter, several models that have been used in reservoir studies to calculate water influx amounts are considered. These models can be generally categorized by a time dependence (i.e., steady-state or unsteady-state) and whether the aquifer is an edge-water or bottom-water drive.

2. STEADY-STATE MODELS

The simplest model we discuss is the Schilthuis steady-state model, in which the rate of water influx, dW_e/dt, is directly proportional to $(p_i - p)$, where the pressure, p, is measured at the original oil-water contact.[5] This model assumes that the pressure at the external boundary of the aquifer is maintained at the initial value p_i and that flow to the reservoir is, by Darcy's Law, proportional to the pressure differential, assuming the water viscosity, average permeability, and aquifer geometry remain constant,

$$W_e = k' \int_0^t (p_i - p)dt \qquad (8.1)$$

$$\frac{dW_e}{dt} = k'(p_i - p) \qquad (8.2)$$

where k' is the water influx constant in barrels per day per pounds per square inch and $(p_i - p)$ is the boundary pressure drop in pounds per square inch. If the value of k' can be found, then the value of the cumulative water influx W_e can be found from Eq. (8.1), from a knowledge of the pressure history of the reservoir. If during any reasonably long period the rate of production and reservoir pressure remain substantially constant, it is obvious that the volumetric withdrawal rate, or *reservoir voidage rate,* must equal the water influx rate, or

$$\frac{dW_e}{dt} = \begin{bmatrix} \text{Rate of active} \\ \text{oil volumetric} \\ \text{voidage} \end{bmatrix} + \begin{bmatrix} \text{Rate of free} \\ \text{gas volumetric} \\ \text{voidage} \end{bmatrix} + \begin{bmatrix} \text{Rate of water} \\ \text{volumetric} \\ \text{voidage} \end{bmatrix}$$

In terms of single-phase oil volume factors

$$\frac{dW_e}{dt} = B_o \frac{dN_p}{dt} + (R - R_{so})\frac{dN_p}{dt} B_g + \frac{dW_p}{dt} B_w \qquad (8.3)$$

where dN_p/dt is the daily oil rate in STB/day and $(R - R_{so})dN_p/dt$ is the daily *free* gas rate in SCF/day. The solution gas-oil ratio R_{so} is subtracted from the *net daily* or *current* gas-oil ratio R because the solution gas R_{so} is accounted for in the oil volume factor B_o of the oil voidage term. Equation (8.3) may be converted to an equivalent one using two-phase volume factors by adding and subtracting the term $R_{soi}B_g dN_p/dt$, and grouping as

$$\frac{dW_e}{dt} = [B_o + (R_{soi} - R_{so})B_g]\frac{dN_p}{dt} + (R - R_{soi})B_g \frac{dN_p}{dt} + B_w \frac{dW_p}{dt}$$

and since $[B_o + (R_{soi} - R_{so})B_g]$ is the two-phase volume factor B_t:

$$\frac{dW_e}{dt} = B_t \frac{dN_p}{dt} + (R - R_{soi})B_g \frac{dN_p}{dt} + B_w \frac{dW_p}{dt} \qquad (8.4)$$

When dW_e/dt has been obtained in terms of the voidage rates by Eqs. (8.3) or

(8.4), then the influx constant k' may be found using Eq. (8.2). Although the influx constant can be obtained in this manner only when the reservoir pressure stabilizes, once it has been found it may be applied to both stabilized and changing reservoir pressures.

Figure 8.3 shows the pressure and production history of the Conroe Field, Texas, and Fig. 8.4 gives the gas and two-phase oil volume factors for the reservoir fluids. Between 33 and 39 months after the start of production, the reservoir pressure stabilized near 2090 psig and the production rate was substantially constant at 44,100 STB/day with a constant gas-oil ratio of 825 SCF/STB. Water production during the period was negligible. Example 8.1 shows the calculation of the water influx constant k' for the Conroe Field from data for this period of stabilized pressure. If the pressure stabilizes and the withdrawal rates are not reasonably constant, the water influx for the period of stabilized pressure may be obtained from the total oil, gas, and water voidages for the period,

$$\Delta W_e = B_t \Delta N_p + (\Delta G_p - R_{soi} \Delta N_p)B_g + B_w \Delta W_p$$

where ΔG_p, ΔN_p, and ΔW_p are the gas, oil, and water produced during the period in surface units. The influx constant is obtained by dividing ΔW_e by the *product* of the days in the interval *and* the stabilized pressure drop $(p_i - p_s)$,

$$k' = \frac{\Delta W_e}{\Delta t(p_i - p_s)}$$

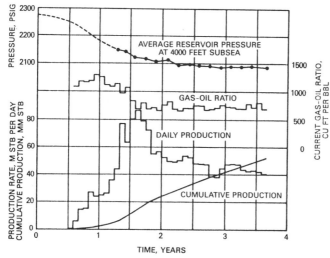

Fig. 8.3. Reservoir pressure and production data, Conroe Field. (*After* Schilthuis.[5])

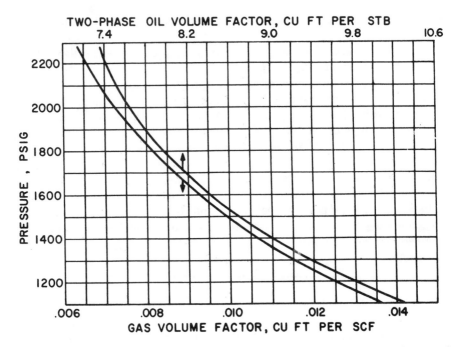

Fig. 8.4. Pressure volume relations for Conroe Field oil and original complement of dissolved gas. (*After* Schilthuis.[5])

Example 8.1. Calculating the water influx constant when reservoir pressure stabilizes.

Given:

The PVT data for Conroe Field, Fig. 8.4:

$$p_i = 2275 \text{ psig}$$

$$p_s = 2090 \text{ psig (stabilized pressure)}$$

$$B_t = 7.520 \text{ cu ft/STB at 2090 psig}$$

$$B_g = 0.00693 \text{ cu ft/SCF at 2090 psig}$$

$$R_{soi} = 600 \text{ SCF/STB (initial solution gas)}$$

$$R = 825 \text{ SCF/STB, from production data}$$

$$dN_p/dt = 44,100 \text{ STB/day, from production data}$$

$$dW_p/dt = 0$$

SOLUTION: At 2090 psig by Eq. (8.4) the daily voidage rate is

$$\frac{dV}{dt} = 7.520 \times 44{,}100 + (825 - 600)\,0.00693 \times 44{,}100 + 0$$

$$= 401{,}000 \text{ cu ft/day}$$

Since this must equal the water influx rate at stabilized pressure conditions, by Eq. (8.2)

$$\frac{dV}{dt} = \frac{dW_e}{dt} = 401{,}000 = k'(2275 - 2090)$$

$$k' = 2170 \text{ cu ft/day/psi}$$

A water influx constant of 2170 cu ft/day/psi means that if the reservoir pressure suddenly drops from an initial pressure of 2275 to, say, 2265 psig (i.e., $\Delta p = 10$ psi) and remains there for 10 days, during this period the water influx will be

$$\Delta W_{e1} = 2170 \times 10 \times 10 = 217{,}000 \text{ cu ft}$$

If at the end of 10 days it drops to, say, 2255 (i.e., $\Delta p = 20$ psi) and remains there for 20 days, the water influx during this second period will be

$$\Delta W_{e2} = 2170 \times 20 \times 20 = 868{,}000 \text{ cu ft}$$

There is four times the influx in the second period because the influx rate was twice as great (because the pressure drop was twice as great) and because the interval was twice as long. The cumulative water influx at the end of 30 days, then, is

$$W_e = k' \int_0^{30} (p_i - p)\, dt = k' \sum_0^{30} (p_i - p)\Delta t$$

$$= 2170[(2275 - 2265) \times 10 + (2275 - 2255) \times 20]$$

$$= 1{,}085{,}000 \text{ cu ft}$$

In Fig. 8.5 the $\int_0^t (p_i - p)dt$ is represented by the area *beneath* the curve of *pressure drop*, $(p_i - p)$, plotted versus time; or it represents the area *above* the curve of *pressure* versus time. The areas may be found by graphical integration.

One of the problems associated with the Schilthuis steady-state model is that as the water is drained from the aquifer, the distance that the water has to travel to the reservoir increases. Hurst suggested a modification to the Schilthuis equation by including a logarithmic term to account for this increasing distance.[6] The Hurst method has met with limited application and is infrequently used.

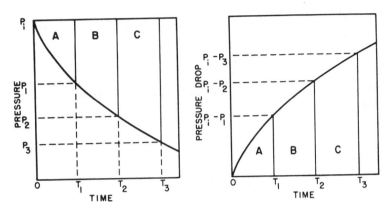

Fig. 8.5. Plot of pressure and pressure drop versus time.

Hurst Method

$$W_e = c' \int_0^t \frac{(p_i - p)dt}{\log at}$$

$$\frac{dW_e}{dt} = \frac{c'(p_i - p)}{\log at}$$

where c' is the water influx constant in barrels per day per pounds per square inch, $(p_i - p)$ is the boundary pressure drop in pounds per square inch, and a is a time conversion constant that depends on the units of the time t.

3. UNSTEADY-STATE MODELS

In nearly all applications, the steady-state models discussed in the previous section are not adequate in describing the water influx. The transient nature of the aquifers suggests that a time-dependent term be included in the calculations for W_e. In the next two sections, unsteady-state models for both edge-water and bottom-water drives are presented. An edge-water drive is defined as water influxing the reservoir from its flanks with negligible flow in the vertical direction. In contrast, a bottom-water drive has significant vertical flow.

3.1. The van Everdingen and Hurst Edge-Water Drive Model

Consider a circular reservoir of radius r_R, as shown in Fig. 8.6, in a horizontal, circular aquifer of radius r_e, which is uniform in thickness, permeability, and porosity, and in rock and water compressibilities. The radial diffusivity equation, Eq. (7.35), expresses the relationship between pressure, radius, and time

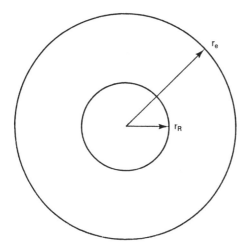

Fig. 8.6. Circular reservoir inside a circular aquifer.

for a radial system such as Fig. 8.6, where the driving potential of the system is the water expandibility and the rock compressibility:

$$\frac{\partial^2 p}{\partial r^2} + \frac{1}{r}\frac{\partial p}{\partial r} = \frac{\phi \mu c_t}{0.0002637k}\frac{\partial p}{\partial t} \qquad (7.35)$$

This equation was solved in Chapter 7 for what is referred to as the *constant terminal rate case*. The constant terminal rate case requires a constant flow rate at the inner boundary, which was the wellbore for the solutions of Chapter 7. This was appropriate for the applications of Chapter 7 since it was desirous to know the pressure behavior at various points in the reservoir because a constant flow of fluid came into the wellbore from the reservoir.

In this chapter, the diffusivity equation is applied to the aquifer where the inner boundary is defined as the interface between the reservoir and the aquifer. With the interface as the inner boundary, it would be more useful to require the pressure at the inner boundary to remain constant and observe the flow rate as it crosses the boundary or as it enters the reservoir from the aquifer. Mathematically, this condition is stated as

$$p = \text{constant} = p_i - \Delta p \text{ at } r = r_R \qquad (8.5)$$

where r_R is a constant and is equal to the outer radius of the reservoir (i.e., the original oil-water contact). The pressure p must be determined at this original oil-water contact. Van Everdingen and Hurst[7] solved the diffusivity equation for this condition, which is referred to as the *constant terminal pressure case,* and the following initial and outer boundary conditions:

Initial condition:

$$p = p_i \text{ for all values of } r$$

Outer boundary condition:

For an infinite aquifer:

$$p = p_i \text{ at } r = \infty$$

For a finite aquifer:

$$\frac{\partial p}{\partial r} = 0 \text{ at } r = r_e$$

At this point, we rewrite the diffusivity equation in terms of the following dimensionless parameters:

$$\text{Dimensionless time: } t_D = 0.0002637 \frac{kt}{\phi\mu c_t r_R^2} \qquad (8.6)$$

$$\text{Dimensionless radius: } r_D = \frac{r}{r_R}$$

$$\text{Dimensionless pressure: } p_D = \frac{p_i - p}{p_i - p_{wf}}$$

where k = average aquifer permeability, md; t = time, hours; ϕ = aquifer porosity, fraction; μ = water viscosity, cp; c_t = aquifer compressibility, psi^{-1}; and r_R = reservoir radius, feet. With these dimensionless parameters, the diffusivity equation becomes:

$$\frac{\partial^2 p_D}{\partial r_D^2} + \frac{1}{r_D}\frac{\partial p_D}{\partial r_D} = \frac{\partial p_D}{\partial t_D} \qquad (8.7)$$

van Everdingen and Hurst converted their solutions to dimensionless, cumulative water influx values and made the results available in a convenient form here given in Tables 8.1 and 8.2 for various ratios of aquifer to reservoir size, expressed by the ratio of their radii, r_e/r_R. Figures 8.7 to 8.10 are plots of some of the tabular values. The data are given in terms of dimensionless time, t_D, and dimensionless water influx, W_{eD}, so that one set of values suffices for all aquifers whose behavior can be represented by the radial form of the diffusivity equation. The water influx is then found by using Eq. (8.8):

$$W_e = B'\Delta p W_{eD} \qquad (8.8)$$

where

$$B' = 1.119\phi c_t r_R^2 h \frac{\theta}{360} \qquad (8.9)$$

B' is the water influx constant in barrels per pounds per square inch and θ is the angle subtended by the reservoir circumference, (i.e., for a full circle, $\theta = 360°$ and for a semicircular reservoir against a fault, $\theta = 180°$). c_t is in psi^{-1} and r_R and h are in feet.

Example 8.1 shows the use of Eq. (8.8) and the values of Tables 8.1 and 8.2 to calculate the cumulative water influx at successive periods for the case of a constant reservoir boundary pressure. The infinite aquifer values may be used for small time values even though the aquifer is limited in size.

Example 8.1. Calculate the water influx after 100 days, 200 days, 400 days, and 800 days into a reservoir the boundary pressure of which is suddenly lowered and held at 2724 psia ($p_i = 2734$ psia).

Given:

$$\phi = 0.20 \qquad\qquad k = 83 \text{ md}$$
$$c_t = 8(10)^{-6} \text{ psi}^{-1} \qquad r_R = 3000 \text{ ft}$$
$$r_e = 30,000 \text{ ft} \qquad\qquad \mu = 0.62 \text{ cp}$$
$$\theta = 360° \qquad\qquad h = 40 \text{ ft}$$

SOLUTION: From Eq. (8.6):

$$t_D = \frac{0.0002637(83)t}{0.20(0.62)[8(10)^{-6}]3000^2} = 0.00245t$$

*remember:
t is in hours*

From Eq. (8.9):

$$B' = 1.119(0.20)[8(10)^{-6}](3000^2)(40)\left(\frac{360}{360}\right) = 644.5$$

At 100 days $t_D = 0.00245(100)(24) = 5.88$ dimensionless time units. From the $r_e/r_R = 10$ curve of Fig. 8.8 find corresponding to $t_D = 5.88$, $W_{eD} = 5.07$ dimensionless influx units. This same value may also be found by interpolation of Table 8.1, since below $t_D = 15$ the aquifer behaves essentially as if it were infinite, and no values are given in Table 8.2. Since $\Delta p = 2734 - 2724 = 10$ psi, and water influx at 100 days from Eq. (8.8) is

$$W_e = B'\Delta p W_{eD} = 644.5(10)(5.07) = 32,680 \text{ bbl}$$

TABLE 8.1.
Infinite aquifer values of dimensionless water influx W_{eD} for values of dimensionless time t_D

Dimensionless time t_D	Fluid influx W_{eD}	Dimensionless time t_D	Fluid influx W_{eD}	Dimensionless time t_D	Fluid influx W_{eD}	Dimensionless time t_D	Fluid influx W_{eD}	Dimensionless time t_D	Fluid influx W_{eD}	Dimensionless time t_D	Fluid influx W_{eD}
0.00	0.000	79	35.697	455	150.249	1190	340.843	3250	816.090	35.000	6780.247
0.01	0.112	80	36.058	460	151.640	1200	343.308	3300	827.088	40.000	7650.096
0.05	0.278	81	36.418	465	153.029	1210	345.770	3350	838.067	50.000	9363.099
0.10	0.404	82	36.777	470	154.416	1220	348.230	3400	849.028	60.000	11,047.299
0.15	0.520	83	37.136	475	155.801	1225	349.460	3450	859.974	70.000	12,708.358
0.20	0.606	84	37.494	480	157.184	1230	350.688	3500	870.903	75.000	13,531.457
0.25	0.689	85	37.851	485	158.565	1240	353.144	3550	881.816	80.000	14,350.121
0.30	0.758	86	38.207	490	159.945	1250	355.597	3600	892.712	90.000	15,975.389
0.40	0.898	87	38.563	495	161.322	1260	358.048	3650	903.594	100.000	17,586.284
0.50	1.020	88	38.919	500	162.698	1270	360.496	3700	914.459	125.000	21,560.732
0.60	1.140	89	39.272	510	165.444	1275	361.720	3750	925.309	$1.5(10)^5$	$2.538(10)^4$
0.70	1.251	90	39.626	520	168.183	1280	362.942	3800	936.144	2.0″	3.308″
0.80	1.359	91	39.979	525	169.549	1290	365.386	3850	946.966	2.5″	4.066″
0.90	1.469	92	40.331	530	170.914	1300	367.828	3900	957.773	3.0″	4.817″
1	1.569	93	40.684	540	173.639	1310	370.267	3950	968.566	4.0″	6.267″
2	2.447	94	41.034	550	176.357	1320	372.704	4000	979.344	5.0″	7.699″
3	3.202	95	41.385	560	179.069	1325	373.922	4050	990.108	6.0″	9.113″
4	3.893	96	41.735	570	181.774	1330	375.139	4100	1000.858	7.0″	$1.051(10)^5$
5	4.539	97	42.084	575	183.124	1340	377.572	4150	1011.595	8.0″	1.189″
6	5.153	98	42.433	580	184.473	1350	380.003	4200	1022.318	9.0″	1.326″
7	5.743	99	42.781	590	187.166	1360	382.432	4250	1033.028	$1.0(10)^6$	1.462″
8	6.314	100	43.129	600	189.852	1370	384.859	4300	1043.724	1.5″	2.126″
9	6.869	105	44.858	610	192.533	1375	386.070	4350	1054.409	2.0″	2.781″
10	7.411	110	46.574	620	195.208	1380	387.283	4400	1065.082	2.5″	3.427″

11	7.940	115	48.277	625	196.544	1390	389.705	4450	1075.743	3.0"	4.064"
12	8.457	120	49.968	630	197.878	1400	392.125	4500	1086.390	4.0"	5.313"
13	8.964	125	51.648	640	200.542	1410	394.543	4550	1097.024	5.0"	6.544"
14	9.461	130	53.317	650	203.201	1420	396.959	4600	1107.646	6.0"	7.761"
15	9.949	135	54.976	660	205.854	1425	398.167	4650	1118.257	7.0"	8.965"
16	10.434	140	56.625	670	208.502	1430	399.373	4700	1128.854	8.0"	$1.016(10)^6$
17	10.913	145	58.265	675	209.825	1440	401.786	4750	1139.439	9.0"	1.134"
18	11.386	150	59.895	680	211.145	1450	404.197	4800	1150.012	$1.0(10)^7$	1.252"
19	11.855	155	61.517	690	213.784	1460	406.606	4850	1160.574	1.5"	1.828"
20	12.319	160	63.131	700	216.417	1470	409.013	4900	1171.125	2.0"	2.398"
21	12.778	165	64.737	710	219.046	1475	410.214	4950	1181.666	2.5"	2.961"
22	13.233	170	66.336	720	221.670	1480	411.418	5000	1192.198	3.0"	3.517"
23	13.684	175	67.928	725	222.980	1490	413.820	5100	1213.222	4.0"	4.610"
24	14.131	180	69.512	730	224.289	1500	416.220	5200	1234.203	5.0"	5.689"
25	14.573	185	71.090	740	226.904	1525	422.214	5300	1255.141	6.0"	6.758"
26	15.013	190	72.661	750	229.514	1550	428.196	5400	1276.037	7.0"	7.816"
27	15.450	195	74.226	760	232.120	1575	434.168	5500	1296.893	80"	8.866"
28	15.883	200	75.785	770	234.721	1600	440.128	5600	1317.709	9.0"	9.911"
29	16.313	205	77.338	775	236.020	1625	446.077	5700	1338.486	$1.0(10)^8$	$1.095(10)^7$
30	16.742	210	78.886	780	237.318	1650	452.016	5800	1359.225	1.5"	1.604"
31	17.167	215	80.428	790	239.912	1675	457.945	5900	1379.927	2.0"	2.108"
32	17.590	220	81.965	800	242.501	1700	463.863	6000	1400.593	2.5"	2.607"
33	18.011	225	83.497	810	245.086	1725	469.771	6100	1421.224	3.0"	3.100"
34	18.429	230	85.023	820	247.668	1750	475.669	6200	1441.820	4.0"	4.071"
35	18.845	235	86.545	825	248.957	1775	481.558	6300	1462.383	5.0"	5.032"
36	19.259	240	88.062	830	250.245	1800	487.437	6400	1482.912	6.0"	5.984"
37	19.671	245	89.575	840	252.819	1825	493.307	6500	1503.408	7.0"	6.928"
38	20.080	250	91.084	850	255.388	1850	499.167	6600	1523.872	8.0"	7.865"
39	20.488	255	92.589	860	257.953	1875	505.019	6700	1544.305	9.0"	8.797"
40	20.894	260	94.090	870	260.515	1900	510.861	6800	1564.706	$1.0(10)^9$	9.725"
41	21.298	265	95.588	875	261.795	1925	516.695	6900	1585.077	1.5"	$1.429(10)^8$

TABLE 8.1.
(Cont'd.)

Dimensionless time t_D	Fluid influx W_{eD}	Dimensionless time t_D	Fluid influx W_{eD}	Dimensionless time t_D	Fluid influx W_{eD}	Dimensionless time t_D	Fluid influx W_{eD}	Dimensionless time t_D	Fluid influx W_{eD}	Dimensionless time t_D	Fluid influx W_{eD}
42	21.701	270	97.081	880	263.073	1950	522.520	7000	1605.418	2.0"	1.880"
43	22.101	275	98.571	890	265.629	1975	528.337	7100	1625.729	2.5"	2.328"
44	22.500	280	100.057	900	268.181	2000	534.145	7200	1646.011	3.0"	2.771"
45	22.897	285	101.540	910	270.729	2025	539.945	7300	1666.265	4.0"	3.645"
46	23.291	290	103.019	920	273.274	2050	545.737	7400	1686.490	5.0"	4.510"
47	23.684	295	104.495	925	274.545	2075	551.522	7500	1706.688	6.0"	5.368"
48	24.076	300	105.968	930	275.815	2100	557.299	7600	1726.859	7.0"	6.220"
49	24.466	305	107.437	940	278.353	2125	563.068	7700	1747.002	8.0"	7.066"
50	24.855	310	108.904	950	280.888	2150	568.830	7800	1767.120	9.0"	7.909"
51	25.244	315	110.367	960	283.420	2175	574.585	7900	1787.212	$1.0(10)^{10}$	8.747"
52	25.633	320	111.827	970	285.948	2200	580.332	8000	1807.278	1.5"	$1.288(10)^9$
53	26.020	325	113.284	975	287.211	2225	586.072	8100	1827.319	2.0"	1.697"
54	26.406	330	114.738	980	288.473	2250	591.806	8200	1847.336	2.5"	2.103"
55	26.791	335	116.189	990	290.995	2275	597.532	8300	1867.329	3.0"	2.505"
56	27.174	340	117.638	1000	293.514	2300	603.252	8400	1887.298	4.0"	3.299"
57	27.555	345	119.083	1010	296.030	2325	608.965	8500	1907.243	5.0"	4.087"
58	27.935	350	120.526	1020	298.543	2350	614.672	8600	1927.166	6.0"	4.868"
59	28.314	355	121.966	1025	299.799	2375	620.372	8700	1947.065	7.0"	5.643"
60	28.691	360	123.403	1030	301.053	2400	626.066	8800	1966.942	8.0"	6.414"
61	29.068	365	124.838	1040	303.560	2425	631.755	8900	1986.796	9.0"	7.183"
62	29.443	370	126.720	1050	306.065	2450	637.437	9000	2006.628	$1.0(10)^{11}$	7.948"
63	29.818	375	127.699	1060	308.567	2475	643.113	9100	2026.438	1.5"	$1.17(10)^{10}$
64	30.192	380	129.126	1070	311.066	2500	648.781	9200	2046.227	2.0"	1.55"

65	30.565	385	130.550	1075	312.314	2550	660.093	9300	2065.996	2.5″	1.92″
66	30.937	390	131.972	1080	313.562	2600	671.379	9400	2085.744	3.0″	2.29″
67	31.308	395	133.391	1090	316.055	2650	682.640	9500	2105.473	4.0″	3.02″
68	31.679	400	134.808	1100	318.545	2700	693.877	9600	2125.184	5.0″	3.75″
69	32.048	405	136.223	1110	321.032	2750	705.090	9700	2144.878	6.0″	4.47″
70	32.417	410	137.635	1120	323.517	2800	716.280	9800	2164.555	7.0″	5.19″
71	32.785	415	139.045	1125	324.760	2850	727.449	9900	2184.216	8.0″	5.89″
72	33.151	420	140.453	1130	326.000	2900	738.598	10,000	2203.861	9.0″	6.58″
73	33.517	425	141.859	1140	328.480	2950	749.725	12,500	2688.967	$1.0(10)^{12}$	7.28″
74	33.883	430	143.262	1150	330.958	3000	760.833	15,000	3164.780	1.5″	$1.08(10)^{11}$
75	34.247	435	144.664	1160	333.433	3050	771.922	17,500	3633.368	2.0″	1.42″
76	34.611	440	146.064	1170	335.906	3100	782.992	20,000	4095.800		
77	34.974	445	147.461	1175	337.142	3150	794.042	25,000	5005.726		
78	35.336	450	148.856	1180	338.376	3200	805.075	30,000	5899.508		

TABLE 8.2.
Limited aquifer values of dimensionless water influx W_{eD} for values of dimensionless time t_D and for several ratios of aquifer-reservoir radii r_e/r_R ⋍ r_{eD}

$r_e/r_R = 1.5$		$r_e/r_R = 2.0$		$r_e/r_R = 2.5$		$r_e/r_R = 3.0$		$r_e/r_R = 3.5$		$r_e/r_R = 4.0$		$r_e/r_R = 4.5$	
Dimensionless time t_D	Fluid influx W_{eD}	Dimensionless time t_D	Fluid flux W_{eD}	Dimensionless time t_D	Fluid influx W_{eD}	Dimensionless time t_D	Fluid influx W_{eD}	Dimensionless time t_D	Fluid influx W_{eD}	Dimensionless time t_D	Fluid flux W_{eD}	Dimensionless time t_D	Fluid influx W_{eD}
$5.0(10)^{-2}$	0.276	$5.0(10)^{-2}$	0.278	$1.0(10)^{-1}$	0.408	$3.0(10)^{-1}$	0.755	1.00	1.571	2.00	2.442	2.5	2.835
6.0″	0.304	7.5″	0.345	1.5″	0.509	4.0″	0.895	1.20	1.761	2.20	2.598	3.0	3.196
7.0″	0.330	$1.0(10)^{-1}$	0.404	2.0″	0.599	5.0″	1.023	1.40	1.940	2.40	2.748	3.5	3.537
8.0″	0.354	1.25″	0.458	2.5″	0.681	6.0″	1.143	1.60	2.111	2.60	2.893	4.0	3.859
9.0″	0.375	1.50″	0.507	3.0″	0.758	7.0″	1.256	1.80	2.273	2.80	3.034	4.5	4.165
$1.0(10)^{-1}$	0.395	1.75″	0.553	3.5″	0.829	8.0″	1.363	2.00	2.427	3.00	3.170	5.0	4.454
1.1″	0.414	2.00″	0.597	4.0″	0.897	9.0″	1.465	2.20	2.574	3.25	3.334	5.5	4.727
1.2″	0.431	2.25″	0.638	4.5″	0.962	1.00	1.563	2.40	2.715	3.50	3.493	6.0	4.986
1.3″	0.446	2.50″	0.678	5.0″	1.024	1.25	1.791	2.60	2.849	3.75	3.645	6.5	5.231
1.4″	0.461	2.75″	0.715	5.5″	1.083	1.50	1.997	2.80	2.976	4.00	3.792	7.0	5.464
1.5″	0.474	3.00″	0.751	6.0″	1.140	1.75	2.184	3.00	3.098	4.25	3.932	7.5	5.684
1.6″	0.486	3.25″	0.785	6.5″	1.195	2.00	2.353	3.25	3.242	4.50	4.068	8.0	5.892
1.7″	0.497	3.50″	0.817	7.0″	1.248	2.25	2.507	3.50	3.379	4.75	4.198	8.5	6.089
1.8″	0.507	3.75″	0.848	7.5″	1.299	2.50	2.646	3.75	3.507	5.00	4.323	9.0	6.276
1.9″	0.517	4.00″	0.877	8.0″	1.348	2.75	2.772	4.00	3.628	5.50	4.560	9.5	6.453
2.0″	0.525	4.25″	0.905	8.5″	1.395	3.00	2.886	4.25	3.742	6.00	4.779	10	6.621
2.1″	0.533	4.50″	0.932	9.0″	1.440	3.25	2.990	4.50	3.850	6.50	4.982	11	6.930

2.2"	0.541	4.75"	0.958	9.5"	1.484	3.50	3.084	4.75	3.951	7.00	5.169	12	7.208
2.3"	0.548	5.00"	0.983	1.0	1.526	3.75	3.170	5.00	4.047	7.50	5.343	13	7.457
2.4"	0.554	5.50"	1.028	1.1	1.605	4.00	3.247	5.50	4.222	8.00	5.504	14	7.680
2.5"	0.559	6.00"	1.070	1.2	1.679	4.25	3.317	6.00	4.378	8.50	5.653	15	7.880
2.6"	0.565	6.50"	1.108	1.3	1.747	4.50	3.381	6.50	4.516	9.00	5.790	16	8.060
2.8"	0.574	7.00"	1.143	1.4	1.811	4.75	3.439	7.00	4.639	9.50	5.917	18	8.365
3.0"	0.582	7.50"	1.174	1.5	1.870	5.00	3.491	7.50	4.749	10	6.035	20	8.611
3.2"	0.588	8.00"	1.203	1.6	1.924	5.50	3.581	8.00	4.846	11	6.246	22	8.809
3.4"	0.594	9.00"	1.253	1.7	1.975	6.00	3.656	8.50	4.932	12	6.425	24	8.968
3.6"	0.599	1.00"	1.295	1.8	2.022	6.50	3.717	9.00	5.009	13	6.580	26	9.097
3.8"	0.603	1.1	1.330	2.0	2.106	7.00	3.767	9.50	5.078	14	6.712	28	9.200
4.0"	0.606	1.2	1.358	2.2	2.178	7.50	3.809	10.00	5.138	15	6.825	30	9.283
4.5"	0.613	1.3	1.382	2.4	2.241	8.00	3.843	11	5.241	16	6.922	34	9.404
5.0"	0.617	1.4	1.402	2.6	2.294	9.00	3.894	12	5.321	17	7.004	38	9.481
6.0"	0.621	1.6	1.432	2.8	2.340	10.00	3.928	13	5.385	18	7.076	42	9.532
7.0"	0.623	1.7	1.444	3.0	2.380	11.00	3.951	14	5.435	20	7.189	46	9.565
8.0"	0.624	1.8	1.453	3.4	2.444	12.00	3.967	15	5.476	22	7.272	50	9.586
		2.0	1.468	3.8	2.491	14.00	3.985	16	5.506	24	7.332	60	9.612
		2.5	1.487	4.2	2.525	16.00	3.993	17	5.531	26	7.377	70	9.621
		3.0	1.495	4.6	2.551	18.00	3.997	18	5.551	30	7.434	80	9.623
		4.0	1.499	5.0	2.570	20.00	3.999	20	5.579	34	7.464	90	9.624
		5.0	1.500	6.0	2.599	22.00	3.999	25	5.611	38	7.481	100	9.625
				7.0	2.613	24.00	4.000	30	5.621	42	7.490		
				8.0	2.619			35	5.624	46	7.494		
				9.0	2.622			40	5.625	50	7.499		
				10.0	2.624								

Use Table 8.1 for $t_{D_{min}} \leq t_D \leq t_{D_{max}}$

289

TABLE 8.2.
(Cont'd.)

$r_e/r_R = 5.0$		$r_e/r_R = 6.0$		$r_e/r_R = 7.0$		$r_e/r_R = 8.0$		$r_e/r_R = 9.0$		$r_e/r_R = 10.0$	
Dimensionless time t_D	Fluid influx W_{eD}	Dimensionless time t_D	Fluid influx W_{eD}	Dimensionless time t_D	Fluid influx W_{eD}	Dimensionless time t_D	Fluid influx W_{eD}	Dimensionless time t_D	Fluid influx W_{eD}	Dimensionless time t_D	Fluid influx W_{eD}
3.0	3.195	6.0	5.148	9.00	6.861	9	6.861	10	7.417	15	9.965
3.5	3.542	6.5	5.440	9.50	7.127	10	7.398	15	9.945	20	12.32
4.0	3.875	7.0	5.724	10	7.389	11	7.920	20	12.26	22	13.22
4.5	4.193	7.5	6.002	11	7.902	12	8.431	22	13.13	24	14.95
5.0	4.499	8.0	6.273	12	8.397	13	8.930	24	13.98	26	14.95
5.5	4.792	8.5	6.537	13	8.876	14	9.418	26	14.79	28	15.78
6.0	5.074	9.0	6.795	14	9.341	15	9.895	28	15.59	30	16.59
6.5	5.345	9.5	7.047	15	9.791	16	10.361	30	16.35	32	17.38
7.0	5.605	10.0	7.293	16	10.23	17	10.82	32	17.10	34	18.16
7.5	5.854	10.5	7.533	17	10.65	18	11.26	34	17.82	36	18.91
8.0	6.094	11	7.767	18	11.06	19	11.70	36	18.52	38	19.65
8.5	6.325	12	8.220	19	11.46	20	12.13	38	19.19	40	20.37
9.0	6.547	13	8.651	20	11.85	22	12.95	40	19.85	42	21.07
9.5	6.760	14	9.063	22	12.58	24	13.74	42	20.48	44	21.76
10	6.965	15	9.456	24	13.27	26	14.50	44	21.09	46	22.42
11	7.350	16	9.829	26	13.92	28	15.23	46	21.69	48	23.07
12	7.706	17	10.19	28	14.53	30	15.92	48	22.26	50	23.71
13	8.035	18	10.53	30	15.11	34	17.22	50	22.82	52	24.33

14	8.339	19	10.85	35	16.39	38	18.41	52	23.36	54	24.94
15	8.620	20	11.16	40	17.49	40	18.97	54	23.89	56	25.53
16	8.879	22	11.74	45	18.43	45	20.26	56	24.39	58	26.11
18	9.338	24	12.26	50	19.24	50	21.42	58	24.88	60	26.67
20	9.731	25	12.50	60	20.51	55	22.46	60	25.36	65	28.02
22	10.07	31	13.74	70	21.45	60	23.40	65	26.48	70	29.29
24	10.35	35	14.40	80	22.13	70	24.98	70	27.52	75	30.49
26	10.59	39	14.93	90	22.63	80	26.26	75	28.48	80	31.61
28	10.80	51	16.05	100	23.00	90	27.28	80	29.36	85	32.67
30	10.98	60	16.56	120	23.47	100	28.11	85	30.18	90	33.66
34	11.26	70	16.91	140	23.71	120	29.31	90	30.93	95	34.60
38	11.46	80	17.14	160	23.85	140	30.08	95	31.63	100	35.48
42	11.61	90	17.27	180	23.92	160	30.58	100	32.27	120	38.51
46	11.71	100	17.36	200	23.96	180	30.91	120	34.39	140	40.89
50	11.79	110	17.41	500	24.00	200	31.12	140	35.92	160	42.75
60	11.91	120	17.45			240	31.34	160	37.04	180	44.21
70	11.96	130	17.46			280	31.43	180	37.85	200	45.36
80	11.98	140	17.48			320	31.47	200	38.44	240	46.95
90	11.99	150	17.49			360	31.49	240	39.17	280	47.94
100	12.00	160	17.49			400	31.50	280	39.56	320	48.54
120	12.00	180	17.50			500	31.50	320	39.77	360	48.91
		200	17.50					360	39.88	400	49.14
		220	17.50					400	39.94	440	49.28
								440	39.97	480	49.36
								480	39.98		

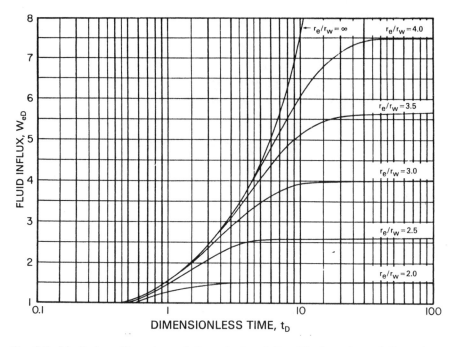

Fig. 8.7. Limited aquifer values of dimensionless influx W_{eD} for values of dimensionless time t_D and aquifer limits given by the ratio r_e/r_R.

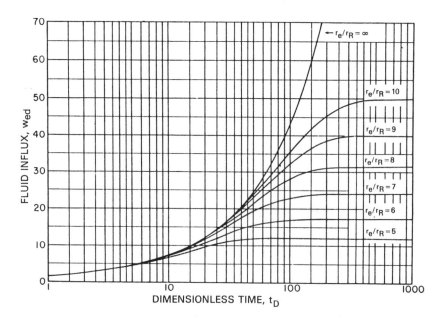

Fig. 8.8. Limited aquifer values of dimensionless influx W_{eD} for values of dimensionless time t_D and aquifer limits given by the ratio r_e/r_R.

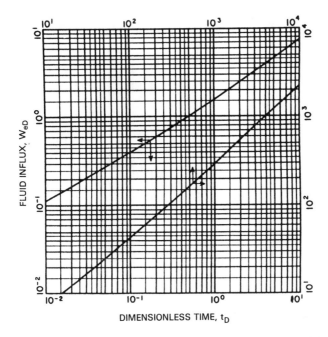

Fig. 8.9. Infinite aquifer values of dimensionless influx W_{eD} for values of dimensionless time t_D.

Similarly at

$t = 100$ days	200 days	400 days	800 days
$t_D = 5.88$	11.76	23.52	47.04
$W_{eD} = 5.07$	8.43	13.90	22.75
$W_e = 32,680$	54,330	89,590	146,600

For aquifers 99 times as large as the reservoirs they surround, or $r_e/r_R = 10$, this means that the effect of the aquifer limits are negligible for dimensionless time values under 15, and that it is some time before the aquifer limits affect the water influx appreciably. This is also illustrated by the coincidence of the curves of Figs. 8.7 and 8.8 with the infinite aquifer curve for the smaller time values. It should also be noted that, unlike a steady-state system, the values of water influx calculated in Ex. 8.1 fail to double for a doubling of the time.

While water is entering the reservoir from the aquifer at a declining rate, in response to the first pressure signal $\Delta p_1 = p_i - p_1$, let a second, sudden pressure drop $\Delta p_2 = p_1 - p_2$ (not $p_i - p_2$) be imposed at the reservoir boundary at a time t_1. This is an application of the principle of superposition, which was discussed in Chapter 7. The total or net effect is the sum of the two as illustrated in Fig. 8.11 where, for simplicity, $\Delta p_1 = \Delta p_2$ and $t_2 = 2t_1$. The upper

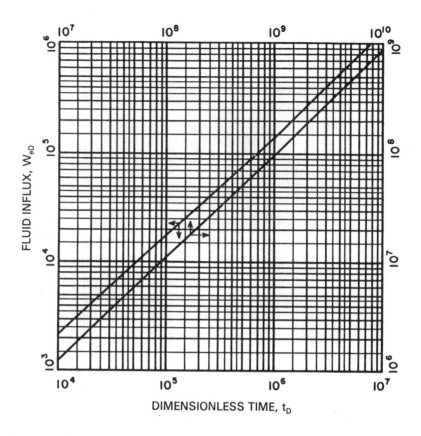

Fig. 8.10. Infinite aquifer values of dimensionless influx W_{eD} for values of dimensionless time t_D.

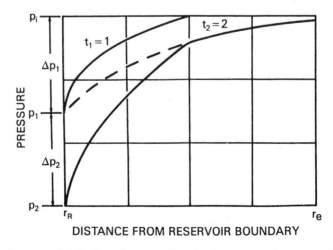

Fig. 8.11. Pressure distributions in an aquifer due to two equal pressure decrements imposed at equal time intervals.

and middle curves represent the pressure distribution in the aquifer in response to the first signal alone, at times t_1 and t_2, respectively. The upper curve may also be used to represent the pressure distribution for the second pressure signal alone at time t_2 because in this simplified case $\Delta p_1 = \Delta p_2$, and $\Delta t_2 = \Delta t_1$. The lower curve, then, is the sum of the upper and middle curves. Mathematically, this means that Eq. (8.10) can be used to calculate the cumulative water influx:

$$W_e = B' \Sigma \Delta p W_{eD} \tag{8.10}$$

This calculation is illustrated in Ex. Prob. 8.2.

Example 8.2. Suppose in Ex. 8.1 at the end of 100 days the reservoir boundary pressure suddenly drops to $p_2 = 2704$ psia (i.e., $\Delta p_2 = p_1 - p_2 = 20$ psi, *not* $p_i - p_2 = 30$ psi). Calculate the water influx at 400 days total time.

The water influx due to the first pressure drop $\Delta p_1 = 10$ psi at 400 days was calculated in Example 8.1 to be 89,590 bbl. This will be the same even though a second pressure drop occurs at 100 days and continues to 400 days. This second drop will have acted for 300 days, or a dimensionless time of $t_D = 0.0588 \times 300 = 17.6$. From Fig. 8.8 or Table 8.2 for $r_e/r_R = 10$ $W_{eD} = 11.14$ for $t_D = 17.6$ and the water influx is

$$\Delta W_{e2} = B' \times \Delta p_2 \times W_{eD2} = 644.5 \times 20 \times 11.14 = 143,600 \text{ bbl}$$

$$W_{e2} = \Delta W_{e1} + \Delta W_{e2} = B' \times \Delta p_1 \times W_{eD1} + B' \times \Delta p_2 \times W_{eD2} = B' \Sigma \Delta p W_{eD}$$

$$= 644.5 \, (10 \times 13.90 + 20 \times 11.14)$$

$$= 89,590 + 143,600 = 233,190 \text{ bbl}$$

Example 8.2 illustrates the calculation of water influx when a second pressure drop occurs 100 days after the first drop in Example 8.1. A continuation of this method may be used to calculate the water influx into reservoirs for which boundary pressure histories are known, and also for which sufficient information is known about the aquifer to calculate the constant B' and the dimensionless time t_D.

The history of the reservoir boundary pressure may be approximated as closely as desired by a series of step-by-step pressure reductions (or increases), as illustrated in Fig. 8.12. The best approximation of the pressure history is made as shown by making the pressure step at any time equal to half of the drop in the previous interval of time plus half of the drop in the succeeding period of time.[8] When reservoir boundary pressures are not known, average reservoir pressures may be substituted with some reduction in the accuracy of the results. In addition, for best accuracy, the average boundary pressure should always be that at the initial rather than the current oil-water contact; otherwise, among other changes, a decreasing value of r_R is unaccounted for.

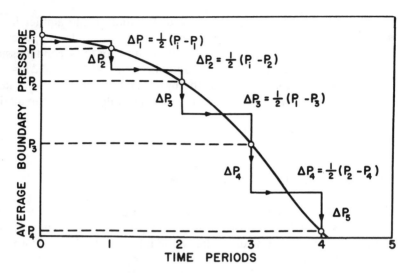

Fig. 8.12. Sketch showing the use of step pressures to approximate the pressure-time curve.

Example 8.3 illustrates the calculation of water influx at two successive time values for the reservoir shown in Fig. 8.13.

Example 8.3. Calculate the water influx at the third and fourth quarter years of production for the reservoir shown in Fig. 8.13. Use $\phi = 0.209$; $k = 275$ md (average reservoir permeability, presumed the same for the aquifer); $\mu = 0.25$ cp; $c_t = 6 \times 10^{-6}$ psi^{-1}; $h = 19.2$ ft; area of reservoir = 1216 ac; estimated area of aquifer = 250,000 ac; $\theta = 180°$.

SOLUTION: Since the reservoir is against a fault $A = \frac{1}{2}\pi r_R^2$ and:

$$r_R^2 = \frac{1216 \times 43,560 \overset{ac}{}\overset{ft^2/ac}{}}{0.5 \times 3.1416\,\pi}$$

$$r_R = 5807 \ ft$$

For $t = 91.3$ days (one-quarter year or one period):

$$6.323 \times 10^{-3} = (24)(0.0002637)$$

$$t_D = \frac{6.323 \times 10^{-3} \times 275 \times 91.3}{0.209 \times 0.25 \times 6 \times 10^{-6} \times (5807)^2} = 15.0$$

$$B' = 1.119 \times 0.209 \times 6 \times 10^{-6} \times (5807)^2 \times 19.2 \times (180°/360°)$$

$$= 455 \ bbl/psi$$

Since the aquifer is 250,000/1216 = 206 times the area of the reservoir, for a considerable time the infinite aquifer values may be used. Table 8.3

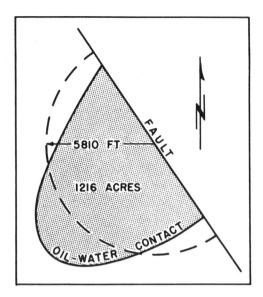

Fig. 8.13. Sketch showing the equivalent radius of a reservoir.

TABLE 8.3.
Boundary step pressures and W_{eD} values for Ex. 8.3

Time Period t	Time in Days t	Dimensionless less Time, t_D	W_{eD}	p, Avg. Reservoir Pressure, psia	p_B, Avg. Boundary Pressure, psia	Δp, Step Pressure, psi
0	0	0	0.0	3793	3793	0.0
1	91.3	15	10.0	3786	3788	2.5
2	182.6	30	16.7	3768	3774	9.5
3	273.9	45	22.9	3739	3748	20.0
4	365.2	60	28.7	3699	3709	32.5
5	456.5	75	34.3	3657	3680	34.0
6	547.8	90	39.6	3613	3643	33.0

[a] Infinite aquifer values from Fig. 8.9 or Table 8.1.

shows the values of boundary step pressures and the W_{eD} values for the first six periods. The calculation of the step pressures Δp is illustrated in Fig. 8.12. For example,

$$\Delta p_3 = \tfrac{1}{2}(p_1 - p_3) = \tfrac{1}{2}(3788 - 3748) = 20.0 \text{ psi}$$

Tables 8.4 and 8.5 show the calculation of $\Sigma \Delta p \times W_{eD}$ at the end of the third

TABLE 8.4.
Water influx at the end of the third quarter for Example 8.3

t_D	W_{eD}	Δp	$\Delta p \times W_{eD}$
45	22.9	2.5	57.3
30	16.7	9.5	158.7
15	10.0	20.0	200.0

$$\Sigma \Delta P = 416$$

TABLE 8.5.
Water influx at the end of the fourth quarter for Example 8.3

t_D	W_{eD}	Δp	$\Delta p \times W_{eD}$
60	28.7	2.5	71.8
45	22.9	9.5	217.6
30	16.7	20.0	334.0
15	10.0	32.5	325.0

$$\Sigma \Delta P = 948$$

and fourth periods, the values being 416.0 and 948.0, respectively. Then the corresponding water influx at the end of these periods is

$$W_e \ (\text{3rd quarter}) = B' \Sigma \Delta p \times W_{eD} = 455 \times 416.0 = 189{,}300 \text{ bbl}$$

$$W_e \ (\text{4th quarter}) = B' \Sigma \Delta p \times W_{eD} = 455 \times 948.4 = 431{,}500 \text{ bbl}$$

In calculating the water influx in Ex. 8.3 at the end of the third quarter, it should be carefully noted in Table 8.4 that since the first pressure drop, $\Delta p_1 = 2.5$ psi, had been operating for the full three quarters ($t_D = 45$), it was multiplied by $W_{eD} = 22.9$, which corresponds to $t_D = 45$. Similarly, for the fourth quarter calculation in Table 8.5, the 2.5 psi was multiplied by $W_{eD} = 28.7$, which is the value for $t_D = 60$. Thus the W_{eD} values are inverted so that the one corresponding to the longest time is multiplied by the first pressure drop, and vice versa. Also, in calculating each successive value of $\Sigma \Delta p \times W_{eD}$, it is not simply a matter of adding a new $\Delta p \times W_{eD}$ term to the former summation, but a complete recalculation, as shown in Tables 8.4 and 8.5.

You should show that the water influx values at the end of the fifth and sixth quarters are 773,100 and 1,201,600 bbl, respectively.

From the previous discussion it is evident that it is possible to calculate water influx independently of material balance calculations from a knowledge of the history of the reservoir, boundary pressure, and the dimensions and physical characteristics of the aquifer, as shown by Chatas.[9] Although strictly speaking, the van Everdingen and Hurst solutions to the diffusivity equation apply only to circular reservoirs surrounded concentrically by horizontal, circular (or infinite) aquifers of constant thickness, porosity, permeability, and effective water compressibility, for many engineering purposes good results may be obtained when the situation is somewhat less than ideal, as it nearly always is. The radius of the reservoir may be approximated by using the radius of a circle equal in area to the area of the reservoir; and where the approximate size of the aquifer is known, the same approximation may be used for the aquifer radius. Where the aquifer is more than approximately 99 times the size (volume) of the reservoir ($r_e/r_R = 10$), the aquifer behaves essentially as if it were infinite for a considerable period so that the values of Table 8.1 may be used. There are to be sure, uncertainties in the permeability, porosity, and thickness of the aquifer which must be estimated from information obtained from wells drilled in the reservoir and whatever wells are drilled in the aquifer. The viscosity of the water can be estimated from the temperature and pressure (Chapter 1, Sect. 7.4) and the water and rock compressibilities from the data of Chapter 1, Sects. 7.3 and 4.2.

Because of the many uncertainties in the dimensions and properties of the aquifer, the calculation of water influx independently of material balance appears somewhat unreliable. For instance in Ex. 8.3 it was assumed that the fault against which the reservoir accumulated was of large (actually infinite) extent; and since the permeability of only the reservoir rock was known, it was assumed that the average permeability of the aquifer was also 275 md. There may be variations in the aquifer thickness and porosity, and the aquifer may contain faults, impermeable areas, and unknown hydrocarbon accumulations, all of which can introduce variations of greater or lesser importance.

As Ex. Probs. 8.1, 8.2, and 8.3 suggest, the calculations for water influx can become long and tedious. The use of the computer with these calculations requires large data files containing the values of W_{eD}, t_D, and r_e/r_R from Tables 8.1 and 8.2 as well as a table lookup routine. Several authors have attempted to develop equations to describe the dimensionless water influx as a function of the dimensionless time and radius ratio.[10, 11, 12] These equations reduce the data storage required when using the computer in calculations of water influx.

3.2. Bottom-Water Drive   *stopped reading here 4/5/99*

The van Everdingen and Hurst model discussed in the previous section is based on the radial diffusivity equation written without a term describing vertical flow from the aquifer. In theory, this model should not be used when there is significant movement of water into the reservoir from a bottom-water drive. To account for the flow of water in a vertical direction, Coats and later Allard and Chen, added a term to Eq. (7.35) to yield the following:

$$\frac{\partial^2 p}{\partial r^2} + \frac{1}{r}\frac{\partial p}{\partial r} + F_k\frac{\partial^2 p}{\partial z^2} = \frac{\phi\mu c_t}{0.0002637k}\frac{\partial p}{\partial t} \tag{8.10}$$

where F_k is the ratio of vertical to horizontal permeability.[13, 14]

Using the definitions of dimensionless time, radius, and pressure and introducing a second dimensionless distance, z_D, Eq. (8.10) becomes Eq. (8.11):

$$z_D = \frac{z}{r_R F_k^{1/2}}$$

$$\frac{\partial^2 p_D}{\partial r_D^2} + \frac{1}{r_D}\frac{\partial p_D}{\partial r_D} + \frac{\partial^2 p_D}{\partial z_D^2} = \frac{\partial p_D}{\partial t_D} \tag{8.11}$$

Coats solved Eq. (8.11) for the terminal rate case for infinite aquifers.[13] Allard and Chen used a numerical simulator to solve the problem for the terminal pressure case.[14] They defined a water influx constant, B', and a dimensionless water influx, W_{eD} analogous to those defined by van Everdingen and Hurst except that B' does not include the angle θ:

$$B' = 1.119\phi h c_t r_R^2 \tag{8.12}$$

The actual values of W_{eD} will be different from those of the van Everdingen and Hurst model because W_{eD} for the bottom-water drive is a function of the vertical permeability. Because of this functionality, the solutions presented by Allard and Chen, found in Tables 8.6 to 8.10, are functions of two dimensionless parameters, r_D' and z_D'.

$$r_D' = \frac{r_e}{r_R} \tag{8.13}$$

$$z_D' = \frac{h}{r_R F_k^{1/2}} \tag{8.14}$$

The method of calculating water influx from the dimensionless values obtained from these tables follows exactly the method illustrated in Exs. 8.1 to 8.3. The procedure is shown in Ex. 8.4, which is a problem taken from Allard and Chen.[14]

Example 8.4. Calculate the water influx as a function of time for the reservoir data and boundary pressure data that follow:

Given:

$r_R = 2000$ ft	$r_e = \infty$
$h = 200$ ft	$k = 50$ md
$F_k = 0.04$	$\phi = 0.10$
$\mu = 0.395$ cp	$c_t = 8 \times 10^{-6}$ psi^{-1}

Time in Days (t)	Average Boundary Pressure, psia (p_B)
0	3000
30	2956
60	2917
90	2877
120	2844
150	2811
180	2791
210	2773
240	2755

SOLUTION:

$$r_D' = \infty$$

$$z_D' = \frac{200}{2000(0.040)^{1/2}} = 0.5$$

$$t_D = \frac{0.0002637(50)}{0.10(0.395)8(10)^{-6}2000^2} = 0.0104t \text{ (where } t \text{ is in hours)}$$

$$B' = 1.119(0.10)(200)8(10)^{-6}2000^2 = 716 \text{ bbl/psi}$$

Time in Days (t)	Dimensionless Time (t_D)	W_{eD}	Average Boundary Pressure, psia (p_B)	Step Pressure (Δp)	Water Influx, M bbl (W_e)
0	0	0	3000	0	0
30	7.5	5.038	2956	22.0	79
60	15.0	8.389	2917	41.5	282
90	22.5	11.414	2877	39.5	572
120	30.0	14.263	2844	36.5	933
150	37.5	16.994	2811	33.0	1353
180	45.0	19.641	2791	26.5	1810
210	52.5	22.214	2773	19.0	2284
240	60.0	24.728	2755	18.0	2782

TABLE 8.6.
Dimensionless influx, W_{eD}, for infinite aquifer for bottom-water drive

t_D				z_D'			
	0.05	0.1	0.3	0.5	0.7	0.9	1.0
0.1	0.700	0.677	0.508	0.349	0.251	0.195	0.176
0.2	0.793	0.786	0.696	0.547	0.416	0.328	0.295
0.3	0.936	0.926	0.834	0.692	0.548	0.440	0.396
0.4	1.051	1.041	0.952	0.812	0.662	0.540	0.486
0.5	1.158	1.155	1.059	0.918	0.764	0.631	0.569
0.6	1.270	1.268	1.167	1.021	0.862	0.721	0.651
0.7	1.384	1.380	1.270	1.116	0.953	0.806	0.729
0.8	1.503	1.499	1.373	1.205	1.039	0.886	0.803
0.9	1.621	1.612	1.477	1.286	1.117	0.959	0.872
1	1.743	1.726	1.581	1.347	1.181	1.020	0.932
2	2.402	2.393	2.288	2.034	1.827	1.622	1.509
3	3.031	3.018	2.895	2.650	2.408	2.164	2.026
4	3.629	3.615	3.477	3.223	2.949	2.669	2.510
5	4.217	4.201	4.048	3.766	3.462	3.150	2.971
6	4.784	4.766	4.601	4.288	3.956	3.614	3.416
7	5.323	5.303	5.128	4.792	4.434	4.063	3.847
8	5.829	5.808	5.625	5.283	4.900	4.501	4.268
9	6.306	6.283	6.094	5.762	5.355	4.929	4.680
10	6.837	6.816	6.583	6.214	5.792	5.344	5.080
11	7.263	7.242	7.040	6.664	6.217	5.745	5.468
12	7.742	7.718	7.495	7.104	6.638	6.143	5.852
13	8.196	8.172	7.943	7.539	7.052	6.536	6.231
14	8.648	8.623	8.385	7.967	7.461	6.923	6.604
15	9.094	9.068	8.821	8.389	7.864	7.305	6.973
16	9.534	9.507	9.253	8.806	8.262	7.682	7.338

17	9.969	9.942	9.679	9.218	8.656	8.056	7.699
18	10.399	10.371	10.100	9.626	9.046	8.426	8.057
19	10.823	10.794	10.516	10.029	9.432	8.793	8.411
20	11.241	11.211	10.929	10.430	9.815	9.156	8.763
21	11.664	11.633	11.339	10.826	10.194	9.516	9.111
22	12.075	12.045	11.744	11.219	10.571	9.874	9.457
23	12.486	12.454	12.147	11.609	10.944	10.229	9.801
24	12.893	12.861	12.546	11.996	11.315	10.581	10.142
25	13.297	13.264	12.942	12.380	11.683	10.931	10.481
26	13.698	13.665	13.336	12.761	12.048	11.279	10.817
27	14.097	14.062	13.726	13.140	12.411	11.625	11.152
28	14.493	14.458	14.115	13.517	12.772	11.968	11.485
29	14.886	14.850	14.501	13.891	13.131	12.310	11.816
30	15.277	15.241	14.884	14.263	13.488	12.650	12.145
31	15.666	15.628	15.266	14.634	13.843	12.990	12.473
32	16.053	16.015	15.645	15.002	14.196	13.324	12.799
33	16.437	16.398	16.023	15.368	14.548	13.659	13.123
34	16.819	16.780	16.398	15.732	14.897	13.992	13.446
35	17.200	17.160	16.772	16.095	15.245	14.324	13.767
36	17.579	17.538	17.143	16.456	15.592	14.654	14.088
37	17.956	17.915	17.513	16.815	15.937	14.983	14.406
38	18.331	18.289	17.882	17.173	16.280	15.311	14.724
39	18.704	18.662	18.249	17.529	16.622	15.637	15.040
40	19.088	19.045	18.620	17.886	16.964	15.963	15.356
41	19.450	19.407	18.982	18.240	17.305	16.288	15.671
42	19.821	19.777	19.344	18.592	17.644	16.611	15.985
43	20.188	20.144	19.706	18.943	17.981	16.933	16.297
44	20.555	20.510	20.065	19.293	18.317	17.253	16.608
45	20.920	20.874	20.424	19.641	18.651	17.573	16.918
46	21.283	21.237	20.781	19.988	18.985	17.891	17.227
47	21.645	21.598	21.137	20.333	19.317	18.208	17.535

TABLE 8.6.
(Cont'd.)

t_D	0.05	0.1	0.3	0.5	0.7	0.9	1.0
				z_D'			
48	22.006	21.958	21.491	20.678	19.648	18.524	17.841
49	22.365	22.317	21.844	21.021	19.978	18.840	18.147
50	22.722	22.674	22.196	21.363	20.307	19.154	18.452
51	23.081	23.032	22.547	21.704	20.635	19.467	18.757
52	23.436	23.387	22.897	22.044	20.962	19.779	19.060
53	23.791	23.741	23.245	22.383	21.288	20.091	19.362
54	24.145	24.094	23.593	22.721	21.613	20.401	19.664
55	24.498	24.446	23.939	23.058	21.937	20.711	19.965
56	24.849	24.797	24.285	23.393	22.260	21.020	20.265
57	25.200	25.147	24.629	23.728	22.583	21.328	20.564
58	25.549	25.496	24.973	24.062	22.904	21.636	20.862
59	25.898	25.844	25.315	24.395	23.225	21.942	21.160
60	26.246	26.191	25.657	24.728	23.545	22.248	21.457
61	26.592	26.537	25.998	25.059	23.864	22.553	21.754
62	26.938	26.883	26.337	25.390	24.182	22.857	22.049
63	27.283	27.227	26.676	25.719	24.499	23.161	22.344
64	27.627	27.570	27.015	26.048	24.816	23.464	22.639
65	27.970	27.913	27.352	26.376	25.132	23.766	22.932
66	28.312	28.255	27.688	26.704	25.447	24.068	23.225
67	28.653	28.596	28.024	27.030	25.762	24.369	23.518
68	28.994	28.936	28.359	27.356	26.075	24.669	23.810
69	29.334	29.275	28.693	27.681	26.389	24.969	24.101
70	29.673	29.614	29.026	28.006	26.701	25.268	24.391
71	30.011	29.951	29.359	28.329	27.013	25.566	24.681
72	30.349	30.288	29.691	28.652	27.324	25.864	24.971

73	30.686	30.625	30.022	28.974	27.634	26.161	25.260
74	31.022	30.960	30.353	29.296	27.944	26.458	25.548
75	31.357	31.295	30.682	29.617	28.254	26.754	25.836
76	31.692	31.629	31.012	29.937	28.562	27.049	26.124
77	32.026	31.963	31.340	30.257	28.870	27.344	26.410
78	32.359	32.296	31.668	30.576	29.178	27.639	26.697
79	32.692	32.628	31.995	30.895	29.485	27.933	26.983
80	33.024	32.959	32.322	31.212	29.791	28.226	27.268
81	33.355	33.290	32.647	31.530	30.097	28.519	27.553
82	33.686	33.621	32.973	31.846	30.402	28.812	27.837
83	34.016	33.950	33.297	32.163	30.707	29.104	28.121
84	34.345	34.279	33.622	32.478	31.011	29.395	28.404
85	34.674	34.608	33.945	32.793	31.315	29.686	28.687
86	35.003	34.935	34.268	33.107	31.618	29.976	28.970
87	35.330	35.263	34.590	33.421	31.921	30.266	29.252
88	35.657	35.589	34.912	33.735	32.223	30.556	29.534
89	35.984	35.915	35.233	34.048	32.525	30.845	29.815
90	36.310	36.241	35.554	34.360	32.826	31.134	30.096
91	36.636	36.566	35.874	34.672	33.127	31.422	30.376
92	36.960	36.890	36.194	34.983	33.427	31.710	30.656
93	37.285	37.214	36.513	35.294	33.727	31.997	30.935
94	37.609	37.538	36.832	35.604	34.026	32.284	31.215
95	37.932	37.861	37.150	35.914	34.325	32.570	31.493
96	38.255	38.183	37.467	36.233	34.623	32.857	31.772
97	38.577	38.505	37.785	36.532	34.921	33.142	32.050
98	38.899	38.826	38.101	36.841	35.219	33.427	32.327
99	39.220	39.147	38.417	37.149	35.516	33.712	32.605
100	39.541	39.467	38.733	37.456	35.813	33.997	32.881
105	41.138	41.062	40.305	38.987	37.290	35.414	34.260
110	42.724	42.645	41.865	40.508	38.758	36.821	35.630
115	44.299	44.218	43.415	42.018	40.216	38.221	36.993

TABLE 8.6.
(Cont'd.)

t_D	z'_D						
	0.05	0.1	0.3	0.5	0.7	0.9	1.0
120	45.864	45.781	44.956	43.520	41.666	39.612	38.347
125	47.420	47.334	46.487	45.012	43.107	40.995	39.694
130	48.966	48.879	48.009	46.497	44.541	42.372	41.035
135	50.504	50.414	49.523	47.973	45.967	43.741	42.368
140	52.033	51.942	51.029	49.441	47.386	45.104	43.696
145	53.555	53.462	52.528	50.903	48.798	46.460	45.017
150	55.070	54.974	54.019	52.357	50.204	47.810	46.333
155	56.577	56.479	55.503	53.805	51.603	49.155	47.643
160	58.077	57.977	56.981	55.246	52.996	50.494	48.947
165	59.570	59.469	58.452	56.681	54.384	51.827	50.247
170	61.058	60.954	59.916	58.110	55.766	53.156	51.542
175	62.539	62.433	61.375	59.534	57.143	54.479	52.832
180	64.014	63.906	62.829	60.952	58.514	55.798	54.118
185	65.484	65.374	64.276	62.365	59.881	57.112	55.399
190	66.948	66.836	65.718	63.773	61.243	58.422	56.676
195	68.406	68.293	67.156	65.175	62.600	59.727	57.949
200	69.860	69.744	68.588	66.573	63.952	61.028	59.217
205	71.309	71.191	70.015	67.967	65.301	62.326	60.482
210	72.752	72.633	71.437	69.355	66.645	63.619	61.744
215	74.191	74.070	72.855	70.740	67.985	64.908	63.001
220	75.626	75.503	74.269	72.120	69.321	66.194	64.255
225	77.056	76.931	75.678	73.496	70.653	67.476	65.506
230	78.482	78.355	77.083	74.868	71.981	68.755	66.753
235	79.903	79.774	78.484	76.236	73.306	70.030	67.997
240	81.321	81.190	79.881	77.601	74.627	71.302	69.238

245	82.734	82.602	81.275	78.962	75.945	72.570	70.476
250	84.144	84.010	82.664	80.319	77.259	73.736	71.711
255	85.550	85.414	84.050	81.672	78.570	75.098	72.943
260	86.952	86.814	85.432	83.023	79.878	76.358	74.172
265	88.351	88.211	86.811	84.369	81.182	77.614	75.398
270	89.746	89.604	88.186	85.713	82.484	78.868	76.621
275	91.138	90.994	89.558	87.053	83.782	80.119	77.842
280	92.526	92.381	90.926	88.391	85.078	81.367	79.060
285	93.911	93.764	92.292	89.725	86.371	82.612	80.276
290	95.293	95.144	93.654	91.056	87.660	83.855	81.489
295	96.672	96.521	95.014	92.385	88.948	85.095	82.700
300	98.048	97.895	96.370	93.710	90.232	86.333	83.908
305	99.420	99.266	97.724	95.033	91.514	87.568	85.114
310	100.79	100.64	99.07	96.35	92.79	88.80	86.32
315	102.16	102.00	100.42	97.67	94.07	90.03	87.52
320	103.52	103.36	101.77	98.99	95.34	91.26	88.72
325	104.88	104.72	103.11	100.30	96.62	92.49	89.92
330	106.24	106.08	104.45	101.61	97.89	93.71	91.11
335	107.60	107.43	105.79	102.91	99.15	94.93	92.30
340	108.95	108.79	107.12	104.22	100.42	96.15	93.49
345	110.30	110.13	108.45	105.52	101.68	97.37	94.68
350	111.65	111.48	109.78	106.82	102.94	98.58	95.87
355	113.00	112.82	111.11	108.12	104.20	99.80	97.06
360	114.34	114.17	112.43	109.41	105.45	101.01	98.24
365	115.68	115.51	113.76	110.71	106.71	102.22	99.42
370	117.02	116.84	115.08	112.00	107.96	103.42	100.60
375	118.36	118.18	116.40	113.29	109.21	104.63	101.78
380	119.69	119.51	117.71	114.57	110.46	105.83	102.95
385	121.02	120.84	119.02	115.86	111.70	107.04	104.13
390	122.35	122.17	120.34	117.14	112.95	108.24	105.30
395	123.68	123.49	121.65	118.42	114.19	109.43	106.47

TABLE 8.6.
(Cont'd.)

t_D	0.05	0.1	0.3	0.5	0.7	0.9	1.0
				z'_D			
400	125.00	124.82	122.94	119.70	115.43	110.63	107.64
405	126.33	126.14	124.26	120.97	116.67	111.82	108.80
410	127.65	127.46	125.56	122.25	117.90	113.02	109.97
415	128.97	128.78	126.86	123.52	119.14	114.21	111.13
420	130.28	130.09	128.16	124.79	120.37	115.40	112.30
425	131.60	131.40	129.46	126.06	121.60	116.59	113.46
430	132.91	132.72	130.75	127.33	122.83	117.77	114.62
435	134.22	134.03	132.05	128.59	124.06	118.96	115.77
440	135.53	135.33	133.34	129.86	125.29	120.14	116.93
445	136.84	136.64	134.63	131.12	126.51	121.32	118.08
450	138.15	137.94	135.92	132.38	127.73	122.50	119.24
455	139.45	139.25	137.20	133.64	128.96	123.68	120.39
460	140.75	140.55	138.49	134.90	130.18	124.86	121.54
465	142.05	141.85	139.77	136.15	131.39	126.04	122.69
470	143.35	143.14	141.05	137.40	132.61	127.21	123.84
475	144.65	144.44	142.33	138.66	133.82	128.38	124.98
480	145.94	145.73	143.61	139.91	135.04	129.55	126.13
485	147.24	147.02	144.89	141.15	136.25	130.72	127.27
490	148.53	148.31	146.16	142.40	137.46	131.89	128.41
495	149.82	149.60	147.43	143.65	138.67	133.06	129.56
500	151.11	150.89	148.71	144.89	139.88	134.23	130.70
510	153.68	153.46	151.24	147.38	142.29	136.56	132.97
520	156.25	156.02	153.78	149.85	144.70	138.88	135.24
530	158.81	158.58	156.30	152.33	147.10	141.20	137.51
540	161.36	161.13	158.82	154.79	149.49	143.51	139.77

550	163.91	163.68	161.34	157.25	151.88	145.82	142.03
560	166.45	166.22	163.85	159.71	154.27	148.12	144.28
570	168.99	168.75	166.35	162.16	156.65	150.42	146.53
580	171.52	171.28	168.85	164.61	159.02	152.72	148.77
590	174.05	173.80	171.34	167.05	161.39	155.01	151.01
600	176.57	176.32	173.83	169.48	163.76	157.29	153.25
610	179.09	178.83	176.32	171.92	166.12	159.58	155.48
620	181.60	181.34	178.80	174.34	168.48	161.85	157.71
630	184.10	183.85	181.27	176.76	170.83	164.13	159.93
640	186.60	186.35	183.74	179.18	173.18	166.40	162.15
650	189.10	188.84	186.20	181.60	175.52	168.66	164.37
660	191.59	191.33	188.66	184.00	177.86	170.92	166.58
670	194.08	193.81	191.12	186.41	180.20	173.18	168.79
680	196.57	196.29	193.57	188.81	182.53	175.44	170.99
690	199.04	198.77	196.02	191.21	184.86	177.69	173.20
700	201.52	201.24	198.46	193.60	187.19	179.94	175.39
710	203.99	203.71	200.90	195.99	189.51	182.18	177.59
720	206.46	206.17	203.34	198.37	191.83	184.42	179.78
730	208.92	208.63	205.77	200.75	194.14	186.66	181.97
740	211.38	211.09	208.19	203.13	196.45	188.89	184.15
750	213.83	213.54	210.62	205.50	198.76	191.12	186.34
760	216.28	215.99	213.04	207.87	201.06	193.35	188.52
770	218.73	218.43	215.45	210.24	203.36	195.57	190.69
780	221.17	220.87	217.86	212.60	205.66	197.80	192.87
790	223.61	223.31	220.27	214.96	207.95	200.01	195.04
800	226.05	225.74	222.68	217.32	210.24	202.23	197.20
810	228.48	228.17	225.08	219.67	212.53	204.44	199.37
820	230.91	230.60	227.48	222.02	214.81	206.65	201.53
830	233.33	233.02	229.87	224.36	217.09	208.86	203.69
840	235.76	235.44	232.26	226.71	219.37	211.06	205.85
850	238.18	237.86	234.65	229.05	221.64	213.26	208.00

TABLE 8.6.
(Cont'd.)

t_D	z'_D						
	0.05	0.1	0.3	0.5	0.7	0.9	1.0
860	240.59	240.27	237.04	231.38	223.92	215.46	210.15
870	243.00	242.68	239.42	233.72	226.19	217.65	212.30
880	245.41	245.08	241.80	236.05	228.45	219.85	214.44
890	247.82	247.49	244.17	238.37	230.72	222.04	216.59
900	250.22	249.89	246.55	240.70	232.98	224.22	218.73
910	252.62	252.28	248.92	243.02	235.23	226.41	220.87
920	255.01	254.68	251.28	245.34	237.49	228.59	223.00
930	257.41	257.07	253.65	247.66	239.74	230.77	225.14
940	259.80	259.46	256.01	249.97	241.99	232.95	227.27
950	262.19	261.84	258.36	252.28	244.24	235.12	229.39
960	264.57	264.22	260.72	254.59	246.48	237.29	231.52
970	266.95	266.60	263.07	256.89	248.72	239.46	233.65
980	269.33	268.98	265.42	259.19	250.96	241.63	235.77
990	271.71	271.35	267.77	261.49	253.20	243.80	237.89
1,000	274.08	273.72	270.11	263.79	255.44	245.96	240.00
1,010	276.35	275.99	272.35	265.99	257.58	248.04	242.04
1,020	278.72	278.35	274.69	268.29	259.81	250.19	244.15
1,030	281.08	280.72	277.03	270.57	262.04	252.35	246.26
1,040	283.44	283.08	279.36	272.86	264.26	254.50	248.37
1,050	285.81	285.43	281.69	275.15	266.49	256.66	250.48
1,060	288.16	287.79	284.02	277.43	268.71	258.81	252.58
1,070	290.52	290.14	286.35	279.71	270.92	260.95	254.69
1,080	292.87	292.49	288.67	281.99	273.14	263.10	256.79
1,090	295.22	294.84	290.99	284.26	275.35	265.24	258.89
1,100	297.57	297.18	293.31	286.54	277.57	267.38	260.98

1,110	263.08	269.52	279.78	288.81	295.63	299.53	299.91
1,120	265.17	271.66	281.98	291.07	297.94	301.87	302.26
1,130	267.26	273.80	284.19	293.34	300.25	304.20	304.60
1,140	269.35	275.93	286.39	295.61	302.56	306.54	306.93
1,150	271.44	278.06	288.59	297.87	304.87	308.87	309.27
1,160	273.52	280.19	290.79	300.13	307.18	311.20	311.60
1,170	275.61	282.32	292.99	302.38	309.48	313.53	313.94
1,180	277.69	284.44	295.19	304.64	311.78	315.86	316.26
1,190	279.77	286.57	297.38	306.89	314.08	318.18	318.59
1,200	281.85	288.69	299.57	309.15	316.38	320.51	320.92
1,210	283.92	290.81	301.76	311.39	318.67	322.83	323.24
1,220	286.00	292.93	303.95	313.64	320.96	325.14	325.56
1,230	288.07	295.05	306.13	315.89	323.25	327.46	327.88
1,240	290.14	297.16	308.32	318.13	325.54	329.77	330.19
1,250	292.21	229.27	310.50	320.37	327.83	332.08	332.51
1,260	294.28	301.38	312.68	322.61	330.11	334.39	334.82
1,270	296.35	303.49	314.85	324.85	332.39	336.70	337.13
1,280	298.41	305.60	317.03	327.08	334.67	339.01	339.44
1,290	300.47	307.71	319.21	329.32	336.95	341.31	341.74
1,300	302.54	309.81	321.38	331.55	339.23	343.61	344.05
1,310	304.60	311.92	323.55	333.78	341.50	345.91	346.35
1,320	306.65	314.02	325.72	336.01	343.77	348.21	348.65
1,330	308.71	316.12	327.89	338.23	346.04	350.50	350.95
1,340	310.77	318.22	330.05	340.46	348.31	352.80	353.24
1,350	312.82	320.31	332.21	342.68	350.58	355.09	355.54
1,360	314.87	322.41	334.38	344.90	352.84	357.38	357.83
1,370	316.92	324.50	336.54	347.12	355.11	359.67	360.12
1,380	318.97	326.59	338.70	349.34	357.37	361.95	362.41
1,390	321.02	328.68	340.85	351.56	359.63	364.24	364.69
1,400	323.06	330.77	343.01	353.77	361.88	366.52	366.98
1,410	325.11	332.86	345.16	355.98	364.14	368.80	369.26

TABLE 8.6.
(Cont'd.)

				z'_D				
t_D	0.05	0.1	0.3	0.5	0.7	0.9	1.0
1,420	371.54	371.08	366.40	358.19	347.32	334.94	327.15
1,430	373.82	373.35	368.65	360.40	349.47	337.03	329.19
1,440	376.10	375.63	370.90	362.61	351.62	339.11	331.23
1,450	378.38	377.90	373.15	364.81	353.76	341.19	333.27
1,460	380.65	380.17	375.39	367.02	355.91	343.27	335.31
1,470	382.92	382.44	377.64	369.22	358.06	345.35	337.35
1,480	385.19	384.71	379.88	371.42	360.20	347.43	339.38
1,490	387.46	386.98	382.13	373.62	362.34	349.50	341.42
1,500	389.73	389.25	384.37	375.82	364.48	351.58	343.45
1,525	395.39	394.90	389.96	381.31	369.82	356.76	348.52
1,550	401.04	400.55	395.55	386.78	375.16	361.93	353.59
1,575	406.68	406.18	401.12	392.25	380.49	367.09	358.65
1,600	412.32	411.81	406.69	397.71	385.80	372.24	363.70
1,625	417.94	417.42	412.24	403.16	391.11	377.39	368.74
1,650	423.55	423.03	417.79	408.60	396.41	382.53	373.77
1,675	429.15	428.63	423.33	414.04	401.70	387.66	378.80
1,700	434.75	434.22	428.85	419.46	406.99	392.78	383.82
1,725	440.33	439.79	434.37	424.87	412.26	397.89	388.83
1,750	445.91	445.37	439.89	430.28	417.53	403.00	393.84
1,775	451.48	450.93	445.39	435.68	422.79	408.10	398.84
1,800	457.04	456.48	450.88	441.07	428.04	413.20	403.83
1,825	462.59	462.03	456.37	446.46	433.29	418.28	408.82
1,850	468.13	467.56	461.85	451.83	438.53	423.36	413.80
1,875	473.67	473.09	467.32	457.20	443.76	428.43	418.77
1,900	479.19	478.61	472.78	462.56	448.98	433.50	423.73

1,925	428.69	438.56	454.20	467.92	478.24	484.13	484.71
1,950	433.64	443.61	459.41	473.26	483.69	489.63	490.22
1,975	438.59	448.66	464.61	478.60	489.13	495.13	495.73
2,000	443.53	453.70	469.81	483.93	494.56	500.62	501.22
2,025	448.47	458.73	475.00	489.26	499.99	506.11	506.71
2,050	453.40	463.76	480.18	494.58	505.41	511.58	512.20
2,075	458.32	468.78	485.36	499.89	510.82	517.05	517.67
2,100	463.24	473.80	490.53	505.19	516.22	522.52	523.14
2,125	468.15	478.81	495.69	510.49	521.62	527.97	528.60
2,150	473.06	483.81	500.85	515.78	527.02	533.42	534.05
2,175	477.96	488.81	506.01	521.07	532.40	538.86	539.50
2,200	482.85	493.81	511.15	526.35	537.78	544.30	544.94
2,225	487.74	498.79	516.29	531.62	543.15	549.73	550.38
2,250	492.63	503.78	521.43	536.89	548.52	555.15	555.81
2,275	497.51	508.75	526.56	542.15	553.88	560.56	561.23
2,300	502.38	513.72	531.68	547.41	559.23	565.97	566.64
2,325	507.25	518.69	536.80	552.66	564.58	571.38	572.05
2,350	512.12	523.65	541.91	557.90	569.92	576.78	577.46
2,375	516.98	528.61	547.02	563.14	575.26	582.17	582.85
2,400	521.83	533.56	552.12	568.37	580.59	587.55	588.24
2,425	526.68	538.50	557.22	573.60	585.91	592.93	593.63
2,450	531.53	543.45	562.31	578.82	591.23	598.31	599.01
2,475	536.37	548.38	567.39	584.04	596.55	603.68	604.38
2,500	541.20	553.31	572.47	589.25	601.85	609.04	609.75
2,550	550.86	563.16	582.62	599.65	612.45	619.75	620.47
2,600	560.50	572.99	592.75	610.04	623.03	630.43	631.17
2,650	570.13	582.80	602.86	620.40	633.59	641.10	641.84
2,700	579.73	592.60	612.95	630.75	644.12	651.74	652.50
2,750	589.32	602.37	623.02	641.07	654.64	662.37	663.13
2,800	598.90	612.13	633.07	651.38	665.14	672.97	673.75
2,850	608.45	621.88	643.11	661.67	675.61	683.56	684.34

TABLE 8.6.
(Cont'd.)

| t_D | \multicolumn{7}{c}{z_D'} |
	0.05	0.1	0.3	0.5	0.7	0.9	1.0
2,900	694.92	694.12	686.07	671.94	653.12	631.60	617.99
2,950	705.48	704.67	696.51	682.19	663.13	641.32	627.52
3,000	716.02	715.20	706.94	692.43	673.11	651.01	637.03
3,050	726.54	725.71	717.34	702.65	683.08	660.69	646.53
3,100	737.04	736.20	727.73	712.85	693.03	670.36	656.01
3,150	747.53	746.68	738.10	723.04	702.97	680.01	665.48
3,200	758.00	757.14	748.45	733.21	712.89	689.64	674.93
3,250	768.45	767.58	758.79	743.36	722.80	699.27	684.37
3,300	778.89	778.01	769.11	753.50	732.69	708.87	693.80
3,350	789.31	788.42	779.42	763.62	742.57	718.47	703.21
3,400	799.71	798.81	789.71	773.73	752.43	728.05	712.62
3,450	810.10	809.19	799.99	783.82	762.28	737.62	722.00
3,500	820.48	819.55	810.25	793.90	772.12	747.17	731.38
3,550	830.83	829.90	820.49	803.97	781.94	756.72	740.74
3,600	841.18	840.24	830.73	814.02	791.75	766.24	750.09
3,650	851.51	850.56	840.94	824.06	801.55	775.76	759.43
3,700	861.83	860.86	851.15	834.08	811.33	785.27	768.76
3,750	872.13	871.15	861.34	844.09	821.10	794.76	778.08
3,800	882.41	881.43	871.51	854.09	830.86	804.24	787.38
3,850	892.69	891.70	881.68	864.08	840.61	813.71	796.68
3,900	902.95	901.95	891.83	874.05	850.34	823.17	805.96
3,950	913.20	912.19	901.96	884.01	860.06	832.62	815.23
4,000	923.43	922.41	912.09	893.96	869.77	842.06	824.49
4,050	933.65	932.62	922.20	903.89	879.47	851.48	833.74
4,100	943.86	942.82	932.30	913.82	889.16	860.90	842.99

4,150	954.06	953.01	942.39	923.73	898.84	870.30	852.22
4,200	964.25	963.19	952.47	933.63	908.50	879.69	861.44
4,250	974.42	973.35	962.53	943.52	918.16	889.08	870.65
4,300	984.58	983.50	972.58	953.40	927.80	898.45	879.85
4,350	994.73	993.64	982.62	963.27	937.42	907.81	889.04
4,400	1,004.9	1,003.8	992.7	973.1	947.1	917.2	898.2
4,450	1,015.0	1,013.9	1,002.7	983.0	956.7	926.5	907.4
4,500	1,025.1	1,024.0	1,012.7	992.8	966.3	935.9	916.6
4,550	1,035.2	1,034.1	1,022.7	1,002.6	975.9	945.2	925.7
4,600	1,045.3	1,044.2	1,032.7	1,012.4	985.5	954.5	934.9
4,650	1,055.4	1,054.2	1,042.6	1,022.2	995.0	963.8	944.0
4,700	1,065.5	1,064.3	1,052.6	1,032.0	1,004.6	973.1	953.1
4,750	1,075.5	1,074.4	1,062.6	1,041.8	1,014.1	982.4	962.2
4,800	1,085.6	1,084.4	1,072.5	1,051.6	1,023.7	991.7	971.4
4,850	1,095.6	1,094.4	1,082.4	1,061.4	1,033.2	1,000.9	980.5
4,900	1,105.6	1,104.5	1,092.4	1,071.1	1,042.8	1,010.2	989.5
4,950	1,115.7	1,114.5	1,102.3	1,080.9	1,052.3	1,019.4	998.6
5,000	1,125.7	1,124.5	1,112.2	1,090.6	1,061.8	1,028.7	1,007.7
5,100	1,145.7	1,144.4	1,132.0	1,110.0	1,080.8	1,047.2	1,025.8
5,200	1,165.6	1,164.4	1,151.7	1,129.4	1,099.7	1,065.6	1,043.9
5,300	1,185.5	1,184.3	1,171.4	1,148.8	1,118.6	1,084.0	1,062.0
5,400	1,205.4	1,204.1	1,191.1	1,168.2	1,137.5	1,102.4	1,080.0
5,500	1,225.3	1,224.0	1,210.7	1,187.5	1,156.4	1,120.7	1,098.0
5,600	1,245.1	1,243.7	1,230.3	1,206.7	1,175.2	1,139.0	1,116.0
5,700	1,264.9	1,263.5	1,249.9	1,226.0	1,194.0	1,157.3	1,134.0
5,800	1,284.6	1,283.2	1,269.4	1,245.2	1,212.8	1,175.5	1,151.9
5,900	1,304.3	1,302.9	1,288.9	1,264.4	1,231.5	1,193.8	1,169.8
6,000	1,324.0	1,322.6	1,308.4	1,283.5	1,250.2	1,211.9	1,187.7
6,100	1,343.6	1,342.2	1,327.9	1,302.6	1,268.9	1,230.1	1,205.5
6,200	1,363.2	1,361.8	1,347.3	1,321.7	1,287.5	1,248.3	1,223.3
6,300	1,382.8	1,381.4	1,366.7	1,340.8	1,306.2	1,266.4	1,241.1

TABLE 8.6.
(Cont'd.)

| | | | | z'_D | | | |
t_D	0.05	0.1	0.3	0.5	0.7	0.9	1.0
6,400	1,402.4	1,400.9	1,386.0	1,359.8	1,324.7	1,284.5	1,258.9
6,500	1,421.9	1,420.4	1,405.3	1,378.8	1,343.3	1,302.5	1,276.6
6,600	1,441.4	1,439.9	1,424.6	1,397.8	1,361.9	1,320.6	1,294.3
6,700	1,460.9	1,459.4	1,443.9	1,416.7	1,380.4	1,338.6	1,312.0
6,800	1,480.3	1,478.8	1,463.1	1,435.6	1,398.9	1,356.6	1,329.7
6,900	1,499.7	1,498.2	1,482.4	1,454.5	1,417.3	1,374.5	1,347.4
7,000	1,519.1	1,517.5	1,501.5	1,473.4	1,435.8	1,392.5	1,365.0
7,100	1,538.5	1,536.9	1,520.7	1,492.3	1,454.2	1,410.4	1,382.6
7,200	1,557.8	1,556.2	1,539.8	1,511.1	1,472.6	1,428.3	1,400.2
7,300	1,577.1	1,575.5	1,559.0	1,529.9	1,491.2	1,446.2	1,417.8
7,400	1,596.4	1,594.8	1,578.1	1,548.6	1,509.3	1,464.1	1,435.3
7,500	1,615.7	1,614.0	1,597.1	1,567.4	1,527.6	1,481.9	1,452.8
7,600	1,634.9	1,633.2	1,616.2	1,586.1	1,545.9	1,499.7	1,470.3
7,700	1,654.1	1,652.4	1,635.2	1,604.8	1,564.2	1,517.5	1,487.8
7,800	1,673.3	1,671.6	1,654.2	1,623.5	1,582.5	1,535.3	1,505.3
7,900	1,692.5	1,690.7	1,673.1	1,642.2	1,600.7	1,553.0	1,522.7
8,000	1,711.6	1,709.9	1,692.1	1,660.8	1,619.0	1,570.8	1,540.1
8,100	1,730.8	1,729.0	1,711.0	1,679.4	1,637.2	1,588.5	1,557.6
8,200	1,749.9	1,748.1	1,729.9	1,698.0	1,655.3	1,606.2	1,574.9
8,300	1,768.9	1,767.1	1,748.8	1,716.6	1,673.5	1,623.9	1,592.3
8,400	1,788.0	1,786.2	1,767.7	1,735.2	1,691.6	1,641.5	1,609.7
8,500	1,807.0	1,805.2	1,786.5	1,753.7	1,709.8	1,659.2	1,627.0
8,600	1,826.0	1,824.2	1,805.4	1,722.2	1,727.9	1,676.8	1,644.3
8,700	1,845.0	1,843.2	1,824.2	1,790.7	1,746.0	1,694.4	1,661.6
8,800	1,864.0	1,862.1	1,842.9	1,809.2	1,764.0	1,712.0	1,678.9

8,900	1,833.0	1,881.1	1,861.7	1,827.7	1,782.1	1,729.6	1,696.2
9,000	1,901.9	1,900.0	1,880.5	1,846.0	1,800.1	1,747.1	1,713.4
9,100	1,920.8	1,918.9	1,889.2	1,864.5	1,818.1	1,764.7	1,730.7
9,200	1,939.7	1,937.4	1,917.9	1,882.9	1,836.1	1,782.2	1,747.9
9,300	1,958.6	1,956.6	1,936.6	1,901.3	1,854.1	1,799.7	1,765.1
9,400	1,977.4	1,975.4	1,955.2	1,919.7	1,872.0	1,817.2	1,782.3
9,500	1,996.3	1,994.3	1,973.9	1,938.0	1,890.0	1,834.7	1,799.4
9,600	2,015.1	2,013.1	1,992.5	1,956.4	1,907.9	1,852.1	1,816.6
9,700	2,033.9	2,031.9	2,011.1	1,974.7	1,925.8	1,869.6	1,833.7
9,800	2,052.7	2,050.6	2,029.7	1,993.0	1,943.7	1,887.0	1,850.9
9,900	2,071.5	2,069.4	2,048.3	2,011.3	1,961.6	1,904.4	1,868.0
1.00×10^4	2.090×10^3	2.088×10^3	2.067×10^3	2.029×10^3	1.979×10^3	1.922×10^3	1.855×10^3
1.25×10^4	2.553×10^3	2.551×10^3	2.526×10^3	2.481×10^3	2.421×10^3	2.352×10^3	2.308×10^3
1.50×10^4	3.009×10^3	3.006×10^3	2.977×10^3	2.925×10^3	2.855×10^3	2.775×10^3	2.724×10^3
1.75×10^4	3.457×10^3	3.454×10^3	3.421×10^3	3.362×10^3	3.284×10^3	3.193×10^3	3.135×10^3
2.00×10^4	3.900×10^3	3.897×10^3	3.860×10^3	3.794×10^3	3.707×10^3	3.605×10^3	3.541×10^3
2.50×10^4	4.773×10^3	4.768×10^3	4.724×10^3	4.646×10^3	4.541×10^3	4.419×10^3	4.341×10^3
3.00×10^4	5.630×10^3	5.625×10^3	5.574×10^3	5.483×10^3	5.361×10^3	5.219×10^3	5.129×10^3
3.50×10^4	6.476×10^3	6.470×10^3	6.412×10^3	6.309×10^3	6.170×10^3	6.009×10^3	5.906×10^3
4.00×10^4	7.312×10^3	7.305×10^3	7.240×10^3	7.125×10^3	6.970×10^3	6.790×10^3	6.675×10^3
4.50×10^4	8.139×10^3	8.132×10^3	8.060×10^3	7.933×10^3	7.762×10^3	7.564×10^3	7.437×10^3
5.00×10^4	8.959×10^3	8.951×10^3	8.872×10^3	8.734×10^3	8.548×10^3	8.331×10^3	8.193×10^3
6.00×10^4	1.057×10^4	1.057×10^4	1.047×10^4	1.031×10^4	1.010×10^4	9.846×10^3	9.684×10^3
7.00×10^4	1.217×10^4	1.217×10^4	1.206×10^4	1.188×10^4	1.163×10^4	1.134×10^4	1.116×10^4
8.00×10^4	1.375×10^4	1.375×10^4	1.363×10^4	1.342×10^4	1.315×10^4	1.283×10^4	1.262×10^4
9.00×10^4	1.532×10^4	1.531×10^4	1.518×10^4	1.496×10^4	1.465×10^4	1.430×10^4	1.407×10^4
1.00×10^5	1.687×10^4	1.686×10^4	1.672×10^4	1.647×10^4	1.614×10^4	1.576×10^4	1.551×10^4
1.25×10^5	2.071×10^4	2.069×10^4	2.052×10^4	2.023×10^4	1.982×10^4	1.936×10^4	1.906×10^4
1.50×10^5	2.448×10^4	2.446×10^4	2.427×10^4	2.392×10^4	2.345×10^4	2.291×10^4	2.256×10^4
2.00×10^5	3.190×10^4	3.188×10^4	3.163×10^4	3.119×10^4	3.059×10^4	2.989×10^4	2.945×10^4
2.50×10^5	3.918×10^4	3.916×10^4	3.885×10^4	3.832×10^4	3.760×10^4	3.676×10^4	3.622×10^4

TABLE 8.6.
(Cont'd.)

t_D	z_D'						
	0.05	0.1	0.3	0.5	0.7	0.9	1.0
3.00×10^5	4.636×10^4	4.633×10^4	4.598×10^4	4.536×10^4	4.452×10^4	4.353×10^4	4.290×10^4
4.00×10^5	6.048×10^4	6.004×10^4	5.999×10^4	5.920×10^4	5.812×10^4	5.687×10^4	5.606×10^4
5.00×10^5	7.436×10^4	7.431×10^4	7.376×10^4	7.280×10^4	7.150×10^4	6.998×10^4	6.900×10^4
6.00×10^5	8.805×10^4	8.798×10^4	8.735×10^4	8.623×10^4	8.471×10^4	8.293×10^4	8.178×10^4
7.00×10^5	1.016×10^5	1.015×10^5	1.008×10^5	9.951×10^4	9.777×10^4	9.573×10^4	9.442×10^4
8.00×10^5	1.150×10^5	1.149×10^5	1.141×10^5	1.127×10^5	1.107×10^5	1.084×10^5	1.070×10^5
9.00×10^5	1.283×10^5	1.282×10^5	1.273×10^5	1.257×10^5	1.235×10^5	1.210×10^5	1.194×10^5
1.00×10^6	1.415×10^5	1.412×10^5	1.404×10^5	1.387×10^5	1.363×10^5	1.335×10^5	1.317×10^5
1.50×10^6	2.059×10^5	2.060×10^5	2.041×10^5	2.016×10^5	1.982×10^5	1.943×10^5	1.918×10^5
2.00×10^6	2.695×10^5	2.695×10^5	2.676×10^5	2.644×10^5	2.601×10^5	2.551×10^5	2.518×10^5
2.50×10^5	3.320×10^5	3.319×10^5	3.296×10^5	3.254×10^5	3.202×10^5	3.141×10^5	3.101×10^5
3.00×10^6	3.937×10^5	3.936×10^5	3.909×10^5	3.864×10^5	3.803×10^5	3.731×10^5	3.684×10^5
4.00×10^6	5.154×10^5	5.152×10^5	5.118×10^5	5.060×10^5	4.981×10^5	4.888×10^5	4.828×10^5
5.00×10^6	6.352×10^5	6.349×10^5	6.308×10^5	6.238×10^5	6.142×10^5	6.029×10^5	5.956×10^5
6.00×10^6	7.536×10^5	7.533×10^5	7.485×10^5	7.402×10^5	7.290×10^5	7.157×10^5	7.072×10^5
7.00×10^6	8.709×10^5	8.705×10^5	8.650×10^5	8.556×10^5	8.427×10^5	8.275×10^5	8.177×10^5
8.00×10^6	9.972×10^5	9.867×10^5	9.806×10^5	9.699×10^5	9.555×10^5	9.384×10^5	9.273×10^5
9.00×10^6	1.103×10^6	1.102×10^6	1.095×10^6	1.084×10^6	1.067×10^6	1.049×10^6	1.036×10^6
1.00×10^7	1.217×10^6	1.217×10^6	1.209×10^6	1.196×10^6	1.179×10^6	1.158×10^6	1.144×10^6
1.50×10^7	1.782×10^6	1.781×10^6	1.771×10^6	1.752×10^6	1.727×10^6	1.697×10^6	1.678×10^6
2.00×10^7	2.337×10^6	2.336×10^6	2.322×10^6	2.298×10^6	2.266×10^6	2.227×10^6	2.202×10^6
2.50×10^7	2.884×10^6	2.882×10^6	2.866×10^6	2.837×10^6	2.797×10^6	2.750×10^6	2.720×10^6
3.00×10^7	3.425×10^6	3.423×10^6	3.404×10^6	3.369×10^6	3.323×10^6	3.268×10^6	3.232×10^6
4.00×10^7	4.493×10^6	4.491×10^6	4.466×10^6	4.422×10^6	4.361×10^6	4.290×10^6	4.244×10^6
5.00×10^7	5.547×10^6	5.544×10^6	5.514×10^6	5.460×10^6	5.386×10^6	5.299×10^6	5.243×10^6

6.00×10^7	6.232×10^6	6.299×10^6	6.401×10^6	6.488×10^6	6.551×10^6	6.587×10^6	6.590×10^6
7.00×10^7	7.213×10^6	7.290×10^6	7.407×10^6	7.507×10^6	7.579×10^6	7.620×10^6	7.624×10^6
8.00×10^7	8.188×10^6	8.274×10^6	8.407×10^6	8.519×10^6	8.600×10^6	8.647×10^6	8.651×10^6
9.00×10^7	9.156×10^6	9.252×10^6	9.400×10^6	9.524×10^6	9.615×10^6	9.666×10^6	9.671×10^6
1.00×10^8	1.012×10^7	1.023×10^7	1.039×10^7	1.052×10^7	1.062×10^7	1.067×10^7	1.069×10^7
1.50×10^8	1.483×10^7	1.499×10^7	1.522×10^7	1.541×10^7	1.555×10^7	1.567×10^7	1.567×10^7
2.00×10^8	1.954×10^7	1.974×10^7	2.004×10^7	2.029×10^7	2.048×10^7	2.059×10^7	2.059×10^7
2.50×10^8	2.415×10^7	2.439×10^7	2.476×10^7	2.507×10^7	2.531×10^7	2.545×10^7	2.546×10^7
3.00×10^8	2.875×10^7	2.904×10^7	2.947×10^7	2.984×10^7	3.010×10^7	3.026×10^7	3.027×10^7
4.00×10^8	3.782×10^7	3.819×10^7	3.875×10^7	3.923×10^7	3.958×10^7	3.978×10^7	3.979×10^7
5.00×10^8	4.679×10^7	4.724×10^7	4.793×10^7	4.851×10^7	4.894×10^7	4.918×10^7	4.920×10^7
6.00×10^8	5.568×10^7	5.621×10^7	5.702×10^7	5.771×10^7	5.821×10^7	5.850×10^7	5.852×10^7
7.00×10^8	6.450×10^7	6.511×10^7	6.605×10^7	6.684×10^7	6.741×10^7	6.774×10^7	6.777×10^7
8.00×10^8	7.327×10^7	7.396×10^7	7.501×10^7	7.590×10^7	7.655×10^7	7.693×10^7	7.700×10^7
9.00×10^8	8.199×10^7	8.275×10^7	8.393×10^7	8.492×10^7	8.564×10^7	8.606×10^7	8.609×10^7
1.00×10^9	9.066×10^7	9.151×10^7	9.281×10^7	9.390×10^7	9.469×10^7	9.515×10^7	9.518×10^7
1.50×10^9	1.336×10^8	1.348×10^8	1.367×10^8	1.382×10^8	1.394×10^8	1.400×10^8	1.401×10^8
2.00×10^9	1.758×10^8	1.774×10^8	1.799×10^8	1.819×10^8	1.834×10^8	1.843×10^8	1.843×10^8
2.50×10^9	2.177×10^8	2.196×10^8	2.226×10^8	2.251×10^8	2.269×10^8	2.280×10^8	2.281×10^8
3.00×10^9	2.592×10^8	2.615×10^8	2.650×10^8	2.680×10^8	2.701×10^8	2.713×10^8	2.714×10^8
4.00×10^9	3.413×10^8	3.443×10^8	3.489×10^8	3.528×10^8	3.556×10^8	3.572×10^8	3.573×10^8
5.00×10^9	4.227×10^8	4.263×10^8	4.320×10^8	4.367×10^8	4.401×10^8	4.421×10^8	4.422×10^8
6.00×10^9	5.033×10^8	5.077×10^8	5.143×10^8	5.199×10^8	5.240×10^8	5.262×10^8	5.265×10^8
7.00×10^9	5.835×10^8	5.885×10^8	5.961×10^8	6.025×10^8	6.072×10^8	6.098×10^8	6.101×10^8
8.00×10^9	6.632×10^8	6.688×10^8	6.775×10^8	6.847×10^8	6.900×10^8	6.930×10^8	6.932×10^8
9.00×10^9	7.424×10^8	7.487×10^8	7.584×10^8	7.664×10^8	7.723×10^8	7.756×10^8	7.760×10^8
1.00×10^{10}	8.214×10^8	8.283×10^8	8.389×10^8	8.478×10^8	8.543×10^8	8.574×10^8	8.583×10^8
1.50×10^{10}	1.209×10^9	1.219×10^9	1.235×10^9	1.247×10^9	1.257×10^9	1.264×10^9	1.263×10^9
2.00×10^{10}	1.596×10^9	1.610×10^9	1.630×10^9	1.646×10^9	1.659×10^9	1.666×10^9	1.666×10^9
2.50×10^{10}	1.977×10^9	1.993×10^9	2.018×10^9	2.038×10^9	2.055×10^9	2.063×10^9	2.065×10^9
3.00×10^{10}	2.357×10^9	2.376×10^9	2.405×10^9	2.430×10^9	2.447×10^9	2.458×10^9	2.458×10^9

TABLE 8.6.
(Cont'd.)

t_D	z'_D						
	0.05	0.1	0.3	0.5	0.7	0.9	1.0
4.00×10^{10}	3.240×10^9	3.239×10^9	3.226×10^9	3.203×10^9	3.171×10^9	3.133×10^9	3.108×10^9
5.00×10^{10}	4.014×10^9	4.013×10^9	3.997×10^9	3.968×10^9	3.929×10^9	3.883×10^9	3.852×10^9
6.00×10^{10}	4.782×10^9	4.781×10^9	4.762×10^9	4.728×10^9	4.682×10^9	4.627×10^9	4.591×10^9
7.00×10^{10}	5.546×10^9	5.544×10^9	5.522×10^9	5.483×10^9	5.430×10^9	5.366×10^9	5.325×10^9
8.00×10^{10}	6.305×10^9	6.303×10^9	6.278×10^9	6.234×10^9	6.174×10^9	6.102×10^9	6.055×10^9
9.00×10^{10}	7.060×10^9	7.058×10^9	7.030×10^9	6.982×10^9	6.914×10^9	6.834×10^9	6.782×10^9
1.00×10^{11}	7.813×10^9	7.810×10^9	7.780×10^9	7.726×10^9	7.652×10^9	7.564×10^9	7.506×10^9
1.50×10^{11}	1.154×10^{10}	1.153×10^{10}	1.149×10^{10}	1.141×10^{10}	1.130×10^{10}	1.118×10^{10}	1.109×10^{10}
2.00×10^{11}	1.522×10^{10}	1.521×10^{10}	1.515×10^{10}	1.505×10^{10}	1.491×10^{10}	1.474×10^{10}	1.463×10^{10}
2.50×10^{11}	1.886×10^{10}	1.885×10^{10}	1.878×10^{10}	1.866×10^{10}	1.849×10^{10}	1.828×10^{10}	1.814×10^{10}
3.00×10^{11}	2.248×10^{10}	2.247×10^{10}	2.239×10^{10}	2.224×10^{10}	2.204×10^{10}	2.179×10^{10}	2.163×10^{10}
4.00×10^{11}	2.965×10^{10}	2.964×10^{10}	2.953×10^{10}	2.934×10^{10}	2.907×10^{10}	2.876×10^{10}	2.855×10^{10}
5.00×10^{11}	3.677×10^{10}	3.675×10^{10}	3.662×10^{10}	3.638×10^{10}	3.605×10^{10}	3.566×10^{10}	3.540×10^{10}
6.00×10^{11}	4.383×10^{10}	4.381×10^{10}	4.365×10^{10}	4.337×10^{10}	4.298×10^{10}	4.252×10^{10}	4.221×10^{10}
7.00×10^{11}	5.085×10^{10}	5.082×10^{10}	5.064×10^{10}	5.032×10^{10}	4.987×10^{10}	4.933×10^{10}	4.898×10^{10}
8.00×10^{11}	5.783×10^{10}	5.781×10^{10}	5.760×10^{10}	5.723×10^{10}	5.673×10^{10}	5.612×10^{10}	5.572×10^{10}
9.00×10^{11}	6.478×10^{10}	6.476×10^{10}	6.453×10^{10}	6.412×10^{10}	6.355×10^{10}	6.288×10^{10}	6.243×10^{10}
1.00×10^{12}	7.171×10^{10}	7.168×10^{10}	7.143×10^{10}	7.098×10^{10}	7.035×10^{10}	6.961×10^{10}	6.912×10^{10}
1.50×10^{12}	1.060×10^{11}	1.060×10^{11}	1.056×10^{11}	1.050×10^{11}	1.041×10^{11}	1.030×10^{11}	1.022×10^{11}
2.00×10^{12}	1.400×10^{11}	1.399×10^{11}	1.394×10^{11}	1.386×10^{11}	1.374×10^{11}	1.359×10^{11}	1.350×10^{11}

TABLE 8.7.
Dimensionless Influx, W_{eD}, for $r_D' = 4$ for bottom-water drive

t_D	z_D'						
	0.05	0.1	0.3	0.5	0.7	0.9	1.0
2	2.398	2.389	2.284	2.031	1.824	1.620	1.507
3	3.006	2.993	2.874	2.629	2.390	2.149	2.012
4	3.552	3.528	3.404	3.158	2.893	2.620	2.466
5	4.053	4.017	3.893	3.627	3.341	3.045	2.876
6	4.490	4.452	4.332	4.047	3.744	3.430	3.249
7	4.867	4.829	4.715	4.420	4.107	3.778	3.587
8	5.191	5.157	5.043	4.757	4.437	4.096	3.898
9	5.464	5.434	5.322	5.060	4.735	4.385	4.184
10	5.767	5.739	5.598	5.319	5.000	4.647	4.443
11	5.964	5.935	5.829	5.561	5.240	4.884	4.681
12	6.188	6.158	6.044	5.780	5.463	5.107	4.903
13	6.380	6.350	6.240	5.983	5.670	5.316	5.113
14	6.559	6.529	6.421	6.171	5.863	5.511	5.309
15	6.725	6.694	6.589	6.345	6.044	5.695	5.495
16	6.876	6.844	6.743	6.506	6.213	5.867	5.671
17	7.014	6.983	6.885	6.656	6.371	6.030	5.838
18	7.140	7.113	7.019	6.792	6.523	6.187	5.999
19	7.261	7.240	7.140	6.913	6.663	6.334	6.153
20	7.376	7.344	7.261	7.028	6.785	6.479	6.302
22	7.518	7.507	7.451	7.227	6.982	6.691	6.524
24	7.618	7.607	7.518	7.361	7.149	6.870	6.714
26	7.697	7.685	7.607	7.473	7.283	7.026	6.881
28	7.752	7.752	7.674	7.563	7.395	7.160	7.026
30	7.808	7.797	7.741	7.641	7.484	7.283	7.160
34	7.864	7.864	7.819	7.741	7.618	7.451	7.350
38	7.909	7.909	7.875	7.808	7.719	7.585	7.496
42	7.931	7.931	7.909	7.864	7.797	7.685	7.618
46	7.942	7.942	7.920	7.898	7.842	7.752	7.697
50	7.954	7.954	7.942	7.920	7.875	7.808	7.764
60	7.968	7.968	7.965	7.954	7.931	7.898	7.864
70	7.976	7.976	7.976	7.968	7.965	7.942	7.920
80	7.982	7.982	7.987	7.976	7.976	7.965	7.954
90	7.987	7.987	7.987	7.984	7.983	7.976	7.965
100	7.987	7.987	7.987	7.987	7.987	7.983	7.976
120	7.987	7.987	7.987	7.987	7.987	7.987	7.987

TABLE 8.8.
Dimensionless Influx, W_{eD}, for $r'_D = 6$ for bottom-water drive

t_D	z'_D						
	0.05	0.1	0.3	0.5	0.7	0.9	1.0
6	4.780	4.762	4.597	4.285	3.953	3.611	3.414
7	5.309	5.289	5.114	4.779	4.422	4.053	3.837
8	5.799	5.778	5.595	5.256	4.875	4.478	4.247
9	6.252	6.229	6.041	5.712	5.310	4.888	4.642
10	6.750	6.729	6.498	6.135	5.719	5.278	5.019
11	7.137	7.116	6.916	6.548	6.110	5.648	5.378
12	7.569	7.545	7.325	6.945	6.491	6.009	5.728
13	7.967	7.916	7.719	7.329	6.858	6.359	6.067
14	8.357	8.334	8.099	7.699	7.214	6.697	6.395
15	8.734	8.709	8.467	8.057	7.557	7.024	6.713
16	9.093	9.067	8.819	8.398	7.884	7.336	7.017
17	9.442	9.416	9.160	8.730	8.204	7.641	7.315
18	9.775	9.749	9.485	9.047	8.510	7.934	7.601
19	10.09	10.06	9.794	9.443	8.802	8.214	7.874
20	10.40	10.37	10.10	9.646	9.087	8.487	8.142
22	10.99	10.96	10.67	10.21	9.631	9.009	8.653
24	11.53	11.50	11.20	10.73	10.13	9.493	9.130
26	12.06	12.03	11.72	11.23	10.62	9.964	9.594
28	12.52	12.49	12.17	11.68	11.06	10.39	10.01
30	12.95	12.92	12.59	12.09	11.46	10.78	10.40
35	13.96	13.93	13.57	13.06	12.41	11.70	11.32
40	14.69	14.66	14.33	13.84	13.23	12.53	12.15
45	15.27	15.24	14.94	14.48	13.90	13.23	12.87
50	15.74	15.71	15.44	15.01	14.47	13.84	13.49
60	16.40	16.38	16.15	15.81	15.34	14.78	14.47
70	16.87	16.85	16.67	16.38	15.99	15.50	15.24
80	17.20	17.18	17.04	16.80	16.48	16.06	15.83
90	17.43	17.42	17.30	17.10	16.85	16.50	16.29
100	17.58	17.58	17.49	17.34	17.12	16.83	16.66
110	17.71	17.69	17.63	17.50	17.34	17.09	16.93
120	17.78	17.78	17.73	17.63	17.49	17.29	17.17
130	17.84	17.84	17.79	17.73	17.62	17.45	17.34
140	17.88	17.88	17.85	17.79	17.71	17.57	17.48
150	17.92	17.91	17.88	17.84	17.77	17.66	17.58
175	17.95	17.95	17.94	17.92	17.87	17.81	17.76
200	17.97	17.97	17.96	17.95	17.93	17.88	17.86
225	17.97	17.97	17.97	17.96	17.95	17.93	17.91
250	17.98	17.98	17.98	17.97	17.96	17.95	17.95
300	17.98	17.98	17.98	17.98	17.98	17.97	17.97
350	17.98	17.98	17.98	17.98	17.98	17.98	17.98
400	17.98	17.98	17.98	17.98	17.98	17.98	17.98
450	17.98	17.98	17.98	17.98	17.98	17.98	17.98
500	17.98	17.98	17.98	17.98	17.98	17.98	17.98

TABLE 8.9.

Dimensionless Influx, W_{eD}, for $r'_D = 8$ for bottom-water drive

t_D	\multicolumn{7}{c}{z'_D}						
	0.05	0.1	0.3	0.5	0.7	0.9	1.0
9	6.301	6.278	6.088	5.756	5.350	4.924	4.675
10	6.828	6.807	6.574	6.205	5.783	5.336	5.072
11	7.250	7.229	7.026	6.650	6.204	5.732	5.456
12	7.725	7.700	7.477	7.086	6.621	6.126	5.836
13	8.173	8.149	7.919	7.515	7.029	6.514	6.210
14	8.619	8.594	8.355	7.937	7.432	6.895	6.578
15	9.058	9.032	8.783	8.351	7.828	7.270	6.940
16	9.485	9.458	9.202	8.755	8.213	7.634	7.293
17	9.907	9.879	9.613	9.153	8.594	7.997	7.642
18	10.32	10.29	10.01	9.537	8.961	8.343	7.979
19	10.72	10.69	10.41	9.920	9.328	8.691	8.315
20	11.12	11.08	10.80	10.30	9.687	9.031	8.645
22	11.89	11.86	11.55	11.02	10.38	9.686	9.280
24	12.63	12.60	12.27	11.72	11.05	10.32	9.896
26	13.36	13.32	12.97	12.40	11.70	10.94	10.49
28	14.06	14.02	13.65	13.06	12.33	11.53	11.07
30	14.73	14.69	14.30	13.68	12.93	12.10	11.62
34	16.01	15.97	15.54	14.88	14.07	13.18	12.67
38	17.21	17.17	16.70	15.99	15.13	14.18	13.65
40	17.80	17.75	17.26	16.52	15.64	14.66	14.12
45	19.15	19.10	18.56	17.76	16.83	15.77	15.21
50	20.42	20.36	19.76	18.91	17.93	16.80	16.24
55	21.46	21.39	20.80	19.96	18.97	17.83	17.24
60	22.40	22.34	21.75	20.91	19.93	18.78	18.19
70	23.97	23.92	23.36	22.55	21.58	20.44	19.86
80	25.29	25.23	24.71	23.94	23.01	21.91	21.32
90	26.39	26.33	25.85	25.12	24.24	23.18	22.61
100	27.30	27.25	26.81	26.13	25.29	24.29	23.74
120	28.61	28.57	28.19	27.63	26.90	26.01	25.51
140	29.55	29.51	29.21	28.74	28.12	27.33	26.90
160	30.23	30.21	29.96	29.57	29.04	28.37	27.99
180	30.73	30.71	30.51	30.18	29.75	29.18	28.84
200	31.07	31.04	30.90	30.63	30.26	29.79	29.51
240	31.50	31.49	31.39	31.22	30.98	30.65	30.45
280	31.72	31.71	31.66	31.56	31.39	31.17	31.03
320	31.85	31.84	31.80	31.74	31.64	31.49	31.39
360	31.90	31.90	31.88	31.85	31.78	31.68	31.61
400	31.94	31.94	31.93	31.90	31.86	31.79	31.75
450	31.96	31.96	31.95	31.94	31.91	31.88	31.85
500	31.97	31.97	31.96	31.96	31.95	31.93	31.90
550	31.97	31.97	31.97	31.96	31.96	31.95	31.94
600	31.97	31.97	31.97	31.97	31.97	31.96	31.95
700	31.97	31.97	31.97	31.97	31.97	31.97	31.97
800	31.97	31.97	31.97	31.97	31.97	31.97	31.97

TABLE 8.10.
Dimensionless Influx, W_{eD}, for $r_D' = 10$ for bottom-water drive

t_D	z_D'						
	0.05	0.1	0.3	0.5	0.7	0.9	1.0
22	12.07	12.04	11.74	11.21	10.56	9.865	9.449
24	12.86	12.83	12.52	11.97	11.29	10.55	10.12
26	13.65	13.62	13.29	12.72	12.01	11.24	10.78
28	14.42	14.39	14.04	13.44	12.70	11.90	11.42
30	15.17	15.13	14.77	14.15	13.38	12.55	12.05
32	15.91	15.87	15.49	14.85	14.05	13.18	12.67
34	16.63	16.59	16.20	15.54	14.71	13.81	13.28
36	17.33	17.29	16.89	16.21	15.35	14.42	13.87
38	18.03	17.99	17.57	16.86	15.98	15.02	14.45
40	18.72	18.68	18.24	17.51	16.60	15.61	15.02
42	19.38	19.33	18.89	18.14	17.21	16.19	15.58
44	20.03	19.99	19.53	18.76	17.80	16.75	16.14
46	20.67	20.62	20.15	19.36	18.38	17.30	16.67
48	21.30	21.25	20.76	19.95	18.95	17.84	17.20
50	21.92	21.87	21.36	20.53	19.51	18.38	17.72
52	22.52	22.47	21.95	21.10	20.05	18.89	18.22
54	23.11	23.06	22.53	21.66	20.59	19.40	18.72
56	23.70	23.64	23.09	22.20	21.11	19.89	19.21
58	24.26	24.21	23.65	22.74	21.63	20.39	19.68
60	24.82	24.77	24.19	23.26	22.13	20.87	20.15
65	26.18	26.12	25.50	24.53	23.34	22.02	21.28
70	27.47	27.41	26.75	25.73	24.50	23.12	22.36
75	28.71	28.55	27.94	26.88	25.60	24.17	23.39
80	29.89	29.82	29.08	27.97	26.65	25.16	24.36
85	31.02	30.95	30.17	29.01	27.65	26.10	25.31
90	32.10	32.03	31.20	30.00	28.60	27.03	26.25
95	33.04	32.96	32.14	30.95	29.54	27.93	27.10
100	33.94	33.85	33.03	31.85	30.44	28.82	27.98
110	35.55	35.46	34.65	33.49	32.08	30.47	29.62
120	36.97	36.90	36.11	34.98	33.58	31.98	31.14
130	38.28	38.19	37.44	36.33	34.96	33.38	32.55
140	39.44	39.37	38.64	37.56	36.23	34.67	33.85
150	40.49	40.42	39.71	38.67	37.38	35.86	35.04
170	42.21	42.15	41.51	40.54	39.33	37.89	37.11
190	43.62	43.55	42.98	42.10	40.97	39.62	38.90
210	44.77	44.72	44.19	43.40	42.36	41.11	40.42
230	45.71	45.67	45.20	44.48	43.54	42.38	41.74
250	46.48	46.44	46.01	45.38	44.53	43.47	42.87
270	47.11	47.06	46.70	46.13	45.36	44.40	43.84
290	47.61	47.58	47.25	46.75	46.07	45.19	44.68
310	48.03	48.00	47.72	47.26	46.66	45.87	45.41
330	48.38	48.35	48.10	47.71	47.16	46.45	46.03
350	48.66	48.64	48.42	48.08	47.59	46.95	46.57

TABLE 8.10.
(Cont'd.)

t_D	z_D'						
	0.05	0.1	0.3	0.5	0.7	0.9	1.0
400	49.15	49.14	48.99	48.74	48.38	47.89	47.60
450	49.46	49.45	49.35	49.17	48.91	48.55	48.31
500	49.65	49.64	49.58	49.45	49.26	48.98	48.82
600	49.84	49.84	49.81	49.74	49.65	49.50	49.41
700	49.91	49.91	49.90	49.87	49.82	49.74	49.69
800	49.94	49.94	49.93	49.92	49.90	49.85	49.83
900	49.96	49.96	49.94	49.94	49.93	49.91	49.90
1,000	49.96	49.96	49.96	49.96	49.94	49.93	49.93
1,200	49.96	49.96	49.96	49.96	49.96	49.96	49.96

4. PSEUDOSTEADY-STATE MODELS

The edge-water and bottom-water, unsteady-state methods discussed in Sect. 3 provide correct procedures for calculating water influx in nearly any reservoir application. However, the calculations tend to be somewhat cumbersome, and therefore there have been various attempts to simplify the calculations. The most popular and seemingly accurate method is one developed by Fetkovich using an aquifer material balance and an equation that describes the flow rate from the aquifer.[15] The equations for flow rate used by Fetkovich are similar to the productivity index equation defined in Chapter 7. The productivity index required pseudosteady-state flow conditions. Thus, this method neglects the effects of the transient period in the calculations of water influx, which will obviously introduce errors into the calculations. However, the method has been found to give results similar to those of the van Everdingen and Hurst model in many applications.

Fetkovich first wrote a material balance equation on the aquifer for constant water and rock compressibilities as

$$\bar{p} = -\left(\frac{p_i}{W_{ei}}\right)W_e + p_i \tag{8.15}$$

where $\bar{p}$ is the average pressure in the aquifer after the removal of W_e bbl of water, p_i is the initial pressure of the aquifer, and W_{ei} is the initial encroachable water in place at the initial pressure, Fetkovich next defined a generalized rate equation as

$$q_w B_w = J(\bar{p} - p_R)^{m_a} \tag{8.16}$$

where $q_w B_w$ is the flow rate of water from the aquifer, J is the productivity index of the aquifer and is a function of the aquifer geometry, p_R is the pressure at the reservoir-aquifer boundary, and m_a is equal to 1 for Darcy flow during the pseudosteady-state flow region. Equations (8.15) and (8.16) can be combined to yield the following equation (see References 15 and 16 for the complete derivation):

$$W_e = \frac{W_{ei}}{p_i}(p_i - p_R)\left(1 - e^{-\frac{Jp_i t}{W_{ei}}}\right) \qquad (8.17)$$

This equation was derived for constant pressures at both the reservoir-aquifer boundary, p_R, and the average pressure in the aquifer, $\bar{p}$. At this point, to apply the equation to a typical reservoir application where both of these pressures are changing with time, it would normally be required to use the principle of superposition. Fetkovich showed that by calculating the water influx for a short time period, Δt, with a corresponding average aquifer pressure, $\bar{p}$, and an average boundary pressure, $\bar{p}_R$, and then starting the calculation over again for a new period and new pressures, superposition was not needed. The following equations are used in the calculation for water influx with this method

$$\Delta W_{en} = \frac{W_{ei}}{p_i}(\bar{p}_{n-1} - \bar{p}_{Rn})\left(1 - e^{-\frac{Jp_i \Delta t_n}{W_{ei}}}\right) \qquad (8.18)$$

$$\bar{p}_{n-1} = p_i\left(1 - \frac{W_e}{W_{ei}}\right) \qquad (8.19)$$

$$\bar{p}_{Rn} = \frac{p_{Rn-1} + p_{Rn}}{2} \qquad (8.20)$$

where n represents a particular interval, $\bar{p}_{n-1}$ is the average aquifer pressure at the end of the $n - 1$ time interval, $\bar{p}_{Rn}$ is the average reservoir-aquifer boundary pressure during interval n, and W_e is the total, or cumulative, water influx and is given by

$$W_e = \Sigma \Delta W_{en} \qquad (8.21)$$

The productivity index, J, used in the calculation procedure is a function of the geometry of the aquifer. Table 8.11 contains several aquifer productivity indexes as presented by Fetkovich.[15] When you use the equations for the condition of a constant pressure outer aquifer boundary, the average aquifer pressure in Eq. (8.18) will always be equal to the initial outer boundary pressure, which is usually p_i. Example 8.5 illustrates the use of the Fetkovich method.

Example 8.5. Repeat the water influx calculations for the reservoir in Ex. 8.3 using the Fetkovich approach.

TABLE 8.11.
Productivity indices for radial and linear aquifers (taken from reference 15)
Fetkovich

Type of Outer Aquifer Boundary	Radial Flow[a]	Linear Flow[b]
Finite—no flow	$J = \dfrac{0.00708kh\left(\dfrac{\theta}{360}\right)}{\mu[\ln(r_e/r_R) - 0.75]}$	$J = \dfrac{0.003381kwh}{\mu L}$
Finite—constant pressure	$J = \dfrac{0.00708kh\left(\dfrac{\theta}{360}\right)}{\mu[\ln(r_e/r_R)]}$	$J = \dfrac{0.001127kwh}{\mu L}$

[a] Units are in normal field units with k in millidarcies.
[b] w is width and L is length of linear aquifer.

SOLUTION:

$$\text{area of aquifer} = \frac{1}{2}\pi r_e^2 \quad \text{or} \quad r_e = \left[\frac{250,000(43560)}{0.5\pi}\right]^{1/2} = 83,263 \text{ ft}$$

$$\text{area of reservoir} = \frac{1}{2}\pi r_R^2 \quad \text{or} \quad r_R = \left[\frac{1216(43560)}{0.5\pi}\right]^{1/2} = 5807 \text{ ft}$$

$$W_{ei} = \frac{c_t\left(\dfrac{\theta}{360}\right)\pi(r_e^2 - r_R^2)h\,\phi p_i}{5.615}$$

$$W_{ei} = \frac{6(10)^{-6}\left(\dfrac{180}{360}\right)\pi(83,263^2 - 5807^2)19.2(0.209)3793}{5.615} = 176.3(10)^6 \text{ bbl}$$

$$J = \frac{0.00708kh\left(\dfrac{\theta}{360}\right)}{\mu[\ln(r_e/r_R) - 0.75]} = \frac{0.00708(275)(19.2)\left(\dfrac{180}{360}\right)}{0.25\left[\ln\left(\dfrac{83,263}{5807}\right) - 0.75\right]} = 39.08$$

$$\Delta W_{en} = \frac{W_{ei}}{p_i}(\bar{p}_{n-1} - \bar{p}_{Rn})\left(1 - e^{-\frac{Jp_i\Delta t_n}{W_{ei}}}\right)$$

$$= \frac{176.3(10)^6}{3793}(\bar{p}_{n-1} - \bar{p}_{Rn})\left(1 - e^{-\frac{39.08(3793)(91.3)}{176.3(10)^6}}\right)$$

$$\Delta W_{en} = 3435(\bar{p}_{n-1} - \bar{p}_{Rn}) \tag{8.22}$$

$$\bar{p}_{n-1} = p_i\left(1 - \frac{\Sigma \Delta W_{en}}{W_{ei}}\right)$$

$$\bar{p}_{n-1} = 3793\left(1 - \frac{\Sigma \Delta W_{en}}{176.3(10)^6}\right) \tag{8.23}$$

Solving Eqs. (8.22) and (8.23), we get Table 8.12.

TABLE 8.12.

Time	p_R	$\bar{p}_{Rn}$	$\bar{p}_{n-1} - \bar{p}_{Rn}$	ΔW_e	W_e	$\bar{p}_n$
0	3793	3793	0	0	0	3793
1	3788	3790.5	2.5	8,600	8,600	3792.8
2	3774	3781	11.8	40,500	49,100	3791.9
3	3748	3761	30.9	106,100	155,200	3789.7
4	3709	3728.5	61.2	210,000	365,300	3785.1
5	3680	3694.5	90.6	311,200	676,500	3778.4
6	3643	3661.5	116.9	401,600	1,078,100	3769.8

The water influx values calculated by the Fetkovich method agree fairly closely with those calculated by the van Everdingen and Hurst method used in Ex. 8.3. The Fetkovich method consistently gives water influx values smaller than the values calculated by the van Everdingen and Hurst method for this problem (Fig. 8.14). This result could be because the Fetkovich method does not apply to an aquifer that remains in the transient time flow. It is apparent from observing the values of p_{n-1}, which are the average pressure values in the aquifer, that the pressure in the aquifer is not dropping very fast, which would indicate that the aquifer is very large and that the water flow from it to the reservoir could be transient in nature.

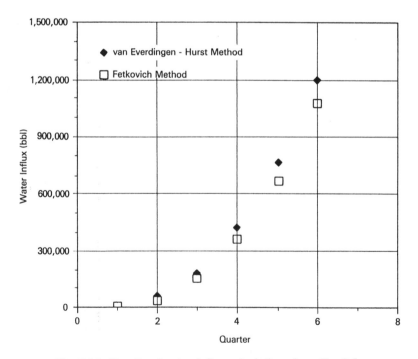

Fig. 8.14. Results of water influx calculations from Ex. 8.5.

PROBLEMS

8.1 Assuming the Schilthuis steady-state water influx model, use the pressure drop history for the Conroe Field given in Fig. 8.15, and a water influx constant, k', of 2170 ft^3/day/psi, to find the cumulative water encroachment at the end of the second and fourth periods by graphical integration for Table 6.1.

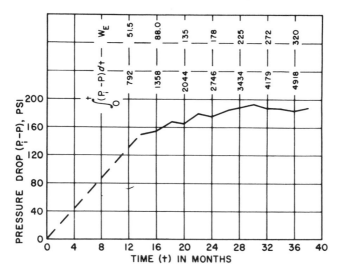

Fig. 8.15. Calculation of quantity of water that has encroached into Conroe Field. (*After* Schilthuis.[5])

8.2 The pressure history for the Peoria Field is given in Fig. 8.16. Between 36 and 48 months, production in the Peoria Field remained substantially constant at 8450 STB/day, at a daily gas-oil ratio of 1052 SCF/STB, and 2550 STB of water per day. The initial solution GOR was 720 SCF/STB. The cumulative produced GOR at 36 months was 830 SCF/STB, and at 48 months it was 920 SCF/STB. The two-phase formation volume factor at 2500 psia was 9.050 ft^3/STB, and the

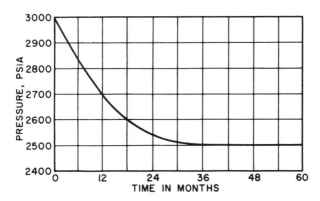

Fig. 8.16. Pressure decline in the Peoria Field.

gas volume factor at the same pressure, 0.00490 ft³/SCF. Calculate the cumulative water influx during the first 36 months.

8.3 During a period of production from a certain reservoir, the average reservoir pressure remained constant at 3200 psia. During the stabilized pressure, the oil and water producing rates were 30,000 STB/day and 5000 STB/day, respectively. Calculate the incremental water influx for a later period when the pressure drops from 3000 to 2800 psia. Assume the following relationship for pressure and time holds:

$$\frac{dp}{dt} = -0.003p, \text{ psia/month}$$

Other data are the following:

$$p_i = 3500 \text{ psia}$$

$$R_{soi} = 750 \text{ SCF/STB}$$

$$B_t = 1.45 \text{ bbl/STB at 3200 psia}$$

$$B_g = 0.002 \text{ bbl/STB at 3200 psia}$$

$$R = 800 \text{ SCF/STB at 3200 psia}$$

$$B_w = 1.04 \text{ bbl/STB at 3200 psia}$$

8.4 The pressure decline in a reservoir from the initial pressure down to a certain pressure, p, was approximately linear at -0.500 psi/day. Assuming the Schilthuis steady-state water influx model and a water influx constant of k', in ft³/day-psia, determine an expression for the water influx as a function of time in bbl.

8.5 An aquifer of 28,850 ac includes a reservoir of 451 ac. The formation has a porosity of 22%, thickness of 60 ft, a compressibility of $4(10)^{-6}$ psi^{-1}, and a permeability of 100 md. The water has a viscosity of 0.30 cp and a compressibility of $3(10)^{-6}$ psi^{-1}. The connate water saturation of the reservoir is 26%, and the reservoir is approximately centered in this closed aquifer. It is exposed to water influx on all of its periphery.

(a) Calculate the effective radii of the aquifer and the reservoir, and their ratio.

(b) Calculate the volume of water the aquifer can supply to the reservoir by rock compaction and water expansion per psi of pressure drop throughout the aquifer.

(c) Calculate the volume of the initial hydrocarbon contents of the reservoir.

(d) Calculate the pressure drop throughout the aquifer required to supply water equivalent to the initial hydrocarbon contents of the reservoir.

(e) Calculate the theoretical time conversion constant for the aquifer.

(f) Calculate the theoretical value of B' for the aquifer.

(g) Calculate the water influx at 100, 200, 400, and 800 days if the reservoir boundary pressure is lowered and maintained at 3450 psia from an initial pressure of 3500 psia.

(h) If the boundary pressure were changed from 3450 psia to 3460 psia after 100 days and maintained there, what would the influx be at 200, 400, and 800 days as measured from the first pressure decrement at time zero?

(i) Calculate the cumulative water influx at 500 days from the following boundary pressure history:

t (days)	0	100	200	300	400	500
p (psia)	3500	3490	3476	3458	3444	3420

(j) Repeat part (i) assuming an infinite aquifer, and again assuming $r_e/r_R = 5.0$

(k) At what time in days do the aquifer limits begin to affect the influx? infinite to finite

(l) From the limiting value of W_{eD} for $r_e/r_R = 8.0$, find the maximum water influx available per psi drop. Compare this result with that calculated in part (b).

8.6 Find the cumulative water influx for the fifth and sixth periods in Ex. 8.3 and Table 8.3.

8.7 The actual pressure history of a reservoir is simulated by the following data, which assume that the pressure at the original oil-water contact is changed instantaneously by a finite amount, Δp.

(a) Use the van Everdingen and Hurst method to calculate the total cumulative water influx.

(b) How much of this water influx occurred in the first two years?

Time in Years	ΔP, psi
0	40
0.5	60
1.0	94
1.5	186
2.0	110
2.5	120
3.0	—

Other reservoir properties include the following:

Reservoir area = 19,600,000 ft^2
Aquifer area = 686,900,000 ft^2
$k = 10.4$ md $\phi = 25\%$
$\mu_w = 1.098$ cp $c_t = 7.01(10)^{-6}$ psi^{-1}
$h = 10$ ft

8.8 An oil reservoir is located between two intersecting faults as shown in the areal view in Fig. 8.17. The reservoir shown is bounded by an aquifer estimated by geologists to have an area of 26,400 ac. Other aquifer data are the following:

$\phi = 0.21$ $k = 275$ md

$h = 30$ ft $c_t = 7(10)^{-6}$ psi^{-1}

$\mu_w = 0.92$ cp

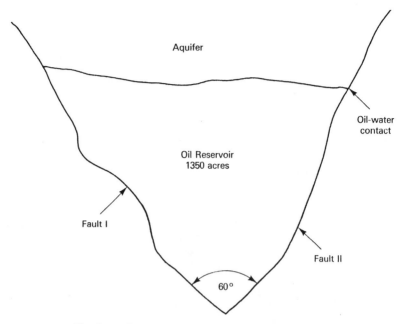

Fig. 8.17. Reservoir between interconnecting faults.

The average reservoir pressure, measured at three-month intervals, is as follows:

Time in Days	p, psia
0	2987
91.3	2962
182.6	2927
273.9	2882
365.2	2837
456.5	2793

Use both the van Everdingen and Hurst and the Fetkovich methods to calculate the water influx that occurred during each of the three-month intervals. Assume that the average reservoir pressure history approximates the oil reservoir–aquifer boundary pressure history.[17]

8.9 For the oil reservoir–aquifer boundary pressure relationship that follows, use the van Everdingen and Hurst method to calculate the cumulative water influx at each quarter (see Fig. 8.18):

$$\phi = 0.20 \qquad k = 200 \text{ md}$$

$$h = 40 \text{ ft} \qquad c_t = 7(10)^{-6} \text{ psi}^{-1}$$

$$\mu_w = 0.80 \text{ cp}$$

Area of oil reservoir = 1000 ac

Area of aquifer = 15,000 ac

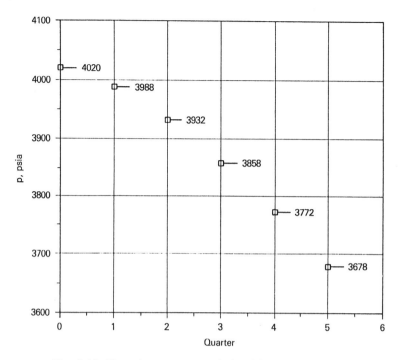

Fig. 8.18. Boundary pressure relationship for Prob. 8.9.

8.10 Repeat Prob. 8.9 using the Fetkovich method, and compare the results with the results of Prob. 8.9.

REFERENCES

[1] W. D. Moore and L. G. Truby, Jr., "Pressure Performance of Five Fields Completed in a Common Aquifer," *Trans.* AIME (1952), **195**, 297.

[2] F. M. Stewart, F. H. Callaway, and R. E. Gladfelter, "Comparison of Methods for Analyzing a Water Drive Field, Torchlight Tensleep Reservoir, Wyoming," *Trans.* AIME (1955), **204**, 197.

[3] D. Havlena and A. S. Odeh, "The Material Balance as an Equation of a Straight Line," *Jour. of Petroleum Technology* (August 1968), 846–900.

[4] D. Havlena and A. S. Odeh, "The Material Balance as an Equation of a Straight Line. Part II—Field Cases," *Jour. of Petroleum Technology* (July 1964), 815–822.

[5] R. J. Schilthuis, "Active Oil and Reservoir Energy," *Trans.* AIME (1936), **118**, 37.

[6] S. J. Pirson, *Elements of Oil Reservoir Engineering,* 2nd ed. New York: McGraw-Hill, 1958, p. 608.

[7] A. F. van Everdingen and W. Hurst, "The Application of the Laplace Transformation to Flow Problems in Reservoirs," *Trans.* AIME (1949), **186**, 305.

[8] A. F. van Everdingen, E. H. Timmerman, and J. J. McMahon, "Application of the Material Balance Equation to a Partial Water-Drive Reservoir," *Trans.* AIME (1953), **198,** 51.

[9] A. T. Chatas, "A Practical Treatment of Nonsteady-State Flow Problems in Reservoir Systems," *Petroleum Engineering* (May 1953), **25,** No. 5, B-42; No. 6, June, p. B-38; No. 8, August, p. B-44.

[10] M. J. Edwardson, et al., "Calculation of Formation Temperature Disturbances Caused by Mud Circulation," *Jour. of Petroleum Technology* (April 1962), 416–25; *Trans.* AIME, 225.

[11] J. R. Fanchi, "Analytical Representation of the van Everdingen-Hurst Influence Functions for Reservoir Simulation," *SPE Jour.* (June 1985), 405–06.

[12] M. A. Klins, A. J. Bouchard, and C. L. Cable, "A Polynomial Approach to the van Everdingen-Hurst Dimensionless Variables for Water Encroachment," SPE *Reservoir Engineering* (Feb. 1988), 320–326.

[13] K. H. Coats, "A Mathematical Model for Water Movement about Bottom-Water-Drive Reservoirs," *SPE Jour.* (March 1962) 44–52; *Trans.* AIME, 225.

[14] D. R. Allard and S. M. Chen, "Calculation of Water Influx for Bottomwater Drive Reservoirs," SPE *Reservoir Engineering* (May 1988), 369–379.

[15] M. J. Fetkovich, "A Simplified Approach to Water Influx Calculations—Finite Aquifer Systems," *Jour. of Petroleum Technology* (July 1971), 814–828.

[16] L. P. Dake, *Fundamentals of Reservoir Engineering,* New York: Elsevier, 1978.

[17] Personal contact with J. T. Smith.

Chapter 9

The Displacement of Oil and Gas

1. INTRODUCTION

This chapter includes a discussion of the displacement of oil and gas both by external flooding processes and by internal displacement processes. It is not meant to be an exhaustive treatise but only an introduction. Several good books cover the material in this chapter more extensively.[*,1-5] The reservoir engineer should be exposed to these concepts because they form the basis for understanding secondary and tertiary flooding techniques as well as some primary recovery mechanisms.

2. RECOVERY EFFICIENCY

The overall recovery efficiency E of any fluid displacement process is given by the product of the macroscopic, or volumetric displacement, efficiency, E_v, and the microscopic displacement efficiency, E_d:

$$E = E_v E_d \tag{9.1}$$

[*] References throughout the text are given at the end of each chapter.

The *macroscopic* displacement efficiency is a measure of how well the displacing fluid has contacted the oil-bearing parts of the reservoir. The *microscopic* displacement efficiency is a measure of how well the displacing fluid mobilizes the residual oil once the fluid has contacted the oil.

The macroscopic displacement efficiency is made up of two other terms: the areal, E_s, sweep efficiency and the vertical, E_i, sweep efficiency.

2.1. Microscopic Displacement Efficiency

The microscopic displacement efficiency is affected by the following factors: interfacial and surface tension forces, wettability, capillary pressure, and relative permeability.

When a drop of one immiscible fluid is immersed in another fluid and comes to rest on a solid surface, the surface area of the drop will take a minimum value owing to the forces acting at the fluid-fluid and rock-fluid interfaces. The forces per unit length acting at the fluid-fluid and rock-fluid interfaces are referred to as *interfacial tensions*. The interfacial tension between two fluids represents the amount of work required to create a new unit of surface area at the interface. The interfacial tension can also be thought of as a measure of the immiscibility of two fluids. Typical values of oil-brine interfacial tensions are on the order of 20 to 30 dynes/cm.

The tendency for a solid to prefer one fluid over another is called *wettability*. Wettability is a function of the chemical composition of both the fluids and the rock. Surfaces can be either oil-wet or water-wet, depending on the chemical composition of the fluids. The degree to which a rock is either oil-wet or water-wet is strongly affected by the absorption or desorption of constituents in the oil phase. Large, polar compounds in the oil phase can absorb onto the solid surface, leaving an oil film that may alter the wettability of the surface.

The concept of wettability leads to another significant factor in the recovery of oil. This factor is *capillary pressure*. To illustrate capillary pressure, consider a capillary tube that contains both oil and brine, the oil having a lower density than the brine. The pressure in the oil phase immediately above the oil-brine interface in the capillary tube will be slightly greater than the pressure in the water phase just below the interface. This difference in pressure is called the capillary pressure, P_c, of the system. The greater pressure will always occur in the nonwetting phase. An expression relating the contact angle, θ, the radius, r_c, of the capillary in feet, the oil-brine interfacial tension, σ_{wo}, in dynes/cm, and the capillary pressure in psi is given by:

$$P_c = \frac{9.519(10)^{-7}\sigma_{wo}\cos\theta}{r_c} \qquad (9.2)$$

This equation suggests that the capillary pressure in a porous medium is a function of the chemical composition of the rock and fluids, the pore size distribution of the sand grains in the rock, and the saturation of the fluids in the pores. Capillary pressures have also been found to be a function of the saturation history, although this dependence is not reflected in Eq. (9.2). For this reason, different values of capillary pressure are obtained during the drainage process (i.e., displacing the wetting phase by the nonwetting phase), then during the imbibition process (i.e., displacing the nonwetting phase with the wetting phase). This historesis phenomenon is exhibited in all rock-fluid systems.

It has been shown that the pressure required to force a nonwetting phase through a small capillary can be very large. For instance, the pressure drop required to force an oil drop through a tapering constriction that has a forward radius of 0.00002 ft, a rearward radius of 0.00005 ft, a contact angle of 0°, and an interfacial tension of 25 dyn/cm is 0.71 psi. If the oil drop were 0.00035 ft long, a pressure gradient of 2029 psi/ft would be required to move the drop through the constriction. Pressure gradients of this magnitude are not realizable in reservoirs. Typical pressure gradients obtained in reservoir systems are of the order of 1 to 2 psi/ft.

Another factor affecting the microscopic displacement efficiency is the fact that when two or more fluid phases are present and flowing, the saturation of one phase affects the permeability of the other(s). The next section discusses in detail the important concept of relative permeability.

2.2. Relative Permeability

Except for gases at low pressures, the permeability of a rock is a property of the rock and not of the fluid that flows through it, provided that the fluid 100% saturates the pore space of the rock. This permeability at 100% saturation of a single fluid is called the *absolute permeability* of the rock. If a core sample 0.00215 ft^2 in cross section and 0.1 ft long flows a 1.0 cp brine with a formation volume factor of 1.0 bbl/STB at the rate of 0.30 STB/day under a 30 psi pressure differential, it has an absolute permeability of

$$k = \frac{q_w B_w \mu_w L}{0.001127 A_c \Delta p} = \frac{0.30(1.0)(0.1)}{0.001127(0.00215)(30)} = 413 \text{ md}$$

If the water is replaced by an oil of 3.0 cp viscosity and 1.2 bbl/STB formation volume factor, under the same pressure differential the flow rate will be 0.0834 STB/day, and again the absolute permeability is

$$k = \frac{q_o B_o \mu_o L}{0.001127 A_c \Delta p} = \frac{0.0834(1.2)(3.0)(0.1)}{0.001127(0.00215)(30)} = 413 \text{ md}$$

If the same core is maintained at 70% water saturation ($S_w = 70\%$) and 30% oil saturation ($S_o = 30\%$), and at these and only these saturations and under the same pressure drop it flows 0.18 STB/day of the brine and 0.01 STB/day of the oil, then the effective permeability to water is

$$k_w = \frac{q_w B_w \mu_w L}{0.001127 A_c \Delta p} = \frac{0.18(1.0)(1.0)(0.1)}{0.001127(0.00215)(30)} = 248 \text{ md}$$

and the effective permeability to oil is

$$k_o = \frac{q_o B_o \mu_o L}{0.001127 A_c \Delta p} = \frac{0.01(1.2)(3.0)(0.1)}{0.001127(0.00215)(30)} = 50 \text{ md}$$

The effective permeability, then, is the permeability of a rock to a particular fluid when that fluid has a pore saturation of less than 100%. As noted in the foregoing example, the sum of the effective permeabilities (i.e., 298 md) is always less than the absolute permeability, 413 md.

When two fluids, such as oil and water, are present, their relative rates of flow are determined by their relative viscosities, their relative formation volume factors, and their relative permeabilities. Relative permeability is the ratio of effective permeability to the absolute permeability. For the previous example, the relative permeabilities to water and to oil are

$$k_{rw} = \frac{k_w}{k} = \frac{248}{413} = 0.60$$

$$k_{ro} = \frac{k_o}{k} = \frac{50}{413} = 0.12$$

The flowing water-oil ratio at reservoir conditions depends on the viscosity ratio and the effective permeability ratio (i.e., on the mobility ratio), or

$$\frac{q_w B_w}{q_o B_o} = \frac{\dfrac{0.001127 k_w A_c \Delta p}{\mu_w L}}{\dfrac{0.001127 k_o A_c \Delta p}{\mu_o L}} = \frac{k_w/\mu_w}{k_o/\mu_o} = \frac{\lambda_w}{\lambda_o} = M$$

For the previous example

$$\frac{q_w B_w}{q_o B_o} = \frac{k_w/\mu_w}{k_o/\mu_o} = \frac{248/1.0}{50/3.0} = 14.9$$

At 70% water saturation and 30% oil saturation, the water is flowing at 14.9 times the oil rate. Relative permeabilities may be substituted for effective permeabilities in the previous calculation because the relative permeability ratio, k_{rw}/k_{ro}, equals the effective permeability ratio, k_w/k_o. The term *relative permeability ratio* is more commonly used. For the previous example:

$$\frac{k_{rw}}{k_{ro}} = \frac{k_w/k}{k_o/k} = \frac{k_w}{k_o} = \frac{248}{50} = \frac{0.60}{0.12} = 5$$

Water flows at 14.9 times the oil rate because of a viscosity ratio of 3 and a relative permeability ratio of 5, both of which favor the water flow. Although the relative permeability ratio varies with the water-oil saturation ratio, in this example 70/30, or 2.33, the relationship is unfortunately far from one of simple proportionality.

Figure 9.1 shows a typical plot of oil and water relative permeability curves for a particular rock as a function of water saturation. Starting at 100% water saturation, the curves show that a decrease in water saturation to 85% (a 15% increase in oil saturation) sharply reduces the relative permeability to water from 100% down to 60%, and at 15% oil saturation the relative permeability to oil is essentially zero. This value of oil saturation, 15% in this case, is called the *critical saturation,* the saturation at which oil first begins to flow as the oil saturation increases. It is also called the *residual saturation,* the value below which the oil saturation cannot be reduced in an oil-water system. This explains why oil recovery by water drive is not 100% efficient. If the initial

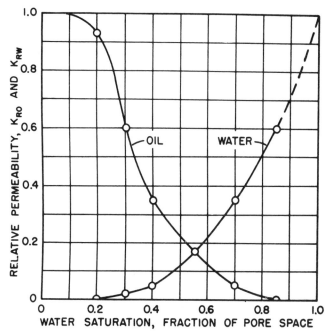

Fig. 9.1. Water-oil relative permeability curves.

connate water saturation is 20% for this particular rock, then the recovery from the portion of the reservoir invaded by high-pressure water influx is

$$\text{Recovery} = \frac{\text{initial-final}}{\text{initial}} = \frac{0.80 - 0.15}{0.80} = 81\%$$

Experiments show that essentially the same relative permeability curves are obtained for a gas-water system as for the oil-water system, which also means that the critical, or residual, gas saturation will be the same. Furthermore, it has been found that if both oil and free gas are present, the residual *hydrocarbon* saturation (oil and gas) will be about the same, in this case 15%. Suppose, then, that the rock is invaded by water at a pressure below saturation pressure so that free gas is present. If, for example, the residual free gas saturation behind the flood front is 10%, then the oil saturation is 5%, and neglecting small changes in the formation volume factors of the oil, the recovery is increased to:

$$\text{Recovery} = \frac{0.80 - 0.05}{0.80} = 94\%$$

Returning to Fig. 9.1, as the water saturation decreases further, the relative permeability to water continues to decrease and the relative permeability to oil increases. At 20% water saturation, the (connate) water is immobile, and the relative permeability to oil is quite high. This explains why some rocks may contain as much as 50% connate water and yet produce water-free oil. Most reservoir rocks are preferentially water wet—that is, the water phase and not the oil phase is next to the walls of the pore spaces. Because of this, at 20% water saturation the *water* occupies the *least favorable* portions of the pore spaces—that is, as thin layers about the sand grains, as thin layers on the walls of the pore cavities, and in the smaller crevices and capillaries. The *oil*, which occupies 80% of the pore space, is in the *most favorable* portions of the pore spaces, which is indicated by a relative permeability of 93%. The curves further indicate that about 10% of the pore spaces contribute nothing to the permeability, for at 10% water saturation, the relative permeability to oil is essentially 100%. Conversely, on the other end of the curves, 15% of the pore spaces contribute 40% of the permeability, for an increase in oil saturation from zero to 15% reduces the relative permeability to water from 100% to 60%.

In describing two-phase flow mathematically, it is always the relative permeability ratio that enters the equations. Figure 9.2 is a plot of the relative permeability ratio, k_o/k_w, versus water saturation for the same data of Fig. 9.1. Because of the wide range of k_o/k_w values, the relative permeability ratio is usually plotted on the log scale of semilog paper. Like many relative permeability ratio curves, the central or main portion of the curve is quite linear. As a straight line on semilog paper, the relative permeability ratio may be expressed as a function of the water saturation by:

$$\frac{k_o}{k_w} = ae^{-bS_w} \tag{9.3}$$

The constants a and b may be determined from the graph shown in Fig. 9.2, or they may be determined from simultaneous equations. At $S_w = 0.30$, $k_o/k_w = 25$, and at $S_w = 0.70$, $k_o/k_w = 0.14$. Then

$$25 = ae^{-0.30b} \qquad \text{and} \qquad 0.14 = ae^{-0.70b}$$

Solving simultaneously, the intercept $a = 1220$, and the slope $b = 13.0$. Equation (9.3) indicates that the relative permeability ratio for a rock is a function of only the relative saturations of the fluids present. Although it is true that the viscosities, the interfacial tensions, and other factors have some effect on

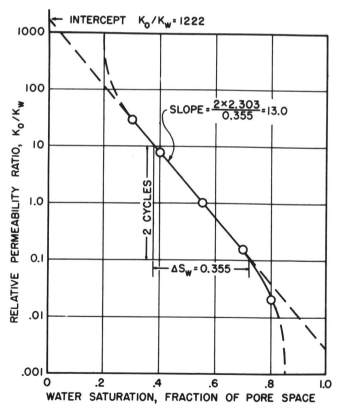

Fig. 9.2. Semilog plot of relative permeability ratio versus saturation.

the relative permeability ratio, for a given rock it is mainly a function of the fluid saturations.

In many rocks there is a transition zone between the water and the oil zones. In the true water zone, the water saturation is essentially 100%, although in some reservoirs a small oil saturation may be found a considerable distance vertically below the oil-water contact. In the oil zone, there is usually present connate water, which is essentially immobile. For the present example, the connate water saturation is 20% and the oil saturation is 80%. Only water will be produced from a well completed in the true water zone, and only oil will be produced from the true oil zone. In the transition zone (Figure 9.3), both oil and water will be produced, and the fraction that is water will depend on the oil and water saturations at the point of completion. If the well in Fig. 9.3 is completed in a uniform sand at a point where $S_o = 60\%$ and $S_w = 40\%$, the fraction of water or water cut may be calculated using Eq. (7.19):

$$q_w B_w = \frac{0.00708 k_w h (p_e - p_w)}{\mu_w \ln(r_e/r_w)}$$

$$q_o B_o = \frac{0.00708 k_o h (p_e - p_w)}{\mu_o \ln(r_e/r_w)}$$

Since water cut, f_w, is defined as

$$f_w = \frac{q_w B_w}{q_w B_w + q_o B_o} \tag{9.4}$$

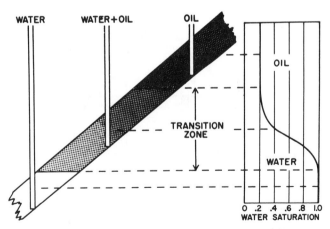

Fig. 9.3. Sketch showing the variation in oil and water saturations in the transition zone.

Combining these equations and cancelling common terms,

$$f_w = \frac{k_w/\mu_w}{k_w/\mu_w + k_o/\mu_o}$$

$$f_w = \frac{1}{1 + \dfrac{k_o \mu_w}{k_w \mu_o}} = \frac{1}{1 + \dfrac{k_{ro} \mu_w}{k_{rw} \mu_o}} \qquad (9.5)$$

Since k_o/k_w is a function of saturation, Eq. (9.3) may be substituted in Eq. (9.5) to give

$$f_w = \frac{1}{1 + \left(\dfrac{\mu_w}{\mu_o}\right) ae^{-bS_w}} \qquad (9.6)$$

Either Eq. (9.5) or (9.6) can be used with the data of Fig. 9.1 or 9.2, respectively, and with viscosity data to calculate the water cut. From Fig. 9.1, at $S_w = 0.40$, $k_{rw} = 0.045$ and $k_{ro} = 0.36$. If $\mu_w = 1.0$ cp and $\mu_o = 3.0$ cp,

$$f_w = \frac{1}{1 + \dfrac{k_{ro} \mu_w}{k_{rw} \mu_o}} = \frac{1}{1 + \dfrac{0.36(1.0)}{0.045(3.0)}} = 0.27$$

If the calculations for the water-cut are repeated at several water saturations, and then the calculated values plotted versus water saturation, Fig. 9.4 will be the result. This plot is referred to as the *fractional flow curve*. The curve shows that the fractional flow of water ranges from 0 (for $S_w \leq$ the connate water saturation) to 1 (for $S_w \geq$ one minus the residual oil saturation).

2.3. Macroscopic Displacement Efficiency

The following factors affect the macroscopic displacement efficiency: heterogeneities and anisotropy, mobility of the displacing fluids compared with the mobility of the displaced fluids, the physical arrangement of injection and production wells, and the type of rock matrix in which the oil or gas exists.

Heterogeneities and anisotropy of a hydrocarbon-bearing formation have a significant effect on the macroscopic displacement efficiency. The movement of fluids through the reservoir will not be uniform if there are large variations in such properties as porosity, permeability, and clay cement. Limestone formations generally have wide fluctuations in porosity and permeability. Also, many formations have a system of microfractures or large macrofractures. Any time a fracture occurs in a reservoir, fluids will try to travel through the fracture because of the high permeability of the fracture, which may lead to substantial bypassing of hydrocarbon.

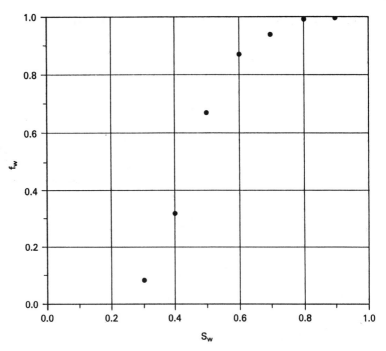

Fig. 9.4. Fractional flow curve for the relative permeability data of Fig. 9.1.

Many producing zones are variable in permeability, both vertically and horizontally, leading to reduced vertical, E_i, and areal, E_s, sweep efficiencies. Zones or strata of higher or lower permeability often exhibit lateral continuity throughout a reservoir or a portion thereof. Where such permeability stratification exists, the displacing water sweeps faster through the more permeable zones so that much of the oil in the less permeable zones must be produced over a long period of time at high water-oil ratios. The situation is the same whether the water comes from natural influx or from injection systems, commonly called *pressure maintenance* when the reservoir pressure is fairly high, and *secondary recovery* when the reservoir pressure is low or depleted.

The areal sweep efficiency is also affected by the type of flow geometry in a reservoir system. As an example, linear displacement occurs in uniform beds of constant cross section where the entire input and outflow ends are open to flow. Under these conditions, the flood front advances as a plane (neglecting gravitational forces), and when it breaks through at the producing end, the sweep efficiency is 100%—that is, 100% of the bed volume has been contacted by the displacing fluid. If the displacing and displaced fluids are injected into and produced from wells located at the input and outflow ends of a uniform linear bed, such as the direct line-drive pattern arrangement shown in Fig. 9.5(a), the flood front is not a plane, and at breakthrough the sweep efficiency is far from 100%, as shown in Figure 9.5(b).

Mobility is a relative measure of how easily a fluid moves thorugh porous media. The apparent mobility has been defined as the ratio of effective permeability to fluid viscosity. Since the effective permeability is a function of fluid saturations, several apparent mobilities can be defined. When a fluid is being injected into a porous medium containing both the injected fluid and a second fluid, the apparent mobility of the displacing phase is usually measured at the average displacing phase saturation when the displacing phase just begins to break through at the production site. The apparent mobility of the non-displacing phase is measured at the displacing phase saturation that occurs just before the beginning of the injection of the displacing phase. Areal sweep efficiencies are a strong function of the mobility ratio. A phenomenon called *viscous fingering* can take place if the mobility of the displacing phase is much greater than the mobility of the displaced phase. Viscous fingering simply refers to the penetration of the much more mobile displacing phase into the phase that is being displaced.

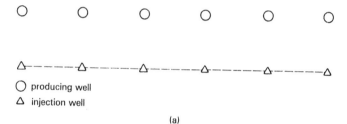

○ producing well
△ injection well

(a)

Fig. 9.5(a). Direct-line-drive flooding network.

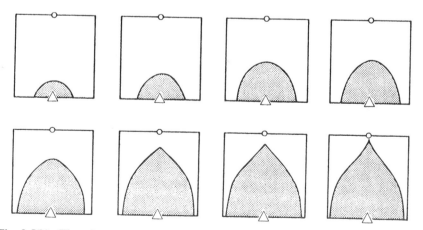

Fig. 9.5(b). The photographic history of a direct-line-drive fluid-injection system, under steady-state conditions, as obtained with a blotting-paper electrolytic model. (*After* Wyckoff, Botset, and Muskat.[6])

Viscous fingering simply refers to the penetration of the much more mobile displacing phase into the phase that is being displaced.

Figure 9.6(b) shows the effect of mobility ratio on areal sweep efficiency at initial breakthrough for a five-spot network (shown in Fig. 9.6(a) obtained using the X-ray shadowgraph. The pattern at breakthrough for a mobility ratio of 1 obtained with an electrolytic model is included for comparison.

The arrangement of injection and production wells depends primarily on the geology of the formation and the size (areal extent) of the reservoir. For a given reservoir, an operator has the option of using the existing well arrange-

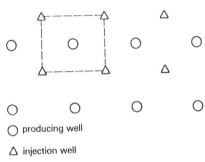

O producing well

△ injection well

Fig. 9.6(a). Five-spot flooding network.

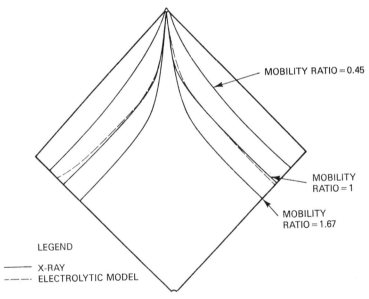

MOBILITY RATIO = 0.45

MOBILITY RATIO = 1

MOBILITY RATIO = 1.67

LEGEND

———— X-RAY
– – – – ELECTROLYTIC MODEL

Fig. 9.6(b). X-ray shadowgraph studies showing the effect of mobility ratio on areal sweep efficiency at breakthrough. (*After* Slobod and Caudle.[7])

ment or drilling new wells in other locations. If the operator opts to use the existing well arrangement, there may be a need to consider converting production wells to injection wells, or vice versa. An operator should also recognize that, when a production well is converted to an injection well, the production capacity of the reservoir will have been reduced. This decision can often lead to major cost items in an overall project and should be given a great deal of consideration. Knowledge of any directional permeability effects and other heterogeneities can aid in the consideration of well arrangements. The presence of faults, fractures, and high-permeability streaks can dictate the shutting in of a well near one of these heterogeneities. Directional permeability trends can lead to a poor sweep efficiency in a developed pattern and can suggest that the pattern be altered in one direction or that a different pattern be used.

Sandstone formations are characterized by a more uniform pore geometry than limestone formations. Limestone formations have large holes (vugs) and can have significant fractures that are often connected. Limestone formations are associated with connate water that can have high levels of divalent ions such as Ca^{2+} and Mg^{2+}. Vugular porosity and high divalent ion content in their connate waters hinders the application of injection processes in limestone reservoirs. A sandstone formation can be composed of such small sand grains and be so tightly packed that fluids will not readily flow through the formation.

3. IMMISCIBLE DISPLACEMENT PROCESSES *Skipped*

3.1. The Buckley-Leverett Displacement Mechanism

Oil is displaced from a rock by water somewhat as fluid is displaced from a cylinder by a leaky piston. Buckley and Leverett developed a theory of displacement based on the relative permeability concept.[8] Their theory is presented here.

Consider a linear bed containing oil and water (Fig. 9.7). Let the total throughput, $q_t' = q_w B_w + q_o B_o$ in reservoir barrels be the same at all cross

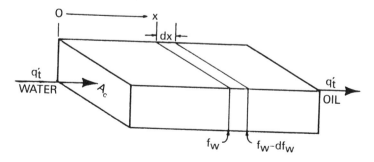

Fig. 9.7.

sections. For the present, we will neglect gravitational and capillary forces that may be acting. Let S_w be the water saturation in any element at time t (days). Then if oil is being displaced from the element, at time $(t + dt)$ the water saturation will be $(S_w + dS_w)$. If ϕ is the total porosity fraction, A_c is the cross section in square feet, and dx is the thickness of the element in feet, then the rate of increase of water in the element at time t in barrels per day is

$$\frac{dW}{dt} = \frac{\phi A_c \, dx}{5.615} \left(\frac{\partial S_w}{\partial t}\right)_x \tag{9.7}$$

The subscript x on the derivative indicates that this derivative is different for each element. If f_w is the fraction of water in the total flow of q_t' barrels per day, then $f_w q_t'$ is the rate of water entering the left-hand face of the element dx. The oil saturation will be slightly higher at the right-hand face, so the fraction of water flowing there will be slightly less, or $f_w - df_w$. Then the rate of water leaving the element is $(f_w - df_w)q_t'$. The net rate of gain of water in the element at any time then is

$$\frac{dW}{dt} = (f_w - df_w)q_t' - f_w q_t' = -q_t' df_w \tag{9.8}$$

Equating (9.7) and (9.8),

$$\left(\frac{\partial S_w}{\partial t}\right)_x = -\frac{5.615 q_t'}{\phi A_c} \left(\frac{\partial f_w}{\partial x}\right)_t \tag{9.9}$$

Now for a given rock, the fraction of water f_w is a function only of the water saturation S_w, as indicated by Eq. (9.5), assuming constant oil and water viscosities. The water saturation, however, is a function of both time and position, x, which may be expressed as $f_w = F(S_w)$ and $S_w = G(t, x)$. Then

$$dS_w = \left(\frac{\partial S_w}{\partial t}\right)_x dt + \left(\frac{\partial S_w}{\partial x}\right)_t dx \tag{9.10}$$

Now, there is interest in determining the rate of advance of a constant saturation plane, or front, $(\partial x/\partial t)_{S_w}$, i.e., where S_w is constant. Then, from Eq. (9.10)

$$\left(\frac{\partial x}{\partial t}\right)_{S_w} = -\frac{(\partial S_w/\partial t)_x}{(\partial S_w/\partial x)_t} \tag{9.11}$$

Substituting Eq. (9.9) in Eq. (9.11),

$$\left(\frac{\partial x}{\partial t}\right)_{S_w} = \frac{5.615 q_t'}{\phi A_c} \frac{(\partial f_w/\partial x)_t}{(\partial S_w/\partial x)_t} \tag{9.12}$$

But

$$\frac{(\partial f_w/\partial x)_t}{(\partial S_w/\partial x)_t} = \left(\frac{\partial f_w}{\partial S_w}\right)_t \tag{9.13}$$

Eq. (9.12) then becomes

$$\left(\frac{\partial x}{\partial t}\right)_{S_w} = \frac{5.615 q_t'}{\phi A_c} \left(\frac{\partial f_w}{\partial S_w}\right)_t \tag{9.14}$$

Because the porosity, area, and throughput are constant and because for any value of S_w, the derivative $\partial f_w/\partial S_w$ is a constant, then the rate dx/dt is constant. This means that the distance a plane of constant saturation, S_w, advances is directly proportional to time and to the value of the derivative $(\partial f_w/\partial S_w)$ at that saturation, or

$$x = \frac{5.615 q_t' t}{\phi A_c} \left(\frac{\partial f_w}{\partial S_w}\right)_{S_w} \tag{9.15}$$

We now apply Eq. (9.15) to a reservoir under active water drive where the walls are located in uniform rows along the strike on 40 ac spacing, as shown in Fig. 9.8. This gives rise to approximate linear flow, and if the daily production of each of the three wells located along the dip is 200 STB of oil per day, then for an active water drive and an oil volume factor of 1.50 bbl/STB, the total reservoir throughput, q_t', will be 900 bbl/day.

The cross-sectional area is the product of the width, 1320 ft, and the true formation thickness, 20 ft, so that for a porosity of 25%, Eq. (9.15) becomes

$$x = \frac{5.615 \times 900 \times t}{0.25 \times 1320 \times 20} \left(\frac{\partial f_w}{\partial S_w}\right)_{S_w}$$

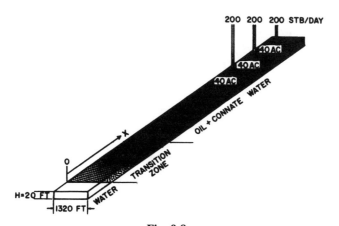

Fig. 9.8.

If we let $x = 0$ at the bottom of the transition zone, as indicated in Fig. 9.8, then the distances the various constant water saturation planes will travel in, say, 60, 120, and 240 days are given by:

$$x_{60} = 46 \ (\partial f_w/\partial S_w)_{S_w}$$

$$x_{120} = 92 \ (\partial f_w/\partial S_w)_{S_w}$$

$$x_{240} = 184 \ (\partial f_w/\partial S_w)_{S_w} \qquad (9.16)$$

The value of the derivative $(\partial f_w/\partial S_w)$ may be obtained for any value of water saturation, S_w, by plotting f_w from Eq. (9.5) versus S_w and graphically taking the slopes at values of S_w. This is shown in Fig. 9.9 at 40% water saturation using the relative permeability ratio data of Table 9.1 and a water-oil viscosity ratio of 0.50. For example, at $S_w = 0.40$, where $k_o/k_w = 5.50$ (Table 9.1),

$$f_w = \frac{1}{1 + 0.50 \times 5.50} = 0.267$$

The slope taken graphically at $S_w = 0.40$ and $f_w = 0.267$ is 2.25, as shown in Fig. 9.9.

The derivative $(\partial f_w/\partial S_w)$ may also be obtained mathematically using

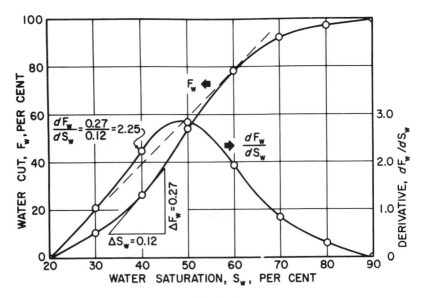

Fig. 9.9.

TABLE 9.1.
Buckley-Leverett frontal advance calculations

(1)	(2)	(3)	(4)	(5)	(6)	(7)
S_w	$\dfrac{k_o}{k_w}$	$\dfrac{\mu_w}{\mu_o}\dfrac{f_w}{}=0.50$ Eq. (9.5)	$\dfrac{\partial f_w}{\partial S_w}$ Eq. (9.17)	$46\dfrac{\partial f_w}{\partial S_w}$ (60 days) Eq. (9.16)	$92\dfrac{\partial f_w}{\partial S_w}$ (120 days) Eq. (9.16)	$184\dfrac{\partial f_w}{\partial S_w}$ (240 days) Eq. (9.16)
0.20	inf.	0.000	0.00	0	0	0
0.30	17.0	0.105	1.08	50	100	200
0.40	5.50	0.267	2.25	104	208	416
0.50	1.70	0.541	2.86	131	262	524
0.60	0.55	0.784	1.95	89	179	358
0.70	0.17	0.922	0.83	38	76	153
0.80	0.0055	0.973	0.30	14	28	55
0.90	0.000	1.000	0.00	0	0	0

Eq. (9.3) to represent the relationship between the relative permeability ratio and the water saturation. Differentiating Eq. (9.16), we obtain

$$\frac{\partial f_w}{\partial S_w} = \frac{(\mu_w/\mu_o)bae^{-bS_w}}{[1 + (\mu_w/\mu_o)ae^{-bS_w}]^2} = \frac{(\mu_w/\mu_o)b(k_o/k_w)}{[1 + (\mu_w/\mu_o)(k_o/k_w)]^2} \tag{9.17}$$

For the k_o/k_w data of Table 9.1, $a = 540$ and $b = 11.5$. Then, at $S_w = 0.40$, for example, by Eq. (9.17).

$$\frac{\partial f_w}{\partial S_w} = \frac{0.50 \times 11.5 \times 5.50}{[1 + 0.50 \times 5.50]^2} = 2.25$$

Figure 9.9 shows the fractional water cut, f_w, and also the derivative $(\partial f_w/\partial S_w)$ plotted against water saturation from the data of Table 9.1. Equation (9.17) was used to determine the values of the derivative. Since Eq. (9.3) does not hold for the very high and for the quite low water saturation ranges, some error is introduced below 30% and above 80% water saturation. Since these are in the regions of the lower values of the derivatives, the overall effect on the calculation is small.

The lowermost curve of Fig. 9.10 represents the initial distribution of water and oil in the linear sand body of Fig. 9.8. Above the transition zone, the connate water saturation is constant at 20%. Equation (9.16) may be used with the values of the derivatives, calculated in Table 9.1 and plotted in Fig. 9.9, to construct the *frontal advance* curves shown in Fig. 9.10 at 60, 120, and 240 days. For example, at 50% water saturation, the value of the derivative is

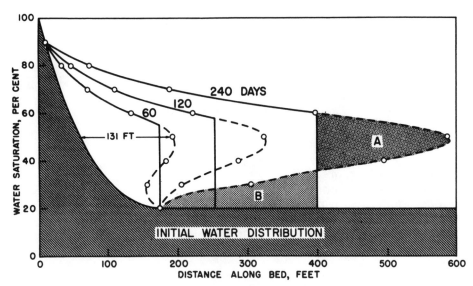

Fig. 9.10. Fluid distributions at initial conditions and at 60, 120, and 240 days.

2.86; so by Eq. (9.16), at 60 days the 50% water saturation plane, or front, will advance a distance of:

$$x = 46 \left(\frac{\partial f_w}{\partial S_w} \right)_{S_w} = 46 \times 2.86 = 131 \text{ feet}$$

This distance is plotted as shown in Fig. 9.10 along with the other distances that have been calculated in Table 9.1 for the other time values and other water saturations. These curves are characteristically double-valued or triple-valued. For example, Fig. 9.10 indicates that the water saturation after 240 days at 400 ft is 20, 36, and 60%. The saturation can be only one value at any place and time. The difficulty is resolved by dropping perpendiculars so that the areas to the right (A) equal the areas to the left (B), as shown in Fig. 9.10.

Figure 9.11 represents the initial water and oil distributions in the reservoir unit and also the distributions after 240 days, provided the flood front has not reached the lowermost well. The area to the right of the *flood front* in Fig. 9.11 is commonly called the *oil bank* and the area to the left is sometimes called the *drag zone*. The area above the 240-day curve and below the 90% water saturation curve represents oil that may yet be recovered, or *dragged* out of the high-water saturation portion of the reservoir by flowing large volumes of water through it. The area above the 90% water saturation represents unrecoverable oil since the critical oil saturation is 10%.

This presentation of the displacement mechanism has assumed that capillary and gravitational forces are negligible. These two forces account for the

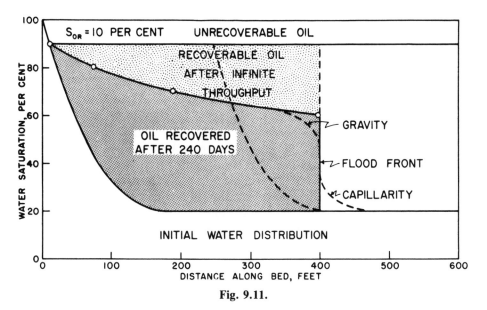

Fig. 9.11.

initial distribution cf oil and water in the reservoir unit, and they also act to modify the sharp flood front in the manner indicated in Fig. 9.11. If production ceases after 240 days, the oil-water distribution will approach one similar to the initial distribution, as shown by the dashed curve in Fig. 9.11.

Figure 9.11 also indicates that a well in this reservoir unit will produce water-free oil until the flood front approaches the well. Thereafter, in a relatively short period, the water cut will rise sharply and be followed by a relatively long period of production at high, and increasingly higher, water cuts. For example, just behind the flood front at 240 days, the water saturation rises from 20% to about 60%—that is, the water cut rises from zero to 78.4% (see Table 9.1). When a producing formation consists of two or more rather definite strata, or stringers, of different permeabilities, the rates of advance in the separate strata will be proportional to their permeabilities, and the overall effect will be a combination of several separate displacements, such as described for a single homogeneous stratum.

3.2. The Displacement of Oil by Gas, with and without Gravitational Segregation

The method discussed in the previous section also applies to the displacement of oil by gas drive. The treatment of oil displacement ᴜy gas in this section considers only gravity drainage along dip. Richardson and Blackwell showed that in some cases there can be a significant vertical component of drainage.[9]

Due to the high oil-gas viscosity ratios and the high gas-oil relative permeability ratios at low gas saturations, the displacement efficiency by gas is generally much lower than that by water, unless the gas displacement is accompanied by substantial gravitational segregation. This is basically the same reason for the low recoveries from reservoirs produced under the dissolved gas-drive mechanism. The effect of gravitational segregation in water-drive oil reservoirs is usually of much less concern because of the higher displacement efficiencies and the lower oil-water density differences, whereas the converse is generally true for gas-oil systems. Welge showed that capillary forces may generally be neglected in both, and he introduced a gravitational term in Eq. (9.5), as will be shown in the following equations. [10] As with water displacement, a linear system is assumed, and a constant gas pressure throughout the system is also assumed so that a constant throughput rate may be used. These assumptions also allow us to eliminate changes caused by gas density, oil density, oil volume factor, and the like. Equation (7.1) may be applied to both the oil and gas flow, assuming the connate water is essentially immobile, so that the fraction of the flowing reservoir fluid volume, which is gas, is

$$f_g = \frac{v_g}{v_t} = -\frac{0.001127k_g}{\mu_q v_t}\left[\left(\frac{dp}{dx}\right)_g - 0.00694 \ \rho_g \cos \alpha\right] \quad (9.18)$$

The total velocity is v_t, which is the total throughput rate q_t' divided by the cross-sectional area A_c. The reservoir gas density, ρ_g, is in lb_m/ft^3. When capillary forces are neglected, as they are in this application, the pressure gradients in the oil and gas phases are equal. Equation (7.1) may be solved for the pressure gradient by applying it to the oil phase, or

$$\left(\frac{dp}{dx}\right)_o = \left(\frac{dp}{dx}\right)_g = -\frac{\mu_o v_o}{0.001127k_o} + 0.00694 \ \rho_o \cos \alpha \quad (9.19)$$

Substituting the pressure gradient of Eq. (9.19) in Eq. (9.18),

$$f_g = -\frac{0.001127k_g}{\mu_g v_t}\left[-\frac{\mu_o v_o}{0.001127k_o} + 0.00694 \ (\rho_o - \rho_g) \cos \alpha\right] \quad (9.20)$$

Expanding and multiplying through by $(k_o/k_g)(\mu_g/\mu_o)$,

$$f_g\left[\frac{k_o\mu_g}{k_g\mu_o}\right] = \frac{v_o}{v_t} - \frac{7.821(10^{-6})k_o(\rho_o - \rho_g)\cos \alpha}{\mu_o v_t} \quad (9.21)$$

But v_o/v_t is the fraction of oil flowing, which equals 1 minus the gas flowing, $(1 - f_g)$. Then, finally,

$$f_g = \frac{1 - \left[\dfrac{7.821(10^{-6})k_o(\rho_o - \rho_g)\cos\alpha}{\mu_o v_t}\right]}{1 + \dfrac{k_o\mu_g}{k_g\mu_o}} \qquad (9.22)$$

The relative permeability ratio (k_{ro}/k_{rg}) may be used for the effective permeability ratio in the denominator of Eq. (9.22); however, the permeability to oil, k_o, in the numerator is the effective permeability and cannot be replaced by the relative permeability. It may, however, be replaced with $(k_{ro}k)$, where k is the absolute permeability. The total velocity, v_t, is the total throughput rate, q_t', divided by the cross-sectional area, A_c. Inserting these equivalents, the fractional gas flow equation with gravitational segregation becomes

$$f_g = \frac{1 - \left[\dfrac{7.821(10^{-6})kA_c(\rho_o - \rho_g)\cos\alpha}{\mu_o}\right]\left(\dfrac{k_{ro}}{q_t'}\right)}{1 + \dfrac{k_o\mu_g}{k_g\mu_o}} \qquad (9.23)$$

If the gravitational forces are small, Eq. (9.23) reduces to the same type of fractional flow equation as Eq. (9.5), or

$$f_g = \frac{1}{1 + \dfrac{k_o\mu_g}{k_g\mu_o}} \qquad (9.24)$$

Although Eq. (9.24) is not rate sensitive (i.e., it does not depend on the throughput rate), Eq. (9.23) includes the throughput velocity q_t'/A_c, and is therefore rate sensitive. Since the total throughput rate, q_t', is in the denominator of the gravitational term of Eq. (9.23), rapid displacement (i.e., large (q_t'/A_c)) reduces the size of the gravitational term, and so causes an increase in the fraction of gas flowing, f_g. A large value of f_g implies low displacement efficiency. If the gravitational term is sufficiently large, f_g becomes zero, or even negative, which indicates counter-current flow of gas updip and oil downdip, resulting in maximum displacement efficiency. In the case of a gas cap that overlies most of an oil zone, the drainage is vertical, and $\cos\alpha = 1.00$; in addition, the cross-sectional area is large. If the vertical *effective* permeability k_o is not reduced to a very low level by low permeability strata, gravitational drainage will substantially improve recovery.

The use of Eq. (9.23) is illustrated using the data given by Welge for the Mile Six Pool, Peru, where advantage was taken of good gravitational segregation characteristics to improve recovery.[10] Pressure maintenance by gas injection has been practiced since 1933 by returning produced gas and other gas to the gas cap so that reservoir pressure has been maintained within 200 psi of its

initial value. Figure 9.12 shows the average relative permeability characteristics of the Mile Six Pool reservoir rock. As is common in gas-oil systems, the
saturations are expressed in percentages of the *hydrocarbon* porosity, and the
connate water, being immobile, is considered as part of the rock. The other
pertinent reservoir rock and fluid data are given in Table 9.2. Substituting
these data in Eq. (9.25),

$$f_g = \frac{1 - \left[\dfrac{7.821(10^{-6})(300)(1.237(10)^6)(48.7 - 5.)\cos 72.5°}{1.32}\right]\left(\dfrac{k_{ro}}{11,600}\right)}{1 + \dfrac{k_o(0.0134)}{k_g(1.32)}}$$

$$f_g = \frac{1 - 2.50 k_{ro}}{1 + 0.0102\left(\dfrac{k_o}{k_g}\right)} \tag{9.25}$$

The values of f_g have been calculated in Table 9.2 for three conditions:
(a) assuming negligible gravitational segregation by using Eq. (9.24); (b) using
the gravitational term equal to 2.50 k_{ro} for the Mile Six Pool, Eq. (9.25); and
(c) assuming the gravitational term equals 1.25 k_{ro}, or half the value at Mile
Six Pool. The values of f_g for these three conditions are shown plotted in Fig.
9.13. The negative values of f_g for the conditions that existed in the Mile Six
Pool indicate counter-current gas flow (i.e., gas updip and oil downdip) in the
range of gas saturations between an assumed critical gas saturation of 5% and
about 17%.

The distance of advance of any gas saturation plane may be calculated

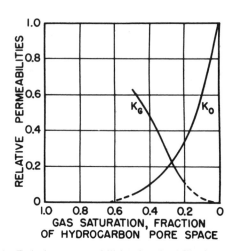

Fig. 9.12. Relative permeabilities for the Mile Six Pool, Peru.

TABLE 9.2.
Mile Six Pool Reservoir data and calculations

Avg. Absolute permeability = 300 md
Avg. hydrocarbon porosity = 0.1625
Avg. connate water = 0.35
Avg. dip angle = 17° 30' (α = 90° – 17°30')
Ave. cross-sectional area = 1,237,000 sq ft
Reservoir oil viscosity = 1.32 cp
Reservoir gas viscosity = 0.0134 cp

Reservoir oil sp. gr. = 0.78 (water = 1)
Reservoir gas sp. gr. = 0.08 (water = 1)
Reservoir temperature = 114°F
Average reservoir pressure = 850 psia
Average throughput = 11,600 reservoir bbl per day
Oil volume factor = 1.25 bbl/STB
Solution gas at 850 psia = 400 SCF/STB
Gas deviation factor = 0.74

S_g	0.05	0.10	0.15	0.20	0.25	0.30	0.35	0.40	0.45	0.50	0.55	0.60
k_o/k_g	inf.	38	8.80	3.10	1.40	0.72	0.364	0.210	0.118	0.072	0.024	0.00
Gravity term = 0												
f_g	0	0.720	0.918	0.969	0.986	0.993	0.996	0.998	0.990	1.00	1.00	1.00
$\partial f_g / \partial S_g$		7.40	1.20	0.60	0.30							
$x = 32\, \partial f_g / \partial S_g$		237	38	19	10							
Gravity term = $2.50 \times k_{ro}$												
k_{ro}	0.77	0.59	0.44	0.34	0.26	0.19	0.14	0.10	0.065	0.040	0.018	0.00
$2.50 \times k_{ro}$	1.92	1.48	1.10	0.85	0.65	0.48	0.35	0.25	0.160	0.10	0.045	0.00
$1 - 2.5\, k_{ro}$	-0.92	-0.48	-0.10	0.15	0.35	0.52	0.65	0.75	0.84	0.90	0.955	1.00
f_g	0	-0.29	-0.092	0.145	0.345	0.516	0.647	0.749	0.840	0.900	0.955	1.00
$\partial f_g / \partial S_g$		3.30	4.40	4.30	3.60	3.00	2.50	1.95	1.60	1.20	0.80	
$32\, \partial f_g / \partial S_g$		106	141	138	115	96	80	62	51	38	26	
Gravity term = $1.25 \times k_{ro}$												
$1.25\, k_{ro}$	0.96	0.74	0.55	0.425	0.325	0.240	0.175	0.125	0.080	0.050	0.023	0.00
$1 - 1.25\, k_{ro}$	0.04	0.26	0.45	0.575	0.675	0.760	0.825	0.875	0.920	0.950	0.977	1.00
f_g		0.190	0.413	0.557	0.666	0.755	0.822	0.920	0.920	0.950	0.977	1.00
$\partial f_g / \partial S_g$		4.00	3.60	2.40	1.90	1.50	1.20	1.00	0.80	0.60		
$32\, \partial f_g / \partial S_g$		128	115	77	61	48	38	32	26	19		

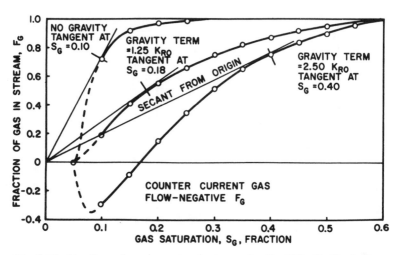

Fig. 9.13. Fraction of gas in reservoir stream for the Mile Six Pool, Peru.

for the Mile Six Pool, using Eq. (9.15), replacing water as the displacing fluid by gas, or

$$x = \frac{5.615 \, q_t' t}{\phi A_c} \left(\frac{\partial f_g}{\partial S_g}\right)_{S_g}$$

In 100 days, then,

$$x = \frac{5.615(11,600)(100)}{0.1625(1,237,000)} \left(\frac{\partial f_g}{\partial S_g}\right)_{S_g}$$

$$x = 32.4 \left(\frac{\partial f_g}{\partial S_g}\right)_{S_g} \tag{9.26}$$

The values of the derivatives $(\partial f_g / \partial S_g)$ given in Table 9.2 have been determined graphically from Fig. 9.13. Figure 9.14 shows the plots of Eq. (9.26) to obtain the gas-oil distributions and the positions of the gas front after 100 days. The shape of the curves will not be altered for any other time. The distribution and fronts at 1000 days, for example, may be obtained by simply changing the scale on the distance axis by a factor of 10.

Welge showed that the position of the front may be obtained by drawing a secant from the origin as shown in Fig. 9.13.[10] For example, the secant is tangent to the lower curve at 40% gas saturation. Then in Fig. 9.14, the front may be found by dropping a perpendicular from the 40% gas saturation as indicated. This will balance the areas of the S-shaped curve, which was done by trial and error in Fig. 9.10 for water displacement. In the case of water displacement, the secant should be drawn, not from the origin, but from the

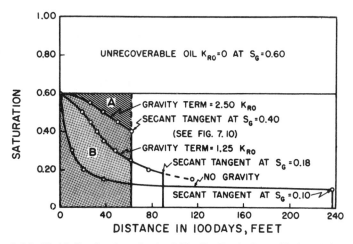

Fig. 9.14. Fluid distributions in the Mile Six Pool after 100 days injection.

connate water saturation, as indicated by the dashed line in Fig. 9.9. This is tangent at a water saturation of 60%. Referring to Fig. 9.10, the 240-day front is seen to occur at 60% water saturation. Owing to the presence of an initial transition zone, the fronts at 60 and 120 days occur at slightly lower values of water saturation.

The much greater displacement efficiency with gravity segregation than without is apparent from Fig. 9.14. Since the permeability to oil is *essentially* zero at 60% gas saturation, the maximum recovery by gas displacement and gravity drainage is 60% of the initial oil in place. Actually some small permeability to oil exists at even very low oil saturations, which explains why some fields may continue to produce at low rates for quite long periods after the pressure has been depleted. The displacement efficiency may be calculated from Fig. 9.14 by the measurement of areas. For example, the displacement efficiency at Mile Six Pool with full gravity segregation is in excess of

$$\text{Recovery} = \frac{\text{Area } B}{\text{Area } A + \text{Area } B} = \frac{32.5}{4.7 + 32.5} = 0.874, \text{ or } 87.4\%$$

If the gravity segregation had been half as effective, the recovery would have been about 60%; without gravity segregation, the recovery would have been only 24%. These recoveries are expressed as percentages of the *recoverable oil*. In terms of the *initial oil* in place, the recoveries are only 60% as large, or 52.4, 36.0, and 14.4%, respectively. Welge, Shreve and Welch, Kern, and others have extended the concepts presented here to the prediction of gas-oil ratios, production rates, and cumulative recoveries, including the treatment of production from wells behind the displacement front.[10, 11, 12] Smith has used the magnitude of the gravity term $[(k_o/\mu_o)(\rho_o - \rho_g) \cos \alpha]$ as a criterion for determining those reservoirs in which gravity segregation is likely to be of considerable importance.[13] The data of Table 9.3 indicate that this gravity term must have a value above about 600 in the units used to be effective. An inspection of Eq. (9.23), however, shows that the throughput velocity (q_t'/A_c) is also of primary importance.

One interesting application of gravity segregation is to the recovery of updip or "attic" oil in active water-drive reservoirs possessing good gravity segregation characteristics. When the structurally highest well(s) has gone to water production, highpressure gas is injected for a period. This gas migrates updip and displaces the oil downdip, where it may be produced from the same well in which the gas was injected. The injected gas is of course, unrecoverable.

It appears from the previous discussions and examples that water is *generally* more efficient than gas in displacing oil from reservoir rocks, mainly because (a) the water viscosity is of the order of 50 times the gas viscosity and (b) the water occupies the less conductive portions of the pore spaces, whereas the gas occupies the more conductive portions. Thus, in water displacement the oil is left to the central and more conductive portions of the pore channels, whereas in gas displacement the gas invades and occupies the more conductive portions first, leaving the oil and water to the less conductive portions. What has been said of water displacement is true for preferentially *water wet* (hydrophilic) rock, which is the case of most reservoir rocks. When the rock is preferentially *oil wet* (hydrophobic), the displacing water will invade the more conductive portions first, just as gas does, resulting in lower displacement efficiencies. In this case, the efficiency by water still exceeds that by gas because of the viscosity advantage the water has over the gas.

3.3 Oil Recovery by Internal Gas Drive

Oil is produced from volumetric, undersaturated reservoirs by expansion of the reservoir fluids. Down to the bubble-point pressure, the production is caused by liquid (oil and connate water) expansion and rock compressibility (see Chapter 5, Sect. 6). Below the bubble point, the expansion of the connate water and the rock compressibility are negligible, and as the oil phase contracts owing to release of gas from solution, production is a result of expansion of the gas phase. When the gas saturation reaches the critical value, free gas begins to flow. At fairly low gas saturations, the gas mobility, k_g/μ_g, becomes large and the oil mobility, k_o/μ_o, small, resulting in high gas-oil ratios and in low oil recoveries, usually in the range of 5 to 25%.

Because the gas orginates internally within the oil, the method described in the previous section for the displacement of oil by external gas drive is not applicable. In addition, constant pressure was assumed in the external displacement so that the gas and oil viscosities and volume factors remained constant during the displacement. With internal gas drive, the pressure drops

TABLE 9.3.

Gravity drainage experience (After R. H. Smith except for Mile Six Pool.[13])

Field and Reservoir	Oil Viscosity, cp	Oil Permeability, From P.I., md	Oil Mobility, md/cp	Dip Angle, deg.	cos α [a]	Density Difference Δρ	Gravity Drainage Term, $\frac{k_o}{\mu_o} \Delta\rho \cos \alpha$	Sand Thickness, ft	Gravity Drainage
Lakeview	17	2000	118	24	0.41	53.7	2,590	100	Yes
Lance Creek, 27B	0.4	80	200	4.5	0.08	39.3	630	Thin Bedding	Yes
Sundance E2–3	1.3	1100	846	22	0.37	40.6	12,710	45	Yes
Oklahoma City	2.1	600	286	36	0.59	34.9	5,880	?	Yes
Kettleman, Temblor	0.8	72	90	30	0.50	35.6	1,600	80	Yes
West Coyote, Emery	1.45	28	19.3	17	0.29	38.1	210	75	?
San Miguelito, First Grubb	1.1	34	30.9	39	0.62	39.3	750	40	Yes
Huntington Beach, Lower Ashton	1.8	125	69	25	0.42	41.8	1,220	50	Yes
Ellwood, Vaqueros	1.5	250	167	32	0.53	43.1	3,810	120	Yes
San Ardo, Campbell	2000	4700	2.35	4	0.07	56.2	10	230	...
Wilmington, Upper Terminal Block V	12.6	284	22.5	4	0.07	52.4	80	40	...
Huntington Beach, Jones	40	600	15	11	0.19	54.3	150	40	No
Mile Six Pool, Peru	1.32	300	224	17.5	0.30	43.7	2,980	635	Yes

[a] α = 90° − Dip angle

361

as production proceeds, and the gas and oil viscosities and volume factors continually change, further complicating the mechanism.

Because of the complexity of the internal gas drive mechanism, a number of simplifying assumptions must be made to keep the mathematical forms reasonably simple. The following assumptions, generally made, do reduce the accuracy of the methods, but in most cases not appreciably:

1. Uniformity of the reservoir at all times regarding porosity, fluid saturations, and relative permeabilities. Studies have shown that the gas and oil saturations about wells are surprisingly uniform at all stages of depletion.

2. Uniform pressure throughout the reservoir in both the gas and oil zones. This means the gas and oil volume factors, the gas and oil viscosities, and the solution gas will be the same throughout the reservoir.

3. Negligible gravity segregation forces.

4. Equilibrium at all times between the gas and the oil phases.

5. A gas liberation mechanism which is the same as that used to determine the fluid properties.

6. No water encroachment and negligible water production.

Several methods appear in the literature for predicting the performance of internal gas drive reservoirs from their rock and fluid properties. Three are discussed in this chapter: (1) Muskat's method, (2) Schilthuis' method, and (3) Tarner's method.[14, 15, 16] These methods relate the pressure decline to the oil recovery and the gas-oil ratio.

You will recall that the material balance is successful in predicting the performance of volumetric reservoirs down to pressures at which free gas begins to flow. In the study of North Snyder, for example, in Chapter 5, Sect. 4, the produced gas-oil ratio was assumed to be equal to the dissolved gas-oil ratio down to the pressure at which the gas saturation reached 10%, the critical gas saturation assumed for that reservoir. Below this pressure (i.e., at higher gas saturations), both gas and oil flow to the wellbores, their relative rates being controlled by their viscosities, which change with pressure and by their relative permeabilities, which change with their saturations. It is not surprising, then, that the material balance principle (static) is combined with the producing gas-oil ratio equation (dynamic) to predict the performance at pressures at which the gas saturation exceeds the critical value.

In the Muskat method, the values of the many variables that affect the production of gas and oil and the values of the rates of changes of these variables with pressure are evaluated at any stage of depletion (pressure). Assuming these values hold for a small drop in pressure, the incremental gas and oil production can be calculated for the small pressure drop. These variables are recalculated at the lower pressure, and the process is continued to any desired abandonment pressure. To derive the Muskat equation, let V_p be the reservoir pore volume in barrels. Then, the stock tank barrels of oil *remaining* N_r at any pressure is given by:

$$N_r = \frac{S_o V_p}{B_o} \text{ stock tank barrels} \tag{9.27}$$

Differentiating with respect to pressure,

$$\frac{dN_r}{dp} = V_p \left(\frac{1}{B_o} \frac{dS_o}{dp} - \frac{S_o}{B_o^2} \frac{dB_o}{dp} \right) \tag{9.28}$$

The gas *remaining* in the reservoir, both free and dissolved, at the same pressure, in standard cubic feet, is

$$G_r = \frac{R_{so} V_p S_o}{B_o} + \frac{(1 - S_o - S_w) V_p}{B_g} \tag{9.29}$$

Differentiating with respect to pressure,

$$\frac{dG_r}{dp} =$$

$$V_p \left[\frac{R_{so}}{B_o} \frac{dS_o}{dp} + \frac{S_o}{B_o} \frac{dR_{so}}{dp} - \frac{R_{so} S_o}{B_o^2} \frac{dB_o}{dp} - \frac{(1 - S_o - S_w)}{B_g^2} \frac{dB_g}{dp} - \frac{1}{B_g} \frac{dS_o}{dp} \right] \tag{9.30}$$

If reservoir pressure is dropping at the rate dp/dt, then the current or producing gas-oil ratio at this pressure is

$$R = \frac{dG_r/dp}{dN_r/dp} \tag{9.31}$$

Substituting Eqs. (9.28) and (9.30) in Eq. (9.31),

$$R = \frac{\dfrac{R_{so}}{B_o} \dfrac{dS_o}{dp} + \dfrac{S_o}{B_o} \dfrac{dR_{so}}{dp} - \dfrac{R_{so} S_o}{B_o^2} \dfrac{dB_o}{dp} - \dfrac{(1 - S_o - S_w)}{B_g^2} \dfrac{dB_g}{dp} - \dfrac{1}{B_g} \dfrac{dS_o}{dp}}{\dfrac{1}{B_o} \dfrac{dS_o}{dp} - \dfrac{S_o}{B_o^2} \dfrac{dB_o}{dp}} \tag{9.32}$$

Equation (9.32) is simply an expression of the material balance for volumetric, undersaturated reservoirs in differential form. The producing gas-oil ratio may also be written as

$$R = R_{so} + \frac{k_g \mu_o B_o}{k_o \mu_g B_g} \tag{9.33}$$

Eq. (9.33) applies both to the flowing free gas and to the solution gas which flows to the wellbore in the oil. These two types of gas make up the total surface producing gas-oil ratio, R in SCF/STB. Equation (9.33) may be equated to Eq. (9.32) and solved for dS_o/dp to give:

$$\frac{dS_o}{dp} = \frac{\dfrac{S_o B_g}{B_o}\dfrac{dR_{so}}{dp} + \dfrac{S_o}{B_o}\dfrac{k_g}{k_o}\dfrac{\mu_o}{\mu_g}\dfrac{dB_o}{dp} - \dfrac{(1 - S_o - S_w)}{B_g}\dfrac{dB_g}{dp}}{1 + \dfrac{k_g}{k_o}\dfrac{\mu_o}{\mu_g}} \tag{9.34}$$

To simplify the handling of Eq. (9.34), the terms in the numerator which are functions of pressure only may be grouped together and given the group symbols $X(p)$, $Y(p)$, and $Z(p)$ as follows:

$$X(p) = \frac{B_g}{B_o}\frac{dR_{so}}{dp} \; ; \; Y(p) = \frac{1}{B_o}\frac{\mu_o}{\mu_g}\frac{dB_o}{dp} \; ; \; Z(p) = \frac{1}{B_g}\frac{dB_g}{dp} \tag{9.35}$$

Using these group symbols and placing Eq. (9.34) in an incremental form,

$$\Delta S_o = \Delta p \left[\frac{S_o X(p) + S_o \dfrac{k_g}{k_o} Y(p) - (1 - S_o - S_w)Z(p)}{1 + \dfrac{k_g}{k_o}\dfrac{\mu_o}{\mu_g}} \right] \tag{9.36}$$

Equation (9.36) gives the change in oil saturation that accompanies a pressure drop, Δp. The functions $X(p)$, $Y(p)$, and $Z(p)$ are obtained from the reservoir fluid properties using Eqs. (9.35). The values of the derivatives dR_{so}/dp, dB_o/dp, and dB_g/dp are found graphically from the plots of R_{so}, B_o, and B_g versus pressure. It has been found that when determining dB_g/dp the numbers are more accurately obtained by plotting $1/B_g$ versus pressure. When this is done, the following substitution is used:

$$\frac{d(1/B_g)}{dp} = -\frac{1}{B_g^2}\frac{dB_g}{dp}$$

$$\frac{dB_g}{dp} = -B_g^2\frac{d(1/B_g)}{dp}$$

or

$$Z(p) = \frac{1}{B_g}\left[-B_g^2\frac{d(1/B_g)}{dp} \right] = -B_g\frac{d(1/B_g)}{dp} \tag{9.37}$$

In calculating ΔS_o for any pressure drop Δp, the values of S_o, $X(p)$, $Y(p)$, $Z(p)$, k_g/k_o, and μ_o/μ_g at the beginning of the interval may be used.

Better results will be obtained, however, if values at the middle of the pressure drop interval are used. The value of S_o at the middle of the interval can be closely estimated from the ΔS_o value for the previous interval, and the value of k_g/k_o used corresponding to the estimated midinterval value of the oil saturation. In addition to Eq. (9.36), the total oil saturation must be calculated. This is done by simply multiplying the value of $\Delta S_o/\Delta p$ by the pressure drop, Δp, and then subtracting the ΔS_o from the oil saturation value that corresponds to the pressure at the beginning of the pressure drop interval, as shown in the following equation:

$$S_{oj} = S_{o(j-1)} - \Delta p \left(\frac{\Delta S_o}{\Delta p}\right) \tag{9.38}$$

where, j corresponds to the pressure at the end of the pressure increment and $j-1$ corresponds to the pressure at the beginning of the pressure increment.

The following procedure is used to solve for the ΔS_o for a given pressure drop Δp:

1. Plot R_{so}, B_o, and B_g or $1/B_g$ versus pressure and determine the slope of each plot.
2. Solve Eq. (9.36) for the $\Delta S_o/\Delta p$ using the oil saturation that corresponds to the initial pressure of the given Δp.
3. Estimate S_{oj} using Eq. (9.38).
4. Solve Eq. (9.36) using the oil saturation from Step 3.
5. Determine an average value for $\Delta S_o/\Delta p$ from the two values calculated in Steps 2 and 4.
6. Using $(\Delta S_o/\Delta p)_{ave}$ solve for S_{oj} using Eq. (9.38). This value of S_{oj} becomes $S_{o(j-1)}$ for the next pressure drop interval.
7. Repeat Steps 2 through 6 for all pressure drops of interest. Muskat Method

The Schilthuis method begins with the general material balance equation, which reduces to the following for a volumetric, undersaturated reservoir, using the single-phase formation volume factor:

$$N = \frac{N_p[B_o + B_g(R_p - R_{so})]}{B_o - B_{oi} + B_g(R_{soi} - R_{so})} \tag{9.39}$$

Notice that this equation contains variables that are a function of only the reservoir pressure, B_t, B_g, R_{soi}, and B_{ti}, and the unknown variables, R_p and N_p. R_p, of course, is the ratio of cumulative oil production, N_p, to cumulative gas production, G_p. To use this equation as a predictive tool for N_p, a method must be developed to estimate R_p. The Schilthuis method uses the total surface producing gas-oil ratio or the instantaneous gas-oil ratio, R, defined previously in Eq. (9.33) as

$$R = R_{so} + \frac{k_g \mu_o B_o}{k_o \mu_g B_g} \qquad (9.33)$$

The first term on the right-hand side of Eq. (9.33) accounts for the production of solution gas, and the second term accounts for the production of free gas in the reservoir. The second term is a ratio of the gas to oil flow equations discussed in Chapter 7. To calculate R with Eq. (9.33), information about the permeabilities to gas and oil is required. This information is usually known from laboratory measurements as a function of fluid saturations and is often available in graphic form (see Fig. 9.15). The fluid saturation equation is also needed:

$$S_L = S_w + (1 - S_w)\left[1 - \frac{N_p}{N}\right]\frac{B_o}{B_{oi}} \qquad (9.40)$$

where

S_L is the total liquid saturation (i.e., $S_L = S_w + S_o$, which also equals $1 - S_g$).

The solution of this set of equations to obtain production values requires a trial-and-error procedure. First, the material balance equation is rearranged to yield the following:

$$\frac{\dfrac{N_p}{N}[B_o + B_g(R_p - R_{so})]}{B_o - B_{oi} + B_g(R_{soi} - R_{so})} - 1 = 0 \qquad (9.41)$$

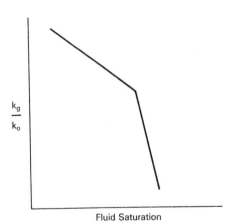

Fig. 9.15. Permeability ratio versus fluid saturation.

All the parameters in Eq. (9.41) are known as functions of pressure from laboratory studies except N_p/N and R_p. When the correct values of these two variables are used in Eq. (9.41) at a given pressure, then the left-hand side of the equation equals zero. The trial-and-error procedure follows this sequence of steps:

1. Guess a value for an incremental oil production ($\Delta N_p/N$) that occurs during a small drop in the average reservoir pressure (Δp).

2. Determine the cumulative oil production to pressure $p_j = p_{j-1} - \Delta p$ by adding all the previous incremental oil productions to the guess during the current pressure drop. The subscript, $j - 1$, refers to the conditions at the beginning of the pressure drop and j to the conditions at the end of the pressure drop.

$$\frac{N_p}{N} = \sum \frac{\Delta N_p}{N} \qquad (9.42)$$

3. Solve the total liquid saturation equation, Eq. (9.40), for S_L at the current pressure of interest.

4. Knowing S_L, determine a value for k_g/k_o from permeability ratio versus saturation information, and then solve Eq. (9.33) for R_j at the current pressure.

5. Calculate the incremental gas production using an average value of the gas-oil ratio over the current pressure drop:

$$R_{ave} = \frac{R_{j-1} + R_j}{2} \qquad (9.43)$$

$$\frac{\Delta G_p}{N} = \frac{\Delta N_p}{N}(R_{ave}) \qquad (9.44)$$

6. Determine the cumulative gas production by adding all previous incremental gas productions in a similar manner to Step 2 in which the cumulative oil was determined.

$$\frac{G_p}{N} = \sum \frac{\Delta G_p}{N} \qquad (9.45)$$

7. Calculate a value for R_p with the cumulative oil and gas amounts.

$$R_p = \frac{G_p/N}{N_p/N} \qquad (9.46)$$

8. With the cumulative oil recovery from Step 2 and the R_p from Step 7, solve

Eq. (9.41) to determine if the left-hand side equals zero. If the left-hand side does not equal zero, then a new incremental recovery should be guessed and the procedure repeated until Eq. (9.41) is satisfied.

Any one of a number of iteration techniques can be used to assist in the trial and error procedure. One that is frequently used is the secant method,[17] which has the following iteration formula:

$$x_{n+1} = x_n - f_n \left[\frac{x_n - x_{n-1}}{f_n - f_{n-1}} \right] \qquad (9.47)$$

To apply the secant method to the foregoing procedure, the left-hand side of Eq. (9.41) becomes the function, f, and the cumulative oil recovery becomes x. The secant method provides the new guess for oil recovery, and the sequence of steps is repeated until the function, f, is zero or within a specified tolerance (e.g., $\pm 10^{-4}$). The solution procedure described earlier is fairly easy to program on a computer.

The Tarner method for predicting reservoir performance by internal gas drive is presented in a form proposed by Tracy.[18] Neglecting the formation and water compressibility terms, the general material balance in terms of the single-phase oil formation volume factor may be written as follows:

$$N = \frac{N_p[B_o - R_{so}B_g] + G_p B_g - (W_e - W_p)}{B_o - B_{oi} + (R_{soi} - R_{so})B_g + \dfrac{mB_{oi}}{B_{gi}}(B_g - B_{gi})} \qquad (9.48)$$

Tracy suggested writing

$$\Phi_n = \frac{B_o - R_{so}B_g}{B_o - B_{oi} + (R_{soi} - R_{so})B_g + \dfrac{mB_{oi}}{B_{gi}}(B_g - B_{gi})} \qquad (9.49)$$

$$\Phi_g = \frac{B_g}{B_o - B_{oi} + (R_{soi} - R_{so})B_g + \dfrac{mB_{oi}}{B_{gi}}(B_g - B_{gi})} \qquad (9.50)$$

$$\Phi_w = \frac{1}{B_o - B_{oi} + (R_{soi} - R_{so})B_g + \dfrac{mB_{oi}}{B_{gi}}(B_g - B_{gi})} \qquad (9.51)$$

where Φ_n, Φ_g, and Φ_w are simply a convenient collection of many terms, all of which are functions of pressure, except the ratio m, the initial free gas-to-oil volume. The general material balance equation may now be written as

$$N = N_p \Phi_n + G_p \Phi_g - (W_e - W_p)\Phi_w \qquad (9.52)$$

Applying this equation to the case of a volumetric, undersaturated reservoir,

$$N = N_p\Phi_n + G_p\Phi_g \qquad (9.53)$$

In progressing from the conditions at any pressure, p_{j-1}, to a lower pressure, p_j, Tracy suggested the estimation of the producing gas-oil ratio, R, at the lower pressure rather than estimating the production ΔN_p during the interval, as we did in the Schilthuis method. The value of R may be estimated by extrapolating the plot of R versus pressure, as calculated at the higher pressure. Then, the estimated average gas-oil ratio between the two pressures is given by Eq. (9.43):

$$R_{ave} = \frac{R_{j-1} + R_j}{2} \qquad (9.43)$$

From this estimated average gas-oil ratio for the Δp interval, the estimated production, ΔN_p, for the interval is made using Eq. (9.53) in the following form:

$$N = (N_{p(j-1)} + \Delta N_p)\Phi_{nj} + (G_{p(j-1)} + R_{ave}(\Delta N_p))\Phi_{gj} \qquad (9.54)$$

From the value of ΔN_p in Eq. (9.58), the value of N_{pj} is found:

$$N_{pj} = N_{p(j-1)} + \Delta N_p \qquad (9.55)$$

In addition to these equations, the total liquid saturation equation is required, Eq. (9.40). The solution procedure becomes as follows:

1. Calculate the values of Φ_n and Φ_g as a function of pressure.
2. Estimate a value for R_j in order to calculate an R_{ave} for a pressure drop of interest, Δp.
3. Calculate ΔN_p by rearranging Eq. (9.54) to give:

$$\Delta N_p = \frac{N - N_{p(j-1)}\Phi_n - G_{p(j-1)}\Phi_g}{\Phi_n + \Phi_g R_{ave}} \qquad (9.56)$$

4. Calculate the total oil recovery from Eq. (9.55).
5. Determine k_g/k_o by calculating the total liquid saturation, S_L from Eq. (9.40) and using k_g/k_o versus saturation information.
6. Calculate a value of R_j by using Eq. (9.33), and compare it with the assumed value in Step 2. If these two values agree within some tolerance, then the ΔN_p calculated in Step 3 is correct for this pressure drop interval. If the value of R_j does not agree with the assumed value in Step 2, then the

TABLE 9.4.
Fluid property data for Example 9.1

Pressure, psia	B_o, bbl/STB	R_{so}, SCF/STB	B_g, bbl/SCF	μ_o, cp	μ_g, cp
2500	1.498	721	0.001048	0.488	0.0170
2300	1.463	669	0.001155	0.539	0.0166
2100	1.429	617	0.001280	0.595	0.0162
1900	1.395	565	0.001440	0.658	0.0158
1700	1.361	513	0.001634	0.726	0.0154
1500	1.327	461	0.001884	0.802	0.0150
1300	1.292	409	0.002206	0.887	0.0146
1100	1.258	357	0.002654	0.981	0.0142
900	1.224	305	0.003300	1.085	0.0138
700	1.190	253	0.004315	1.199	0.0134
500	1.156	201	0.006163	1.324	0.0130
300	1.121	149	0.010469	1.464	0.0126
100	1.087	97	0.032032	1.617	0.0122

calculated value should be used as the new guess and Steps 2 through 6 repeated.

As a further check, the value of R_{ave} can be recalculated and Eq. (9.56) solved for ΔN_p. Again, if the new value agrees with what was previously calculated in Step 3 within some tolerance, it can be assumed that the oil recovery is correct.

The three methods are illustrated in Ex. Prob. 9.1.

Example 9.1. Calculating oil recovery as a function of pressure using the (a) Muskat, (b) Schilthuis, and (c) Tarner methods for an undersaturated, volumetric reservoir. The recovery is calculated for the first 200 psi pressure increment from the initial pressure down to a pressure of 2300 psia.

Given:

Initial reservoir pressure = 2500 psia
Initial reservoir temperature = 180°F
Initial oil in place = 56×10^6 STB
Initial water saturation = 0.20
Fluid property data are given in Table 9.4.
Permeability ratio data are plotted in Fig. 9.16.

SOLUTION: (a) The Muskat method involves the following sequence of steps:

1. R_{so}, B_o, and $1/B_g$ were plotted versus pressure to determine the slopes. Although the plots are not shown, the following values were determined:

$$\frac{dR_{so}}{dp} = 0.26 \qquad \frac{dB_o}{dp} = 0.000171 \qquad \frac{d(1/B_g)}{dp} = 0.433$$

The values of $X(p)$, $Y(p)$, and $Z(p)$ are tabulated as follows as a function of pressure:

Pressure	$X(p)$	$Y(p)$	$Z(p)$
2500 psia	0.000182	0.003277	0.000454
2300 psia	0.000205	0.003795	0.000500

2. Calculate $\Delta S_o/\Delta p$ using $X(p)$, $Y(p)$, and $Z(p)$ at 2500 psia:

$$\frac{\Delta S_o}{\Delta p} = \frac{0.20(0.000182) + 0 + 0}{1 + 0} = 0.000146$$

3. Estimate S_{oj}:

$$S_{oj} = 0.80 - 200(0.000146) = 0.7709$$

4. Calculate $\Delta S_o/\Delta p$ using the S_{oj} from Step 3 and $X(p)$, and $Y(p)$, and $Z(p)$ at 2300 psia

$$\frac{\Delta S_o}{\Delta p} = \frac{\begin{array}{c} 0.7709(0.000205) + 0.7709(0.00001)0.003795 \\ + (1.0 - 0.2 - 0.7709)0.000500 \end{array}}{1 + \dfrac{0.539}{0.0166}(0.00001)}$$

$$\frac{\Delta S_o}{\Delta p} = 0.000173$$

5. Calculate the average $\Delta S_o/\Delta p$.

$$\left(\frac{\Delta S_o}{\Delta p}\right)_{ave} = \frac{0.000146 + 0.000173}{2} = 0.000159$$

6. Calculate S_{oj} using $(\Delta S_o \Delta p)_{ave}$ from Step 5:

$$S_{oj} = 0.8 - 0.000159(200) = 0.7682$$

This value of S_o can now be used to calculate the oil recovery that has occurred down to a pressure of 2300 psia:

$$N_p = N \left(1.0 - \left[\frac{S_o}{1 - S_w}\right]\frac{B_{oi}}{B_o}\right) \tag{9.57}$$

$$N_p = 56(10)^6 \left(1.0 - \left[\frac{0.7682}{1 - 0.2}\right]\frac{1.498}{1.463}\right) = 939,500 \text{ STB}$$

(b) Because the Schilthuis method involves a trail-and-error procedure, the secant iteration formula is used to assist in the solution. For this problem, x in the secant formula becomes N_p/N and f becomes the value of the left-hand side of Eq. (9.41). The secant method requires two values of N_p/N to begin the iteration process.

1. Assume incremental oil recovery:

$$\frac{\Delta N_{p1}}{N} = 0.01 \qquad \text{and} \qquad \frac{\Delta N_{p2}}{N} = 0.02$$

2. Calculate cumulative oil recovery:

$$\frac{N_{p1}}{N} = \frac{\Delta N_{p1}}{N} = 0.01 \qquad \text{and} \qquad \frac{N_{p2}}{N} = \frac{\Delta N_{p2}}{N} = 0.02$$

3. Calculate S_L:

$$S_{L1} = 0.2 + (1 - 0.2)(1 - 0.01)\frac{1.463}{1.498} = .9735$$

$$S_{L2} = 0.2 + (1 - 0.2)(1 - 0.02)\frac{1.463}{1.498} = 0.9657$$

4. Determine k_g/k_o from Fig. 9.16 and calculate R:

$$\frac{k_g}{k_{o1}} = \frac{k_g}{k_{o2}} = 0.00001$$

$$R_{1,1} = R_{1,2} = 721$$

$$R_{2,1} = R_{2,2} = 669 + 0.00001\left(\frac{0.539}{0.0166}\right)\left(\frac{1.463}{0.001155}\right) = 669.4$$

where

$R_{j,k}$ = instantaneous gas-oil ratio
j = current pressure counter
k = guess or iteration counter

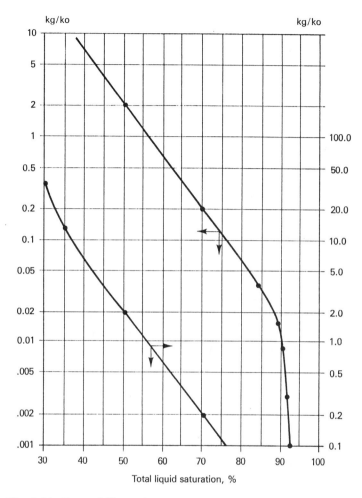

Fig. 9.16. Permeability ratio relationship for Example Problem 9.1.

5. Calculate incremental gas recovery:

$$R_{\text{ave1}} = R_{\text{ave2}} = \frac{721 + 669.4}{2} = 695.2$$

$$\frac{\Delta G_{p1}}{N} = 0.01(695.2) = 6.952$$

$$\frac{\Delta G_{p2}}{N} = 0.02(695.2) = 13.904$$

6. Calculate cumulative gas recovery:

$$\frac{G_{p1}}{N} = \frac{\Delta G_{p1}}{N} = 6.952 \qquad \text{and} \qquad \frac{G_{p2}}{N} = \frac{\Delta G_{p2}}{N} = 13.904$$

7. Calculate R_p:

$$R_{p1} = \frac{6.952}{0.01} = 695.2 \qquad \text{and} \qquad R_{p2} = \frac{13.904}{0.02} = 695.2$$

8. Solve Eq. (9.41) to determine if the left-hand side equals zero or is within some chosen tolerance, say, $\pm 10^{-2}$:

$$f_1 = \frac{0.01[1.523 + 0.001155(695.2 - 721)]}{(1.523 - 1.498)} - 1 = -.4027$$

$$f_2 = \frac{.02[1.523 + 0.001155(695.2 - 721)]}{(1.523 - 1.498)} - 1 = .1946$$

Neither value is within the chosen tolerance, so the two initial guesses of the incremental oil recovery for the first pressure increment are wrong. The secant iteration formula is used to determine a new guess for the incremental oil recovery:

$$\frac{N_{p3}}{N} = 0.02 - 0.1946 \left[\frac{0.02 - 0.01}{0.1946 - (-0.4027)} \right] = 0.0167$$

With the new guess, Steps 2 to 8 are repeated to determine if the tolerance is met for Eq. (9.41). The left-hand side of Eq. (9.41) is found to be -0.0025, which is within the tolerance. Thus, the correct value of fractional oil recovery down to 2300 psia is 0.0167.

To compare with the Muskat method, the recovery ratio must be multiplied by the initial oil in place, 56 M STB, to yield the total cumulative recovery:

$$N_p = 56(10)^6 (0.0167) = 935,200 \text{ STB}$$

(c) The Tarner method requires the following steps:

1. Calculate Φ_n and Φ_g at 2300 psia:

$$\Phi_n = \frac{1.463 - 669(0.001155)}{1.463 - 1.498 + (721 - 669)0.001155} = 27.546$$

$$\Phi_g = \frac{0.001155}{1.463 - 1.498 + (721 - 669)0.001155} = 0.04609$$

2. Assume $R_j = 670$ SCF/STB, which is just slightly larger then R_{so}, suggesting that only a very small amount of gas is flowing to the wellbore and is being produced:

$$R_{ave} = \frac{721 + 670}{2} = 695.5$$

3. Calculate ΔN_p:

$$\Delta N_p = \frac{56,000,000 - 0 - 0}{27.546 + 0.04609(695.9)} = 939,300 \text{ STB}$$

4. Calculate N_p:

$$N_p = \Delta N_p = 939,300 \text{ STB}$$

5. Determine k_g/k_o:

$$S_L = 0.2 + (1 - 0.2)\left[1 - \frac{939,300}{56,000,000}\right]\frac{1.463}{1.498} = 0.968$$

From this value of S_L, the permeability ratio, k_g/k_o, can be obtained from Fig. 9.16. Since the curve is off the plot, a very small value of $k_g/k_o = 0.00001$ is estimated.

6. Calculate R_j and compare it with the assumed value in Step 2:

$$R_j = 669 + 0.00001\left(\frac{0.539}{0.0166}\right)\left(\frac{1.463}{0.001155}\right) = 669.4$$

This value agrees very well with the value of 670 that was assumed in Step 2. For the numbers of Ex. Prob. 9.1, the three methods of calculation yielded values of N_p that are within 0.5%. This suggests that any one of the three methods may be used to predict oil and gas recovery, especially considering that many of the parameters used in the equations could be in error more than 0.5%.

4. INTRODUCTION TO WATERFLOODING

Waterflooding is the use of injection water to produce reservoir oil. The process was discovered quite by accident more than 100 years ago when water from a shallow water-bearing horizon leaked around a packer and entered an oil column in a well. The oil production from the well was curtailed, but production from surrounding wells increased. Over the years, the use of waterflooding has slowly grown until it is now the dominant fluid injection recovery technique. The next several sections provide an overview of this process. You are referred to several good books on the subject that provide detailed design criteria.[1,2,3] In the following sections, the characteristics of good waterflood candidates, the location of injectors and producers in a waterflood, and ways to estimate the recovery of a waterflood are briefly discussed.

4.1. Waterflood Candidates

Several factors lend an oil reservoir to a successful waterflood. They can be generalized in two categories: reservoir characteristics and fluid characteristics.

The main reservoir characteristics that affect a waterflood are depth, structure, homogeneity, and petrophysical properties such as porosity, saturation, and average permeability. The depth of the reservoir affects the waterflood in two ways. First, investment and operating costs generally increase as the depth increases as a result of the increase in drilling and lifting costs. Second, the reservoir must be deep enough for the injection pressure to be less than the fracture pressure of the reservoir. Otherwise, fractures induced by high water injection rates could lead to poor sweep efficiencies if the injected water channels through the reservoir to the producing wells. If the reservoir has a dipped structure, gravity effects can often be used to increase sweep efficiencies. The homogeneity of a reservoir plays an important role in the effectiveness of a waterflood. The presence of faults, permeability trends, and the like effect the location of new injection wells because good communication is required between injection and production wells. However, if serious channeling exists, as in some reservoirs that are significantly heterogeneous, then much of the reservoir oil will be bypassed and the water injection will be rendered useless. If a reservoir has insufficient porosity and oil saturation, then a waterflood may not be economically justified on the basis that not enough oil will be produced to offset investment and operating costs. The average reservoir permeability should be high enough to allow sufficient fluid injection without parting or fracturing the reservoir.

The principal fluid characteristic is the viscosity of the oil compared to that of the injected water. The important variable to consider is actually the mobility ratio, which was defined earlier in Chapter 7 and includes not only the viscosity ratio but a ratio of the relative permeabilities to each fluid phase:

$$M = \frac{k_w/\mu_w}{k_o/\mu_o}$$

A good waterflood has a mobility ratio around 1. If the reservoir oil is extremely viscous, then the mobility ratio will likely be much greater than 1. When this is the case, the water will "finger" through the reservoir and bypass much of the oil.

4.2. Location of Injectors and Producers

The injection and production wells in a waterflood should be placed to accomplish the following: (a) provide the desired oil productivity and the necessary water injection rate to yield this oil productivity; and (b) take advantage of the reservoir characteristics such as dip, faults, fractures, and permeability trends. In general, two kinds of flooding patterns are used: peripheral flooding and pattern flooding.

Pattern flooding is used in reservoirs having a small dip and a large surface area. Some of the more common patterns are shown in Fig. 9.17. Table 9.5 lists the ratio of producing wells to injection wells in the patterns shown in Fig. 9.17. If the reservoir characteristics yield lower injection rates than those desired, the operator should consider using either a seven- or a nine-spot pattern where there are more injection wells per pattern than producing wells. A similar argument can be made for using a four-spot pattern in a reservoir with low flow rates in the production wells.

TABLE 9.5.
Ratio of producing wells to injection wells for several pattern arrangements

Pattern	Ratio of Producing Wells to Injection Wells
Four-spot	2
Five-spot	1
Seven-spot	1/2
Nine-spot	1/3
Direct-line-drive	1
Staggered-line-drive	1

The direct-line-drive and staggered-line-drive patterns are frequently used because they usually involve the lowest investment. Some of the economic factors to consider include the cost of drilling new wells, the cost of switching existing wells to a different type (i.e., a producer to an injector), and the loss of revenue from the production when making a switch from a producer to an injector.

In peripheral flooding, the injectors are grouped together unlike in pattern floods where the injectors are interspersed with the producers. Figure 9.18 illustrates two cases in which peripheral floods are sometimes used. In Fig. 9.18(a), a schematic of an anticlinal reservoir with an underlying aquifer is shown. The injectors are placed so that the injected water either enters the aquifer or is near the aquifer-reservoir interface. The pattern of wells on the surface, shown in Fig. 9.18(a), is a ring of injectors surrounding the producers. A monoclinal reservoir with an underlying aquifer is shown in Fig. 9.18(b). In this case, the injectors are again placed so that the injected water either

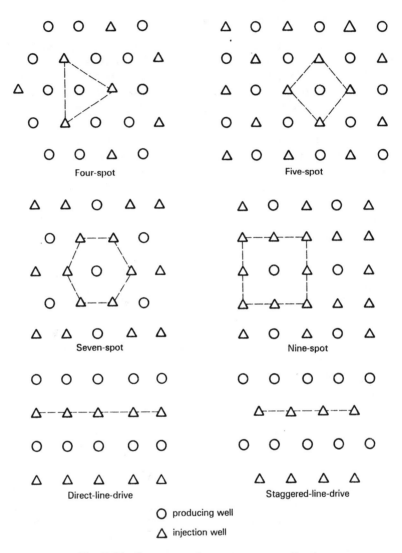

Fig. 9.17. Geometry of common pattern floods.

enters the aquifer or enters near the aquifer-reservoir interface. When this is done, the well arrangement shown in Fig. 9.18(b), where all the injectors are grouped together, is obtained.

4.3. Estimation of Waterflood Recovery Efficiency

Equation (9.1) is an expression for the overall recovery efficiency for any fluid displacement process:

$$E = E_v E_d \qquad (9.1)$$

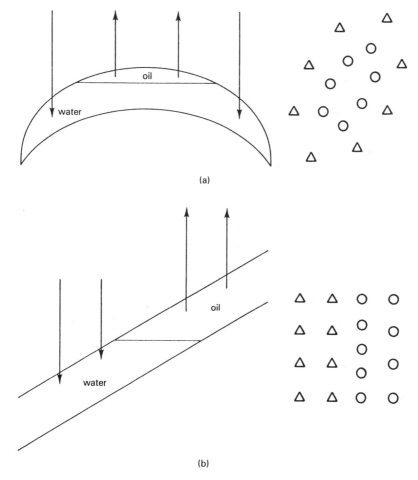

Fig. 9.18. Well arrangements for anticlinal (a) and monoclinal (b) reservoirs with underlying aquifers.

where,

$$E = \text{overall recovery efficiency}$$
$$E_v = \text{volumetric displacement efficiency}$$
$$E_d = \text{microscopic displacement efficiency}$$

The volumetric displacement efficiency is made up of the areal displacement efficiency, E_s, and the vertical displacement efficiency, E_i. To estimate the overall recovery efficiency, values for E_s, E_i, and E_d must be estimated. Methods of estimating these terms are discussed in waterflood textbooks and are too lengthy to present in detail here.[1,2,3] However, some brief, general comments concerning each of the displacement efficiencies can be made.

There are several methods of obtaining estimates for the microscopic displacement efficiency. The basis for one method was presented in Sect. 3.1 in this chapter. The areal displacement, or sweep, efficiency is largely a function of pattern type and mobility ratio. The vertical displacement efficiency is primarily a function of reservoir heterogeneities and thickness of the reservoir formation.

Waterflooding is an important process for the reservoir engineer to understand. It has made and will continue to make large contributions to the recovery of reservoir oil.

5. INTRODUCTION TO ENHANCED OIL RECOVERY PROCESSES

Enhanced oil recovery (EOR) refers to the process of producing liquid hydrocarbons by methods other than the conventional use of reservoir energy and reservoir repressurizing schemes with gas or water.[19] On the average, conventional methods of production produce about one-third of the initial oil in place in a given reservoir. The remaining oil, nearly two-thirds of the initial resource, is a large and attractive target for EOR methods. The next few sections provide an introduction to this important topic in reservoir engineering.

5.1. Mobilization of Residual Oil

During the early stages of a waterflood in a water-wet reservoir system, the brine exists as a film around the sand grains and the oil fills the remaining pore space. At a time intermediate during the flood, the oil saturation has been decreased and exists partly as a continuous phase in some pore channels but as discontinuous droplets in other channels. At the end of the flood, when the oil has been reduced to residual oil saturation, S_{or}, the oil exists primarily as a discontinuous phase of droplets or globules that have been isolated and trapped by the displacing brine.

The waterflooding of oil in an oil-wet system yields a different fluid distribution at S_{or}. Early in the waterflood, the brine forms continuous flow paths through the center portions of some of the pore channels. The brine enters more and more of the pore channels as the waterflood progresses. At residual oil saturation, the brine has entered a sufficient number of pore channels to shut off the oil flow. The residual oil exists as a film around the sand grains. In the smaller flow channels, this film may occupy the entire void space.

The mobilization of the residual oil saturation in a water-wet system requires that the discontinuous globules be connected to form a continuous flow channel that leads to a producing well. In an oil-wet porous medium, the film of oil around the sand grains must be displaced to large pore channels and be connected in a continuous phase before it can be mobilized. The mobilization of oil is governed by the viscous forces (pressure gradients) and the interfacial tension forces that exist in the sand grain-oil-water system.

There have been several investigations of the effect of viscous forces and interfacial tension forces on the trapping and mobilization of residual oil.[3,20,21] From these studies, correlations between a dimensionless parameter called the *capillary number*, N_{vc}, and fraction of oil recovered have been developed. The capillary number is the ratio of viscous force to interfacial tension force and is defined by Eq. (9.58).

$$N_{vc} = (\text{constant})\frac{\upsilon\mu_w}{\sigma_{ow}} = (\text{constant})\frac{k_o\Delta p}{\phi\sigma_{ow}L} \qquad (9.58)$$

where υ is velocity, μ_w is the viscosity of the displacing fluid, σ_{ow} is the interfacial tension between the displaced and displacing fluids, k_o is the effective permeability to the displaced phase, ϕ is the porosity, and $\Delta p/L$ is the pressure drop associated with the velocity.

Figure 9.19 is a schematic representation of the capillary number correlation in which the capillary number is plotted on the abscissa, and the ratio of residual oil saturation (after conducting an EOR process to the residual oil saturation before the EOR process) is plotted as the vertical coordinate. The capillary number increases as the viscous force increases or as the interfacial tension force decreases. The EOR methods that have been developed and applied to reservoir situations are designed either to increase the viscous force associated with the injected fluid or to decrease the interfacial tension force between the injected fluid and the reservoir oil. The next three sections discuss

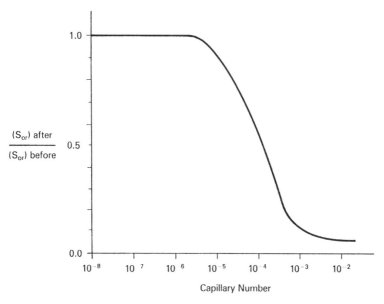

Fig. 9.19. Capillary number correlation.

the three general types of EOR processes: miscible flooding, chemical flooding, and thermal flooding.

5.2. Miscible Injection Processes

In Sect. 2.1 of this chapter, it was noted that microscopic displacement efficiency is largely a function of interfacial forces acting among the oil, rock, and displacing fluid. If the interfacial tension between the trapped oil and displacing fluid could be lowered to 10^{-2} to 10^{-3} dyn/cm, the oil droplets could be deformed so that they would squeeze through the pore constrictions and combine with other droplets to yield a continuous oil phase. A miscible process is one in which the interfacial tension is zero; that is, the displacing fluid and residual oil mix to form one phase. If the interfacial tension is zero, the capillary number becomes infinite and the microscopic displacement efficiency is maximized.

Figure 9.20 is a schematic of a miscible process. Fluid A is injected into the formation. When the fluid contacts residual oil droplets, an oil bank is formed at the leading edge of the injected material. A mixing zone develops between fluid A and the oil bank and grows as a result of molecular diffusion and other processes. Fluid A is then followed by fluid B, which is miscible with fluid A but not generally miscible with the oil and which is much cheaper than fluid A. A mixing zone is also created at the fluid A–fluid B interface. It is important that the amount of fluid A injected be large enough that the two mixing zones do not come in contact but still be small enough to avoid large costs for injected chemicals.

Consider a miscible process with n-decane as the residual oil, propane as fluid A, and methane as fluid B. The system pressure and temperature are

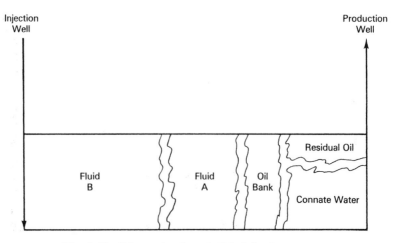

Fig. 9.20. Schematic of a miscible injection process.

2000 psia and 100°F, respectively. At these conditions, both n-decane and propane are liquids and are therefore miscible in all proportions. The system temperature and pressure indicate that any mixture of methane and propane would be in the gas state; therefore, the methane and propane would be miscible in all proportions. However, the methane and n-decane would not be miscible in all proportions because one is a liquid and the other a gas. Notice that, in this example, the propane appears to act as a liquid when in the presence of n-decane and as a gas when in contact with methane. It is this unique capacity of propane and other intermediate gases that leads to the miscible process.

There are, in general, two types of miscible processes. One is referred to as the *single-contact* miscible process and involves such injection fluids as liquified petroleum gases (LPG) and alcohols. The injected fluids are miscible with residual oil immediately on contact. The second type is the *multiple-contact*, or *dynamic*, miscible process. The injected fluids in this case are usually methane, inert fluids, or an enriched methane gas supplemented with a C_2-C_6 fraction; this fraction of alkanes has the unique ability of behaving like a liquid or a gas at many reservoir conditions. The injected fluid and oil are usually not miscible on first contact but rely on a process of chemical exchange of the intermediate hydrocarbons between phases to achieve miscibility. These processes are discussed in great detail elsewhere.[4,5,22]

5.3. Chemical Injection Processes

Chemical flooding relies on the addition of one or more chemical compounds to an injected fluid either to reduce the interfacial tension between the reservoir oil and injected fluid or to improve the sweep efficiency of the injected fluid by making it more viscous, thereby improving the mobility ratio. Both mechanisms are designed to increase the capillary number.

Three general methods are used in chemical flooding technology.[4,5] The first is *polymer flooding,* in which a large macromolecule is used to increase the displacing fluid viscosity. This process leads to improved sweep efficiency in the reservoir of the injected fluid. The remaining two methods, *micellar-polymer flooding* and *alkaline flooding,* make use of chemicals that reduce the interfacial tension between an oil and a displacing fluid.

The addition of molecules of large molecular weight (i.e., polymers) to an injected water can often increase the effectiveness of a conventional waterflood. Polymers are usually added to the water in concentrations ranging from 250 to 2000 parts per million (ppm). A polymer solution is more viscous than a brine without polymer. In a flooding application, the increased viscosity alters the mobility ratio between the injected fluid and the reservoir oil. The improved mobility ratio leads to better vertical and areal sweep efficiencies and thus higher oil recoveries. Polymers have also been used to alter gross permeability variations in some reservoirs. In this application, polymers form a gel-like material by cross-linking with other chemical species. The polymer gel sets up in high permeability streaks and diverts the flow of subsequently injected fluids to a different location.

The improvement in oil recovery when using polymers is a result of an improved sweep efficiency over what is obtained during a conventional water-flood. Typical oil recoveries from polymer flooding are in the range of 1 to 5% of the initial oil in place. It has been found that a polymer flood is more likely to be successful if it is started early in the producing life of the reservoir.

The micellar-polymer process uses a surfactant to lower the interfacial tension between the injected fluid and the reservoir oil. A surfactant is a surface-active agent that contains a hydrophobic ("dislikes" water) part to the molecule and a hydrophilic ("likes" water) part. The surfactant migrates to the interface between the oil and water phases and helps make the two phases more miscible. Interfacial tensions can be reduced from ~30 dyn/cm, found in typical waterflooding applications, to 10^{-4} dyn/cm with the addition of as little as 0.1 to 5.0 weight % surfactant to water-oil systems. Soaps and detergents used in the cleaning industry are examples of surfactants. The same principles involved in washing soiled linen or greasy hands are used in "washing" residual oil off of rock formations. As the interfacial tension between an oil phase and a water phase is reduced, the capacity of the aqueous phase to displace the trapped oil phase from the pores of the rock matrix increases. The reduction of interfacial tension results in a shifting of the relative permeability curves such that the oil will flow more readily at lower oil saturations.

When an alkaline solution is mixed with certain crude oils, surfactant molecules are formed by chemical reactions between the alkaline solution and the oil. When the formation of surfactant molecules occurs in situ, the interfacial tension between the brine and oil phases can be significantly reduced. The reduction of interfacial tension causes the microscopic displacement efficiency to increase, thereby increasing oil recovery.

5.4. Thermal Processes

Primary and secondary production from reservoirs containing heavy, low-gravity crude oils is usually a very small fraction of the initial oil in place. These types of oils are very thick and viscous and as a result do not migrate readily to producing wells. It is not uncommon for viscosities of certain heavy crudes to decrease by several orders of magnitude with an increase of temperature of 100 to 200°F. This suggests that if the temperature of a crude oil in the reservoir can be raised by 100 to 200°F over the normal reservoir temperature, the oil viscosity will be reduced significantly and will flow much more easily to a producing well. The temperature of a reservoir can be raised by injecting a hot fluid or by generating thermal energy in situ by combusting the oil. Hot water or steam can be injected as the hot fluid. Three types of processes are generally used in the industry: (1) the continuous injection of hot fluids, such as hot water or steam; (2) the intermittent injection of steam, referred to as steam cycling; and (3) the injection of air or oxygen-enriched air to aid in the combustion of reservoir oil.[4,5,23]

The continuous injection of hot fluids is usually accomplished by injecting either hot water or steam and is much like a conventional waterflood. When steam is injected into the formation, the thermal energy is used to heat the reservoir oil. Unfortunately, the energy also heats the entire environment, such as formation rock and water. Some energy is also lost to the underburden and overburden. Once the oil viscosity is reduced by the increased temperature, the oil can flow more readily to the producing wells. The steam moves through the reservoir and comes in contact with cold oil, rock, and water. As the steam contacts the cold environment, it condenses. A hot water bank is formed and acts as a waterflood, pushing additional oil to the producing wells.

Several mechanisms responsible for the production of oil from a steam injection process have been identified. These include thermal expansion of the crude oil, viscosity reduction of the crude oil, changes in surface forces as the reservoir temperature increases, and steam distillation of the lighter portions of the crude oil.

The intermittent injection of steam, referred to as the steam stimulation process or the cyclic steam process, begins with the injection of steam for a period of days to weeks. The well is then shut in, and the steam is allowed to soak the area around the injection well. This soak period is fairly short, usually from one to five days. The injection well is then placed on production. The length of the production period is dictated by the oil production rate, and it can last from several months to a year or more. The cycle is repeated as many times as economically feasible. The oil production decreases with each new cycle.

Mechanisms of oil recovery that result from this process include (1) reduction of flow resistance near the wellbore by reducing the crude oil viscosity and (2) enchancement of the solution gas drive mechanism by decreasing the gas solubility in an oil as temperature increases.

In heavy oil reservoirs, the steam stimulation process is often applied to develop injectivity around an injection well so that a continuous steam injection process can be conducted.

The injection of air or oxygen-enriched air is referred to as the *in situ combustion* process. Early attempts to apply the combustion process involved what is called the *forward dry combustion* process. The crude oil was ignited downhole, and then a stream of air or oxygen-enriched air was injected into the well where the combustion was originated. The flame front was then propagated through the reservoir. Large portions of heat energy were lost to the surroundings with this process. To reduce the heat losses, a reverse combustion process was conceived. In reverse combustion, the oil is ignited as in forward combustion, but the air stream is injected into a different well. The air is then "pushed" through the flame front as the flame front moves in the opposite direction. Researchers found the process to work in the laboratory, but when it was tried in the field on a pilot scale, it was never successful. In the field, the flame would shut off because there was no oxygen supply. When oxygen was injected, the oil would often self-ignite. The whole process would then revert to a forward combustion process.

When the reverse combustion process failed, a new technique called the *forward wet combustion* process was introduced. This process begins as a forward dry combustion, but once the flame front has been established, the oxygen stream is replaced with water. As the water comes in contact with the hot zone left by the combustion front, it flashes to steam, using energy that otherwise would have been wasted. The steam moves through the reservoir and aids in the displacement of oil. The wet combustion process has become the primary method of conducting combustion processes.

Not all crude oils are amenable to the combustion process. For the combustion process to function properly, the crude oil must contain enough heavy components to serve as the source of fuel for the combustion. This usually requires an oil of low API gravity.

Most of the oil that has been produced by EOR methods to date has been a result of applying thermal processes. There is a practical reason for this, as well as several technical reasons. To produce more than 1 to 2% of the initial oil in place from a heavy oil reservoir, thermal methods had to be employed. As a result, thermal methods were investigated much earlier than either miscible or chemical methods, and the resulting technology was developed much more rapidly.

PROBLEMS

9.1 (a) A rock 10 cm long and 2 sq cm in cross section flows 0.0080 cu cm/sec of a 2.5 cp oil under a 1.5 atm pressure drop. If the oil saturates the rock 100%, what is its absolute permeability?

(b) What will be the rate of 0.75 cp brine in the same core under a 2.5 atm pressure drop if the brine saturates the rock 100%?

(c) Is the rock more permeable to the oil at 100% oil saturation or to the brine at 100% brine saturation?

(d) The same core is maintained at 40% water saturation and 60% oil saturation. Under a 2.0 atm pressure drop, the oil flow is 0.0030 cu cm/sec and the water flow is 0.004 cu cm/sec. What are the effective permeabilities to water and to oil at these saturations?

(e) Explain why the sum of the two effective permeabilities is less than the absolute permeability.

(f) What are the relative permeabilities to oil and water at 40% water saturation?

(g) What is the relative permeability ratio k_o/k_w at 40% water saturation?

(h) Show that the effective permeability ratio is equal to the relative permeability ratio.

9.2 The following permeability data were measured on a sandstone as a function of its water saturation:

S_w	0	10	20	30*	40	50	60	70	75*	80	90	100
k_{ro}	1.0	1.0	1.0	0.94	0.80	0.44	0.16	0.045	0	0	0	0
k_{rw}	0	0	0	0	0.04	0.11	0.20	0.30	0.36	0.44	0.68	1.0

* Critical saturations for oil and water.

(a) Plot the relative permeabilities to oil and water versus water saturation on Cartesian coordinate paper.

(b) Plot the relative permeability ratio versus water saturation on semilog paper.

(c) Find the constants a and b in Eq. (9.3) from the slope and intercept of your graph. Also find a and b by substituting two sets of data in Eq. (9.3) and solving simultaneous equations.

(d) If $\mu_o = 3.4$ cp, $\mu_w = 0.68$ cp, $B_o = 1.50$ bbl/STB, and $B_w = 1.05$ bbl/STB, what is the surface water cut of a well completed in the transition zone where the water saturation is 50%?

(e) What is the reservoir cut in part (d)?

(f) What percentage of recovery will be realized from this sandstone under high-pressure water drive from that portion of the reservoir above the transition zone invaded by water? The initial water saturation above the transition zone is 30%.

(g) If water drive occurs at a pressure below saturation pressure such that the average gas saturation is 15% in the invaded portion, what percentage of recovery will be realized? The average oil volume factor at the lower pressure is 1.35 bbl/STB and the initial oil volume factor is 1.50 bbl/STB.

(h) What fraction of the absolute permeability of this sandstone is due to the least permeable pore channels that make up 20% of the pore volume? What fraction is due to the most permeable pore channels that make up 25% of the pore volume?

9.3 Given the following reservoir data:

Throughput rate = 1000 bbl/day
Average porosity = 18%
Initial water saturation = 20%
Cross-sectional area = 50,000 ft^2
Water viscosity = 0.62 cp
Oil viscosity = 2.48 cp
k_o/k_w versus S_w data in Fig. 9.1 and 9.2

Assume zero transition zone:

(a) Calculate f_w and plot versus S_w.

(b) Graphically determine $\partial f_w/\partial S_w$ at a number of points, and plot versus S_w.

(c) Calculate $\partial f_w / \partial S_w$ at several values of S_w using Eq. (9.17), and compare with the graphical values of part (b).

(d) Calculate the distances of advance of the constant saturation fronts at 100, 200, and 400 days. Plot on Cartesian coordinate paper versus S_w. Equalize the areas within and without the flood front lines to locate the position of the flood fronts.

(e) Draw a secant line from $S_w = 0.20$ tangent to the f_w versus S_w curve in part (b), and show that the value of S_w at the point of tangency is also the point at which the flood front lines are drawn.

(f) Calculate the fractional recovery when the flood front first intercepts a well, using the areas of the graph of part (d). Express the recovery in terms of (1) the initial oil in place and (2) the recoverable oil in place (i.e., recoverable after infinite throughput).

(g) To what surface water cut will a well rather suddenly rise when it is just enveloped by the flood fronts? Use $B_o = 1.50$ bbl/STB and $B_w = 1.05$ bbl/STB.

(h) Do the answers to parts (f) and (g) depend on how far the front has travelled? Explain.

9.4 Show that for radial displacement where $r_w \ll r$

$$r = \left[\frac{5.615 q'_t}{\pi \phi h} \left(\frac{\partial f_w}{\partial S_w} \right) \right]^{1/2}$$

where r is the distance a constant saturation front has travelled.

9.5 Given the following reservoir data:

S_g	10*	15	20	25	30	35	40	45	50	62*
$k_g k_o$	0	0.08	0.20	0.40	0.85	1.60	3.00	5.50	10.0	
k_{ro}	0.70	0.52	0.38	0.28	0.20	0.14	0.11	0.07	0.04	0

* Critical saturations for gas and oil.

  Absolute permeability = 400 md
  Hydrocarbon porosity = 15%
  Connate water = 28%
  Dip angle = 20°
  Cross-sectional area = 750,000 ft^2
  Oil viscosity = 1.42 cp
  Gas viscosity = 0.015 cp
  Reservoir oil specific gravity = 0.75
  Reservoir gas specific gravity = 0.15 (water = 1)
  Reservoir throughput at constant pressure = 10,000 bbl/day

(a) Calculate and plot the fraction of gas, f_g, versus gas saturation similar to Fig. 9.13 both with and without the gravity segregation term.

(b) Plot the gas saturation versus distance after 100 days of gas injection both with and without the gravity segregation term.

(c) Using the areas of part (b), calculate the recoveries behind the flood front with and without gravity segregation in terms of both initial oil and recoverable oil.

9.6 Derive an equation including a gravity term similar to Eq. (9.23) for water displacing oil.

9.7 Rework the water displacement calculation of Table 9.1, and include a gravity segregation term. Assume an absolute permeability of 500 md, a dip angle of 45°, a density difference of 20% between the reservoir oil and water and an oil viscosity of 1.6 cp. Plot water saturation versus distance after 240 days, and compare with Figure 9.11.

S_w	20	30	40	50	60	70	80	90
k_{ro}	0.93	0.60	0.35	0.22	0.12	0.05	0.01	0

9.8 Continue the calculations of Ex. Prob. 9.1 down to a reservoir pressure of 100 psia using:

(a) Muskat method

(b) Schilthuis method

(c) Tarner method

REFERENCES

[1] C. R. Smith, *Mechanics of Secondary Oil Recovery*. Huntington, NY: Robert E. Krieger Publishing, 1966.

[2] G. P Willhite, *Waterflooding*. Dallas, TX: Society of Petroleum Engineers, 1986.

[3] F. F. Craig, "The Reservoir Engineering Aspects of Waterflooding," Monograph 3, Society of Petroleum Engineers, Dallas, TX, 1971.

[4] H. K. van Poollen and Associates, Inc., *Enhanced Oil Recovery*. Tulsa, OK: Penn-Well Publishing, 1980.

[5] L. W. Lake, *Enhanced Oil Recovery*. Englewood Cliffs, NJ: Prentice Hall, 1989.

[6] R. D. Wycoff, H. G. Botset, and M. Muskat, *Trans.* AIME (1933), **103**, 219.

[7] R. L. Slobod and B. H. Candle, *Trans.* AIME (1952), **195**, 265.

[8] S. E. Buckley and M. C. Leverett, "Mechanism of Fluid Displacement in Sands," *Trans.* AIME (1942), **146**, 107.

[9] J. G. Richardson and R. J. Blackwell, "Use of Simple Mathematical Models for Predicting Reservoir Behavior," *Jour. Of Petroleum Technology,* (September 1971), 1145.

[10] H. J. Welge, "A Simplified Method for Computing Oil Recoveries by Gas or Water Drive," *Trans.* AIME (1952), **195,** 91.

[11] D. R. Shreve and L. W. Welch, Jr., "Gas Drive and Gravity Drainage Analysis for Pressure Maintenance Operations," *Trans.* AIME (1956), **207,** 136.

[12] L. R. Kern, "Displacement Mechanism in Multi-Well Systems," *Trans.* AIME (1952), **195,** 39.

[13] R. H. Smith reported by J. A. Klotz, "The Gravity Drainage Mechanism," *Jour. of Petroleum Technology,* (April 1953), **V,** 19.

[14] M. Muskat, "The Production Histories of Oil Producing Gas-drive Reservoirs," *Jour. of Applied Physics* (1945), **16,** 147.

[15] *Oil and Gas Journal,* "Practical Reservoir Engineering," No. 2, (Petroleum Publishing Co.), Tulsa, OK.

[16] J. Tarner, "How Different Size Gas Caps and Pressure Maintenance Programs Affect Amount of Recoverable Oil," *Oil Weekly,* June 12, 1944, **144,** 32–34.

[17] J. B. Riggs, *An Introduction to Numerical Methods for Chemical Engineers.* Lubbock, TX: Texas Tech. University Press, 1988.

[18] G. W. Tracy, "Simplified Form of the Material Balance Equation," *Trans.* AIME (1955), **204,** 243.

[19] R. E. Terry, "Enhanced Oil Recovery," *Encyclopedia of Physical Science and Technology,* Vol. 5. New York: Academic Press, 1987.

[20] J. J. Taber, "Dynamic and Static Forces Required to Remove a Discontinuous Oil Phase from Porous Media Containing Both Oil and Water," *Soc. Pet. Engr. Jour.* (March 1969), 3.

[21] G. L. Stegemeier, "Mechanisms of Entrapment and Mobilization of Oil in Porous Media," in *Improved Oil Recovery by Surfactant and Polymer Flooding,* D. O. Shah and R. S. Schechter, ed. New York: Academic Press, 1977.

[22] F. I. Stalkup Jr., "Miscible Displacement," Monograph 8, Society of Petroleum Engineers, Dallas, TX, 1983.

[23] M. Prats, "Thermal Recovery," Monograph 7, Society of Petroleum Engineers, Dallas, TX, 1983.

Chapter 10

History Matching

1. INTRODUCTION

One of the most important job functions of the reservoir engineer is the prediction of future production rates from a given reservoir or specific well. Over the years, engineers have developed several methods to accomplish this task. The methods range from simple decline curve analysis techniques to sophisticated multidimensional, multiflow reservoir simulators.[*,1,2,3] Whether a simple or complex method is used, the general approach taken to predict production rates is first to calculate producing rates for a period for which the engineer already has production information. If the calculated rates match the actual rates, the calculation is assumed to be correct and can then be used to make future predictions. If the calculated rates do *not* match the existing production data, some of the process parameters are modified and the calculation repeated. The process of modifying these parameters to match the calculated rates with the actual observed rates is referred to as *history matching*.

The calculational method along with the necessary data used to conduct the history match is often referred to as a mathematical model or simulator. When decline curve analysis is used as the calculational method, the engineer

---

[*] References throughout the text are given at the end of each chapter.

is doing little more than curve fitting, and the only data that are necessary are the existing production data. However, when the calculational technique involves multidimensional mass and energy balance equations and multiflow equations, a large amount of data is required as well as a computer to conduct the calculations. With this complex model, the reservoir is usually divided into a grid. This allows the engineer to use varying input data, such as porosity, permeability, and saturation, in different grid blocks. This often requires estimating much of the data since the engineer usually knows data only at specific coring sites that occur much less frequently then the grid blocks used in the calculational procedure.

Since the purpose of this chapter is to discuss the concept of history matching, a model of complexity in between that of the simple decline curve analysis and the complex multidimensional, multiflow simulator is used. The model presented here uses the simple zero-dimensional Schilthuis material balance equation discussed in earlier chapters.

2. DEVELOPMENT OF THE MODEL

The material balance equations presented in Chapters 2 to 6 and Sect. 3.3 in Chapter 9 do not yield information on future production rates because the equations do not have a time dimension associated with them. These equations simply relate average reservoir pressure to cumulative production. To obtain rate information, a method is needed whereby time can be related to either the average reservoir pressure or cumulative production. In Chapter 7, single-phase flow in porous media was discussed and equations were developed for several situations that related flow rate to average reservoir pressure. It should be possible, then, to combine the material balance equations of Chapters 2 to 6 and 9 with the flow equations from Chapter 7 in a model or simulator that would provide a relationship for flow rates as a function of time. The model will require accurate fluid and rock property data and past production data. Once a model has been tested for a particular well or reservoir system and found to reproduce actual past production data, it can be used to predict future production rates. The importance of the data used in the model cannot be over emphasized. If the data are correct, the prediction of production rates will be fairly accurate.

2.1. The Material Balance Part of the Model

The problem we consider in this chapter involves a volumetric, internal gas-drive reservoir. In Sect. 3.3 of Chapter 9, several different methods to calculate the oil recovery as a function of reservoir pressure for this type of reservoir were presented. For the example in this chapter, the Schilthuis method is used. You will remember that the Schilthuis method requires permeability ratio versus saturation information and the solution of Eqs. (9.33), (9.40), and (9.41) (written with the two-phase formation volume factor):

$$R = R_{so} + \frac{k_g \mu_o B_o}{k_o \mu_g B_g} \tag{9.33}$$

$$S_L = S_w + (1 - S_w)\left[1 - \frac{N_p}{N}\right]\frac{B_o}{B_{oi}} \tag{9.40}$$

$$\frac{\frac{N_p}{N}[B_t + B_g(R_p - R_{soi})]}{B_t - B_{ti}} - 1 = 0 \tag{9.41}$$

2.2. Incorporating a Flow Equation into the Model

The procedure mentioned in the previous section yields oil and gas production as a function of the average reservoir pressure, but it does not give any indication of the time required to produce the oil and gas. To calculate the time and rate at which the oil and gas are produced, a flow equation is needed. It was found in Chapter 7 that most wells reach the pseudosteady-state after flowing for a few hours to a few days. An assumption will be made that the well in this chapter's problem has been produced for a time long enough for the pseudosteady-state flow to have been reached. For this case, Eq. (7.50) can be used to describe the oil flow rate into the wellbore:

$$q_o = \frac{0.00708 k_o h}{\mu_o B_o}\left[\frac{\bar{p} - p_{wf}}{\ln\left(\dfrac{r_e}{r_w}\right) - 0.75}\right] \tag{7.50}$$

This equation assumes pseudosteady-state, radial geometry for an incompressible fluid. The subscript, o, refers to oil, and the average reservoir pressure, $\bar{p}$, is the pressure used to determine the production, N_p, in the Schilthuis material balance equation. The incremental time required to produce an increment of oil for a given pressure drop is found by simply dividing the incremental oil recovery by the rate computed from Eq. (7.50) at the corresponding average pressure:

$$\Delta t = \frac{\Delta N_p}{q_o} \tag{10.1}$$

The total time that corresponds to a particular average reservoir pressure can be determined by summing the incremental times for each of the incremental pressure drops until the average reservoir pressure of interest is reached.

Since Eq. (10.1) requires ΔN_p and the Schilthuis equation determines

$\Delta N_p/N$, N, the initial oil in place, must be estimated. In Sect. 2 of Chapter 5, it was shown that the initial oil in place could be estimated from the volumetric approach by the use of Eq. (10.2):

$$N = \frac{7758 A h \phi (1 - S_{wi})}{B_{oi}} \qquad (10.2)$$

Combining these equations with the solution of the Schilthuis material balance equation yields the necessary production rates of both oil and gas.

3. AN EXAMPLE OF A HISTORY MATCH

We now apply the reservoir model developed in the previous two sections to history match production data from a well in a volumetric, internal gas-drive reservoir. The data for the problem were obtained from personnel at the University of Kansas and is used here by permission.[4]

The well is located in a reservoir that is a sandstone and is produced from two zones separated by a thin shale section of approximately 1 to 2 ft in thickness. The reservoir is classified as a stratigraphic trap. The two producing zones decrease in thickness and permeability in directions where it is believed

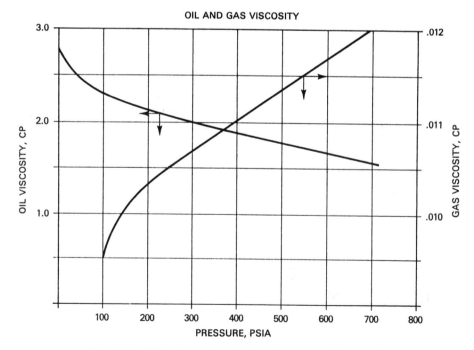

Fig. 10.1. Oil and gas viscosity for history matching problem.

that a pinch-out occurs. Permeability and porosity decrease to unproductive limits both above and below the producing formation. The initial reservoir pressure was 620 psia. The average porosity and initial water saturation values were 21.5% and 37%, respectively. The area drained by the well is 40 ac. Average thicknesses and absolute permeabilities were reported to be 17 ft and 9.6 md for zone 1 and 14 ft and 7.2 md for zone 2. Laboratory data for fluid viscosities, formation volume factors, solution gas-oil ratio, oil relative permeability, and the gas-to-oil permeability ratio are plotted in Figs. 10.1 to 10.5.

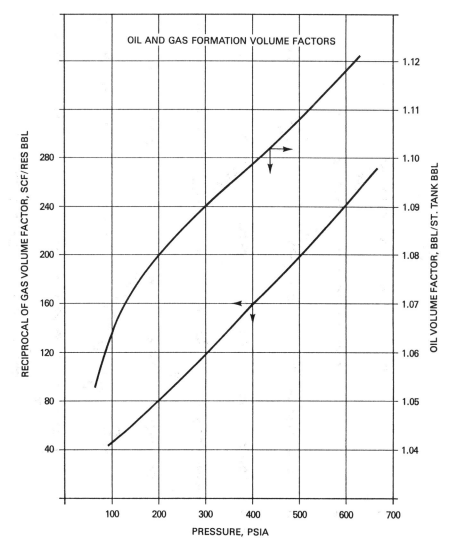

Fig. 10.2. Oil and gas formation volume factor for history matching problem.

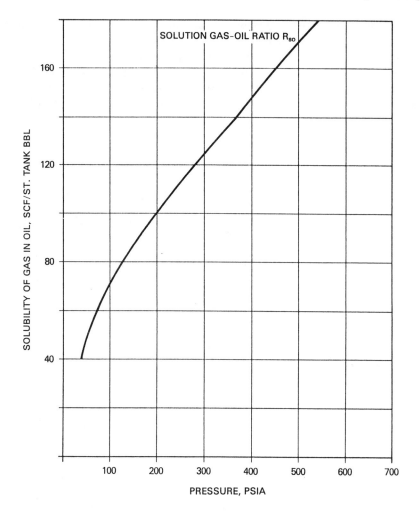

Fig. 10.3. Solution gas-oil ratio for history matching problem.

Actual oil production and instantaneous gas-oil ratios for the first three years of the life of the well are plotted in Fig. 10.6.

3.1. Solution Procedure

One of the first steps in attempting to perform the history match is to convert the fluid property data provided in Figs. 10.1 to 10.3 to a more usuable form. This is done by simply regressing the data to obtain equations for each parameter—for example, oil and gas viscosity, as a function of pressure. The resulting equations can be found in the program listing in Table 10.1. The two permeability relationships also need to be regressed to be used in the program.

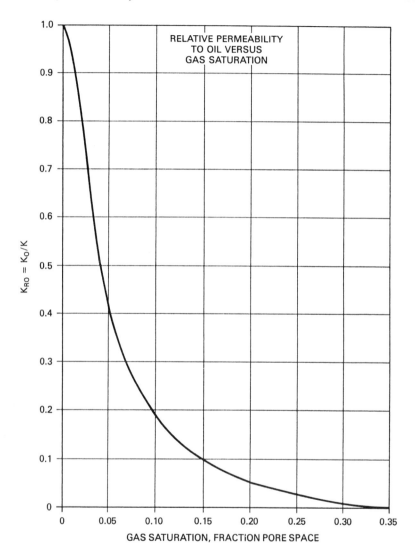

Fig. 10.4. Oil relative permeability for history matching problem.

Both the relative permeability to oil and the permeability ratio can be expressed as functions of gas saturation. These equations are input into the program in a different manner from the fluid property equations. The constants for the regressed equations are placed in a data file to be read into the program at the time of program execution. General expressions are written in the program which use these constants. These relationships are handled this way to facilitate modifications to the equations used in the program if necessary.

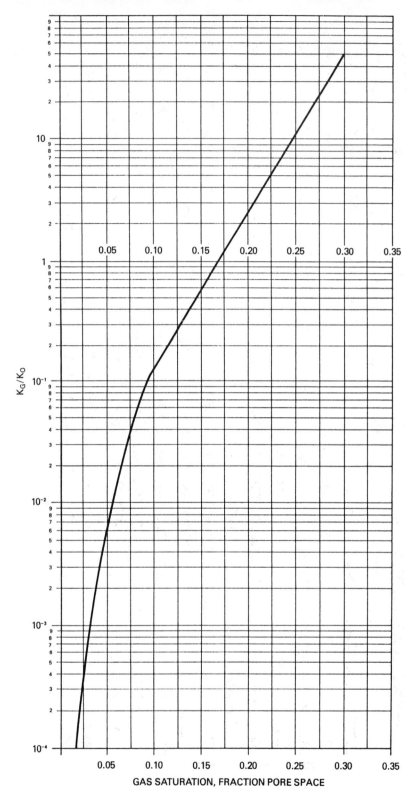

Fig. 10.5. Permeability ratio for history matching problem.

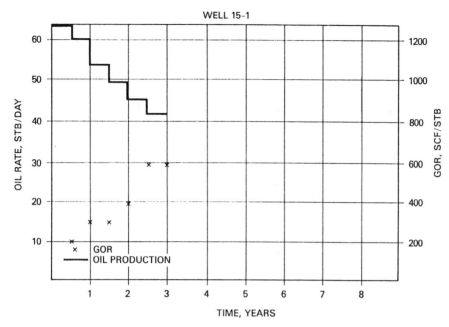

Fig. 10.6. Actual production and instantaneous GOR for history matching problem.

TABLE 10.1.
FORTRAN listing of program for history matching problem

```
C
C  THIS PROGRAM CONDUCTS A HISTORY MATCH
C  THE PROGRAM IS UNIQUE FOR THE PROBLEM IN CHAPTER 10
C  OF THE REVISED CRAFT AND HAWKINS RESERVOIR ENGINEERING
C  TEXT BY RONALD E. TERRY
C
          CHARACTER*6 DUMMYF
          REAL K1,K2,KRO(60),N,NGP(60),NOP(60),NP(60)
          REAL BG(60),BO(60),BT(60),TOP(60),TGP(60),
        1   DT(60),DMB(100),R(60),RP(60),RS(60),QO(60),
        2   VG(60),VO(60),P(60),TOT(60),A0(5),A1(5),
        3   SGL(5),SGU(5),C0(10),C1(10),GASL(10),GASU(10)
C
C  INPUT DATA
C
          DATA PI,DP,PHI,SW,ACRE/620.,5.,.215,.37,40./
          DATA DOP,RW,RE,EPS,PW/1.E-4,.25,660.,1.E-5,25.0/
C
C  GET INPUT VARIABLES FROM FILE NAMED BY USER
C
```

TABLE 10.1.
(Continued)

```
        WRITE(*,598)
 598    FORMAT(1X,'ENTER DATA FILE NAME')
        READ(*,599)DUMMYF
 599    FORMAT(A6)
        OPEN(5,FILE=DUMMYF,STATUS='OLD')
C
C   NKGKO = NUMBER OF KG/KO CURVES
C   SGL(I) = LOWER GAS SATURATION FOR Ith CURVE
C   SGU(I) = UPPER GAS SATURATION FOR Ith CURVE
C   A0 (I) AND A1 (I) = CONSTANTS IN REGRESSED EQUATION
C   log (KGKO) = A0 + A1*SG
C
        READ(5,600)NKGKO
 600    FORMAT(I5)
        DO 4 I=1,NKGKO
          READ(5,605)SGL(I),SGU(I),A0(I),A1(I)
   4    CONTINUE
 605    FORMAT(5F10.0)
C
C   NKRO = NUMBER OF KRO CURVES
C   GASL(I), GASU(I) = LOWER AND UPPER GAS SATURATION FOR
C   Ith CURVE
C   C0(I) AND C1(I) = CONSTANTS IN REGRESSED EQUATION
C   KRO = C0 + C1*SG
C
        READ(5,600)NKRO
        DO 41 I = 1,NKRO
          READ(5,605),GASL(I),GASU(I),C0(I),C1(I)
  41    CONTINUE
        READ(5,605) H1,H2
        READ(5,605), K1,K2
        CLOSE(5)
C
C   OPEN OUTPUT FILE NAMED MATCH
C
        OPEN(6,FILE='MATCH')
C
C   CALCULATE FLUID PROPERTIES
C   FROM REGRESSED EQUATIONS
C
        DO 10 I=1,60
          P(I)=PI-DP*(I-1)
          IF(P(I).LT.200.) GOTO 11
          VO(I)=2.354-0.001134*P(I)
          VG(I)=.009684+.3267E-5*P(I)+.1116E-9*P(I)**2
          RS(I)=52.45+.2475*P(I)-.2097E-4*P(I)**2
```

TABLE 10.1.
(Continued)

```
              BO(I)=1.062+.9257E−4*P(I)
              BG(I)=7.446+.3625*P(I)+.4532E−4*P(I)**2
              BG(I)=1./BG(I)
              GO TO 10
   11         VO(I)=2.670−.004566*P(I)+.9461E−5*P(I)**2
              VG(I)=.007580+.2497E−4*P(I)−.5622E−7*P(I)**2
              RS(I)=16.99+.6565*P(I)−.001198*P(I)**2
              BO(I)=1.033+.3829E−3*P(I)−.7262E−6*P(I)**2
              BG(I)=8.874+.3676*P(I)
              BG(I)=1./BG(1)
   10      CONTINUE
C
C  BEGIN SCHILTHUIS MATERIAL BALANCE CALCULATION
C
           NP(1)=0.
           R(1)=RS(1)
           BT(1)=BO(1)
           NOP(1)=0.
           NGP(1)=0.
           TOT(1)=0.
           DO 20 I=2,60
              BT(I)=BO(I)+(RS(1)−RS(I))*BG(I)
              TOP(1)=NOP(I−1)
              DMB(1)=0.01
              TOP(2)=TOP(1)+DOP
              DO 30 J=2,100
                 SG=(1−(1−TOP(J))*BO(I)/BO(1))*(1−SW)
                 IF(SG.LT.0.0)SG=0.0
C
C  DETERMINE PROPER KG/KO VALUE
C
                 DO 50 KK=1,NKGKO
                    IF(SG.GT.SGU(KK)) GOTO 50
                    A=A0(KK)
                    B=A1(KK)
                    GO TO 55
   50            CONTINUE
   55            CONTINUE
C
C  DETERMINE PROPER KRO VALUE
C
                 DO 42 KI=1,NKRO
                    IF(SG.GT.GASU(KI)) GO TO 42
                    AZ=C0(KI)
                    BZ=C1(KI)
                    GO TO 43
```

TABLE 10.1.
(Continued)

```
     42                CONTINUE
     43                CONTINUE
                      KRO(I)=AZ+BZ*SG
                      IF(KRO(I).LT.0.0) KRO (I)=0.0
                      IF(KRO(I).GT.1.0) KRO(I)=1.0
                      RKGKO=10.**(A+B*SG)
                      R(I)=RS(I)+VO(I)*BO(I)/VG(I)/BG(I)*RKGKO
                      RAV=(R(I-1)+R(I))/2.
                      DGP=DOP*RAV
                      TGP(J)=NGP(I-1)+DGP
                      RP(I)=TGP(J)/TOP(J)
                      DMB(J)=TOP(J)*(BT(I)+BG(I)*(RP(I)-RS(1)))/
        1             (BT(I)-BT(1))-1.
                      TOP(J+1)=TOP(J)-DMB(J)*(TOP(J)-TOP(J-1))/
        1             (DMB(J)-DMB(J-1))
                      IF(ABS(DMB(J)).GT.EPS) GO TO 35
                      NOP(I)=TOP(J)
                      NGP(I)=TOP(J)*RP(I)
                      GOTO 20
     35                DOP=TOP(J+1)-NOP(I-1)
     30                CONTINUE
     20                CONTINUE
C
C   WRITE OUT RESULTS FOR NP/N AND PRESSURE
C
                      WRITE(6,101)
                      WRITE(6,102) P(1),NOP(1),R(1)
                      DO 202 I=2,40,2
                        WRITE(6,102) P(I),NOP(I)R(I)
    202                CONTINUE
C
C   CALCULATE THE INITIAL OIL-IN-PLACE
C
                      HTOT=H1+H2
                      CAPAC=H1*K1+H2*K2
                      N=7758.*ACRE*HTOT*PHI*(1.-SW)/BO(1)
                      WRITE(6,501)
    501                FORMAT ('1',3X, THE HISTORY MATCH FOR WELL 1')
                      WRITE(6,502)
    502                FORMAT(//,1X,' PRESSURE,PSIA ',2X,'OIL PROD. RATE, BPD',
        1             2X, 'INSTANT. GOR, SCF/STB',2X,' TOTAL TIME, YEARS')
C
C   CALCULATED FLOW RATE AND TIME
C
                      DO 40 I=2,40
                        NP(I)=NOP(I)*N
```

TABLE 10.1.
(Continued)

```
              QO(I)=0.00708/BO(I)/VO(I)/(ALOG(RE/RW)-.75)*(P(I)-PW)
       1        *KRO(I)*CAPAC
              DT(I)=(NP(I)-NP(I-1))/QO(I)/365.
              TOT(I)=TOT(I-1)+DT(I)
              IF(MOD(I,2).EQ.0)THEN
C
C  WRITE OUT PRESSURE, FLOW RATE, INSTANTANEOUS GOR, AND TIME
C
              WRITE(6,503) P(I),QO(I),R(I),TOT(I)
      503       FORMAT(1X,F18.6,2X,F18.6,2X,F20.6,2X,F16.6)
              ENDIF
      40      CONTINUE
      101     FORMAT('1',5X,'PRESSURE',10X,'NP/N',10X,'INSTANT. GOR')
      102     FORMAT(8X,F5.0,10X,F7.6,12X,F6.0)
              CLOSE(6)
              STOP
              END
```

With the data of Figs. 10.1 to 10.5 expressed in equation format, the program is now ready to be executed. A listing of the FORTRAN code for the program is presented in Table 10.1. This version of the program was compiled and executed on a personal computer using a commercially available FORTRAN compiler. The program begins by declaring variables and initializing some of the data with DATA statements. Next, the program prompts the user for a name of an input file that contains the constants for the permeability relationships, the zone thicknesses, and absolute permeabilities. An example of an input file that, contains the initial data given in the problem statement is listed in Table 10.2. The first data entry in the file, the number 4, refers to the number of straight-line segments used to describe the original permeability ratio data in Fig. 10.5. The next four lines of data contain the lower and upper gas saturation limits and the constants for the four equations that are in the following general format:

$$\log\left(\frac{K_g}{K_o}\right) = A0 + A1*SG \tag{10.3}$$

The next several lines of data provide similar information for the relative permeability to oil curve in Fig. 10.4. Seven straight-line segments are used to describe this curve, and the saturation and constant values are then listed. The general format for the equations representing the relative permeability to oil data is

$$K_{ro} = C0 + C1*SG \tag{10.4}$$

TABLE 10.2.
Example of data input file for history matching problem

4			
.0000	.0430	−4.7981	54.6520
.0430	.0810	−3.8450	32.0170
.0810	.1090	−2.6308	16.9890
.1090	.3500	−2.1119	12.2410
7			
.0000	.0156	1.0000	−8.2400
.0156	.0450	1.0980	−14.4000
.0450	.0688	.7500	−6.6133
.0688	.1008	.5427	−3.5467
.1008	.1520	.3583	−1.7333
.1520	.2088	.2215	−.8300
.2088	.3325	.1262	−.3760
17.0	14.0		
9.6	7.2		

The last two lines of input represent the zone thicknesses and the absolute permeabilities, respectively.

Returning to the main program, the program now opens up an output file called MATCH. Fluid properties are then calculated as a function of pressure, beginning at the initial pressure of 620 psia and at subsequent pressures determined by deincrementing the pressure by 5 psi. Once the fluid properties are calculated, the program is ready to begin the Schilthuis material balance calculation. After the material balance calculation has generated fractional oil recovery and instantaneous GOR values, these are written to the output file. An example of the output file is shown in Table 10.3. The initial oil in place is then calculated so that oil production values can be determined from the fractional oil recovery numbers. Oil production rates and corresponding times are then calculated, and the average pressures, oil production rates, instantaneous GOR values, and times are written to the output file. Program execution is terminated at this point.

3.2. Discussion of History Matching Results

When the program is executed using the original data given in the problem statement as input, the oil production rate and R, or instantaneous GOR, values obtained result in the plots shown in Fig. 10.7. The oil production rate is plotted in Figure 10.7(a) and compared with the actual data. Notice that the calculated rates begin higher than the actual rates and decrease faster with time or with a greater slope. The calculated instantaneous GOR values are compared with the actual GOR values in Fig. 10.7(b) and found to be much lower than the actual values.

TABLE 10.3.

Example of output file for history matching problem

PRESSURE	NP/N	INSTANT. GOR
620.	.000000	198.
615.	.003574	198.
605.	.010883	196.
595.	.018410	195.
585.	.026131	196.
575.	.033988	200.
565.	.041858	209.
555.	.049524	228.
545.	.056676	264.
535.	.063061	311.
525.	.068856	353.
515.	.074112	405.
505.	.078835	465.
495.	.083060	535.
485.	.086836	611.
475.	.090217	695.
465.	.093254	786.
455.	.095994	882.
445.	.098478	983.
435.	.100740	1089.
425.	.102811	1200.

THE HISTORY MATCH FOR WELL 1

PRESSURE, PSIA	OIL. PROD. RATE, BPD	INSTANT. GOR, SCF/STB	TOTAL TIME, YEARS
615.000000	81.731461	197.584778	.139468
605.000000	76.426636	196.124374	.439406
595.000000	71.218536	195.383209	.770589
585.000000	65.003090	196.084351	1.139370
575.000000	57.519310	199.663467	1.561816
565.000000	50.337765	208.837708	2.043982
555.000000	43.595257	228.218506	2.584777
545.000000	37.472816	264.157867	3.170556
535.000000	32.096237	310.664001	3.780574
525.000000	29.040087	353.154480	4.403254
515.000000	26.670746	404.841736	5.018327
505.000000	24.536453	465.493835	5.619408
495.000000	22.615231	534.584167	6.203073
485.000000	20.882790	611.443237	6.768256
475.000000	19.315432	695.369751	7.315524
465.000000	17.891640	785.692627	7.846456
455.000000	16.592621	881.798767	8.363103
445.000000	15.842712	983.164307	8.856463
435.000000	15.035938	1089.271851	9.330001
425.000000	14.276489	1199.706299	9.786566

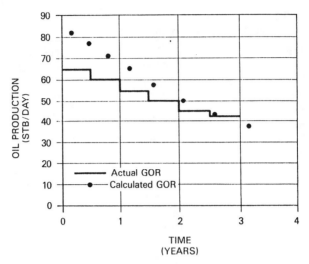

(a) Oil production rate

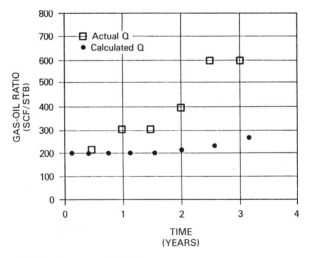

(b) Instantaneous GOR

Fig. 10.7. History match using original data.

At this point, it is necessary to ask how the calculated instantaneous GOR values could be raised in order for them to match the actual values. An examination of Eq. (9.33) suggests that R is a function of fluid property data and the ratio of gas-to-oil permeabilities. To calculate higher values for R, either the fluid property data or the permeability ratio data must be modified.

Because fluid property data are readily and accurately obtained and the permeabilities could change significantly in the reservoir owing to different rock environments, it seems justified to modify the permeability ratio data. It is often the case when conducting a history match, an engineer will find differences between laboratory measured permeability ratios and field measured permeability ratios. Mueller, Warren, and West showed that one of the main reasons for the discrepancy between laboratory k_g/k_o values and field-measured values can be explained by the unequal stages of depletion in the reservoir.[5] For the same reason, field instantaneous GOR values seldom show the slight decline predicted in the early stages of depletion and conversely usually show a rise in gas-oil ratio at an earlier stage of depletion than the prediction. Whereas the theoretical predictions assume a negligible (actually zero) pressure drawdown, so that the saturations are therefore uniform throughout the reservoir, actual well pressure drawdowns will deplete the reservoir in the vicinity of the wellbore in advance of areas further removed. In development programs, too, some wells are often completed years before other wells, and depletion is naturally further advanced in the area of the older wells, which will have gas-oil ratios considerably higher than the newer wells. And even when all wells are completed within a short period, when the formation thickness varies and all wells produce at the same rate, the reservoir will be depleted faster when the formation is thinner. Finally, when the reservoir comprises two or more strata of different specific permeabilities, even if their relative permeability characteristics are the same, the strata with higher permeabilities will be depleted before those with lower permeabilities. Since all of these effects are minimized in high-capacity formations, closer agreement between field and laboratory data can be expected for higher capacity formations. On the other hand, high capacity formations tend to favor gravity segregation. When gravity segregation occurs and advantage is taken of it by shutting in the high-ratio wells or working over wells to reduce their ratios, the field-measured k_g/k_o values will be lower than the laboratory values. Thus the laboratory k_g/k_o values may apply at every point in a reservoir without gravity segregation, and yet the field k_g/k_o values will be higher owing to the unequal depletion of the various portions of the reservoir.

The following procedure is used to generate new permeability ratio values from the actual production data:

1. Plot the actual R values versus time and determine a relationship between R and time.
2. Choose a pressure and determine the fluid property data at that pressure. From the chosen pressure and the output data in Table 10.3, find the time that corresponds with the chosen pressure.
3. From the relationship found in Step 1, calculate R for the time found in Step 2.

4. With the value of R found in Step 3 and the fluid property data found in Step 2, rearrange Eq. (9.33) and calculate a value of the permeability ratio.

5. From the pressure chosen in Step 2 and from the N_p values in the output data in Table 10.3, calculate the value of the gas saturation that corresponds with the calculated value of the permeability ratio.

6. Repeat Steps 2 through 5 for several pressures. The result will be a new permeability ratio–gas saturation relationship.

You should realize that in Steps 2 and 5 the original permeability ratio was used to generate the data of Table 10.3. This suggests that the new permeability ratio–gas saturation relationship could be in error because it is based on the data of Table 10.3 and that to have a more correct relationship, it might be necessary to repeat the procedure. The quality of the history match obtained with the new permeability ratio values will dictate whether this iterative procedure should be used in generating the new permeability ratio–gas saturation relationship. The new permeability ratios determined from the foregoing six-step procedure are plotted with the original permeability ratios in Fig. 10.8.

It is now necessary to regress the new permeability ratio–gas saturation relationship and input the new data into the program before the program can be executed again to obtain a new history match. When this is done, the program yields the results plotted in Fig. 10.9.

The new permeability ratio data has significantly improved the match of the instantaneous GOR values, as can be seen in Fig. 10.9(b). However, the oil production rates are still not a good match. In fact, the new permeability ratio data have yielded a steeper slope for the calculated oil rates, as shown in Fig. 10.9(a), than what is observed in Fig. 10.7(a) from the original data. A look at the calculation scheme helps to understand the effect of the new permeability ratio data.

Because the new values of instantaneous GOR were calculated with the new permeability ratio data, which in turn were determined by using Eq. (9.33) and the actual GOR values, it should be expected that the calculated GOR values will match the actual GOR values. The flow rate calculation, which involves Eq. (7.50), does not use the permeability ratio, so the magnitude of the flow rates would not be expected to be affected by the new permeability ratio data. However, the time calculation does involve N_p, which is a function of the permeability ratio in the Schilthuis material balance calculation. Therefore, the rate at which the flow rates decline will be altered with the new permeability ratio data.

To obtain a more accurate match of oil production rates, it is necessary to modify additional data. This raises the question: What other data can be justifiably changed? We have argued that it was not justifiable to modify the fluid property data. However, the fluid property data and/or equations should be carefully checked for possible errors. In this case, the equations were

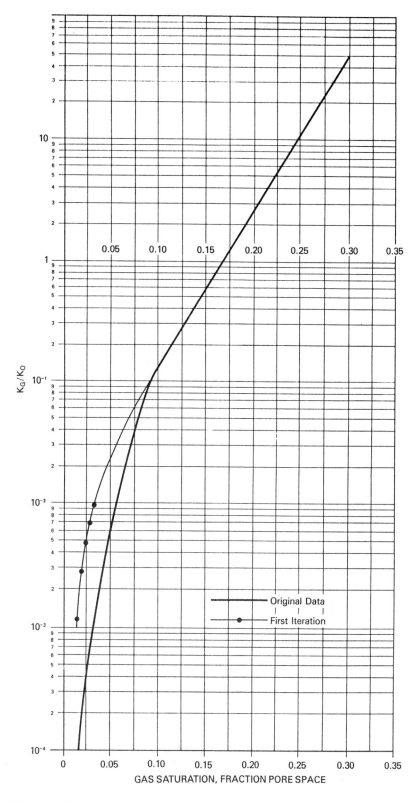

Fig. 10.8. First iteration of permeability ratio for history matching problem.

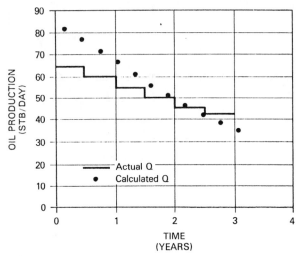

(a) Oil production rate

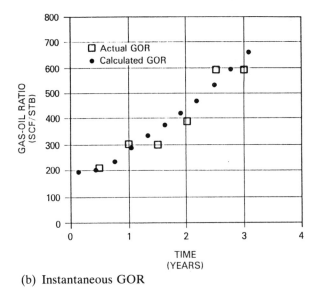

(b) Instantaneous GOR

Fig. 10.9. History match after modifying permeability ratio values.

checked by calculating values of B_o, B_g, R_{so}, μ_o, and μ_g at several pressures and comparing them with the original data. The fluid property equations were found to be correct and accurate. Other assumed reservoir properties that could be in error include the zone thicknesses and absolute permeabilities.

The thicknesses are determined from logging and coring operations from which an isopach map is created. Absolute permeabilities are measured from a small sample of a core taken from a limited number of locations in the reservoir. The number of coring locations is limited largely because of the costs involved in performing the coring operations. Although the actual measurement of both the thickness and permeability from coring material is highly accurate, errors are introduced when one tries to extrapolate the measured information to the entire drainage area of a particular well. For instance, when constructing the isopach map for the zone thickness, you need to make assumptions regarding the continuity of the zone in between coring locations. These assumptions may or may not be correct. Because of the possible errors introduced in determining average values for the thickness and permeability for the well drainage area, varying these parameters and observing the effect of our history match is justified. In the remainder of this section, the effect of changing these parameters on the history matching process is examined. Table 10.4 summarizes the cases that are discussed.

TABLE 10.4.
Description of cases

Case Number	Parameter Varied From Original Data
1	None
2	Permeability ratio
3	Same as Case 2 plus zone thickness
4	Same as Case 2 plus absolute permeability
5	Same as Case 2 plus zone thickness and absolute permeability
6	A second iteration on the permeability ratio data plus zone thickness and absolute permeability

In Case 3, the thicknesses of both zones were adjusted to determine the effect on the history match. Since the calculated flow rates are higher than the actual flow rates, the thicknesses were reduced. Figure 10.10 shows the effect on oil producing rate and instantaneous GOR when the thicknesses are reduced by about 20%.

By reducing the thicknesses, the calculated oil production rates are shifted downward, as shown in Fig. 10.10(a). This yields a good match with the early data but not with the later data because the calculated values decline at a much more rapid rate than do the actual data. The calculated instantaneous GOR values still closely match the actual GOR values. These observations can be supported by noting that the zone thickness enters into the calculation scheme in two places. One is in the calculation for N, the initial oil in place, which is performed by using Eq. (10.2). Then N is multiplied by each of the $\Delta N_p / N$ values determined in the Schilthuis balance. The second place the thickness is used is in the flow equation, Eq. (7.50), which is used to calculate

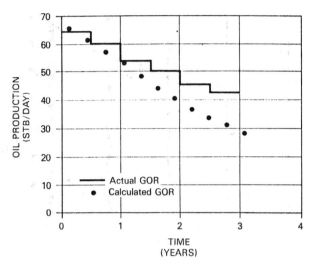

(a) Oil production rate

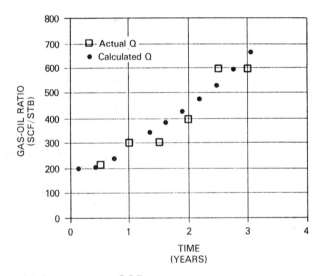

(b) Instantaneous GOR

Fig. 10.10. History match of Case 3. Case 3 used the new permeability ratio data and reduced zone thicknesses.

q_o. The instantaneous GOR values are not affected because neither the calculation for N nor the calculation for q_o is used in the calculation for instantaneous GOR or R. However, the oil flow rate is directly proportional to the thickness, so as the thickness is reduced, the flow rate is also reduced. At first glance, it appears that the decline rate of the flow rate would be altered. But upon further study, it is found that although the flow rate is obviously a function of the thickness, the time is not. To calculate the time, an incremental ΔN_p is divided by the flow rate corresponding with that incremental production. Since both N_p and q_o are directly proportional to the thickness, the thickness cancels out, thereby making the time independent of the thickness. In summary, the net result of reducing the thickness is as follows: (1) the magnitude of the oil flow rate is reduced, (2) the slopes of the oil production and instantaneous GOR curves are not altered, and (3) the instantaneous GOR values are not altered.

To determine the effect on the history match of varying the absolute permeabilities, the permeabilities were reduced by about 20% in Case 4. Figure 10.11 shows the oil production rates and the instantaneous GOR plots for this new case. The quality of the match of oil production rates has improved, but the quality of the match of the instantaneous GOR values has decreased. Again, if the equations involved are examined, an understanding of how changing the absolute permeabilities have affected the history match can be obtained.

Equation (7.50) suggests that the oil flow rate is directly proportional to the effective permeability to oil, k_o. Equation (10.5)

$$k_o = k_{ro}k \qquad (10.5)$$

shows the relationship between the effective permeability to oil and the absolute permeability, k. Combining Eqs. (7.50) and (10.5), it can be seen that the oil flow rate is directly proportional to the absolute permeability. Therefore, when the absolute permeability is reduced, the oil flow rate is also reduced. Since the time values are a function of q_o, the time values are also affected. The magnitude of the instantaneous GOR values are not a function of the absolute permeability since neither the effective nor the absolute permeabilities are used in the Schilthuis material balance calculation. However, the time values are modified, so the slope of both the oil production rate and the instantaneous GOR curves are altered. This is exactly what should happen in order to obtain a better history match of the oil production values. However, although it has improved the oil production history match, the instantaneous GOR match has been made worse. By reducing the absolute permeabilities, it has been found that (1) the magnitude of the oil flow rates are reduced, (2) the magnitude of the instantaneous GOR values are not changed, and (3) the slopes of both the oil production and instantaneous GOR curves are altered.

By modifying the zone thicknesses and absolute permeabilities, the magnitude of the oil flow rates and the slope of the oil flow rate curve can be modified. Also, while adjusting the oil flow rate, slight changes in the slope of the instantaneous GOR curve are obtained. In Case 5, both the zone thickness and the absolute permeability are changed in addition to using the new permeability ratio data. Figure 10.12 contains the history match for Case 5. As can be seen in Figure 10.12(a), the calculated oil flow rates are an excellent match

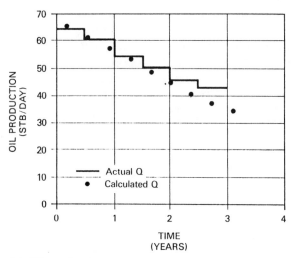

(a) Oil production rate

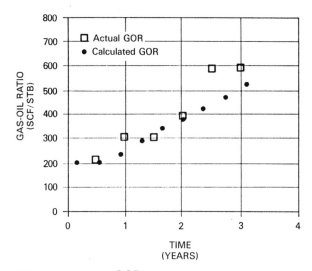

(b) Instantaneous GOR

Fig. 10.11. History match of Case 4. Case 4 used the new permeability ratio data and reduced absolute permeabilities.

to the actual field oil production values. The match of instantaneous GOR values has worsened from Cases 2 to 4 but is still much improved over the match in Case 1, which was obtained by using the original permeability ratio data.

We have said that a second iteration of the permeability ratio values may be necessary, depending on the quality of the final history match that is obtained. This is because the procedure used to obtain the new permeability

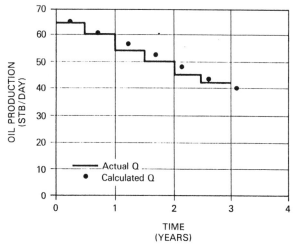

(a) Oil production rate

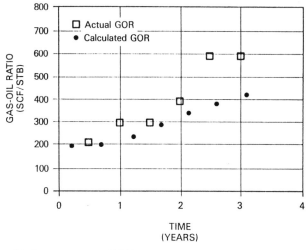

(b) Instantaneous GOR

Fig. 10.12. History match of Case 5. Case 5 used the new permeability ratio data and modified zone thicknesses and absolute permeabilities.

ratio data involves using the old permeability ratio data. The calculated instantaneous GOR values do not match the actual field GOR values very well, so a second iteration of the permeability ratio values is warranted. Following the procedure of obtaining new permeability ratio data in conjunction with the results of Case 5, a second set of new permeability ratios is obtained. This second set is plotted in Fig. 10.13 along with the original data and the first set used in Cases 2 to 5.

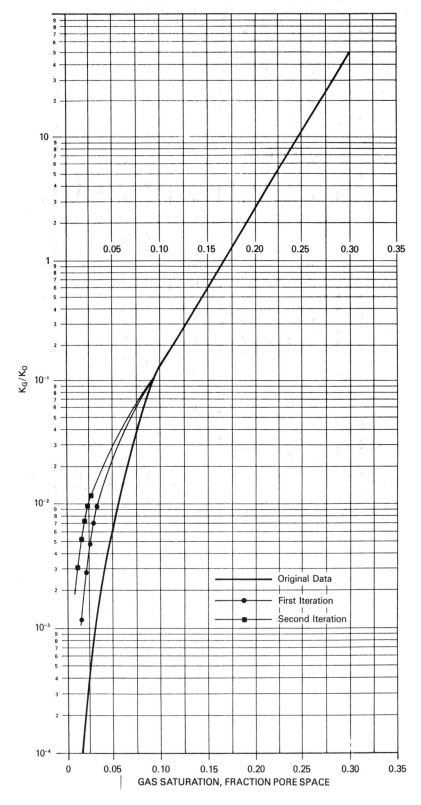

Fig. 10.13. Second iteration of permeability ratios for history matching problem.

By using the permeability ratio data from the second iteration and by adjusting the zone thicknesses and absolute permeabilities as needed, the results shown in Fig. 10.14 are obtained. It can be seen that the quality of the history match for both the oil production rate and the instantaneous GOR values is very good. When a history match is obtained that matches both the

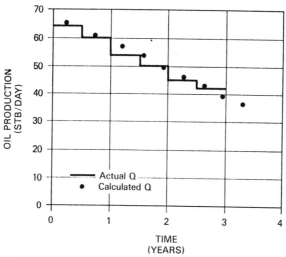

(a) Oil production rate

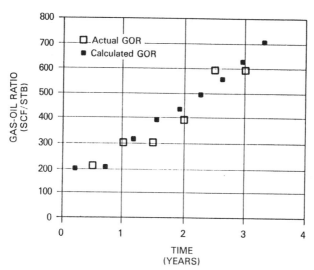

(b) Instantaneous GOR

Fig. 10.14. History match of Case 6. Case 6 used permeability ratio data from a second iteration and modified zone thicknesses and absolute permeabilities.

oil production and instantaneous GOR curves this well, the model can be used
with confidence to predict future production information.

3.3. Summary Comments Concerning History Matching Example

Now that a model has been obtained that matches the available production
data and that can be used to predict future oil and gas production rates, an
assessment of the modifications to the data that were performed during the
history match should be made. Table 10.5 contains information concerning the
data that were varied in the six cases we have discussed.

TABLE 10.5.
Input data for history matching example

Case	Permeability Ratio Data	Absolute Permeability, md		Zone Thickness, ft	
		Zone 1	Zone 2	Zone 1	Zone 2
1	Original data	9.6	7.2	17.0	14.0
2	First iteration	9.6	7.2	17.0	14.0
3	First iteration	9.6	7.2	13.6	11.2
4	First iteration	7.7	5.7	17.0	14.0
5	First iteration	5.9	4.4	22.0	18.2
6	Second iteration	5.9	4.4	22.0	18.2

All other input data were held constant at the original values for all six
cases. The first and second iterations of the gas-to-oil permeability ratio data
led to values higher than the original data, as can be seen in Fig. 10.13. In the
last section, several reasons for a discrepancy between laboratory-measured
permeability ratio values and field-measured values were discussed. These
reasons included greater drawdown in areas closer to the wellbore than in
areas some distance away from the wellbore, completion and placing on production of some wells before others, two or more strata of varying permeabilities, and gravity drainage effects. All these phenomena lead to unequal
stages of depletion within the reservoir. The unequal stages of depletion cause
varying saturations throughout the reservoir and hence varying effective permeabilities. The data plotted in Fig. 10.13 suggest that the discrepancy between the laboratory-measured permeability ratio values and the field-
determined values is not great and that the use of the modified permeability
ratio values in the matching process is justified.

For the final match in Case 6, the zone thicknesses were increased by
about 30% over the original data of Case 1, and the absolute permeabilities
were reduced by about 39%. At first glance, these modifications in zone thickness and absolute permeability might seem excessive. However, remember
that the values of zone thickness and absolute permeability were determined

in the laboratory on a core sample, approximately 6 in. in diameter. Although the techniques used in the laboratory are very accurate in the actual measurement of these parameters, to perform the history match, it was necessary to assume that the measured values would be used as average values over the entire 40 ac of the drainage area of the well. This is a very large extrapolation, and there could be significant error in this assumption. If the small magnitudes of the original values are considered, the adjustments made during the history matching process are not large in magnitude. The adjustments were only 2.8 to 3.7 md and 4.2 to 5.0 ft. These numbers are large relative to the initial values but are certainly not large in magnitude.

We can conclude that the model developed to perform the history match for the well in question is reasonable and defendable. More sophisticated equations could have been developed, but for this particular example the Schilthuis material balance coupled with a flow equation were quite adequate. As long as the simple approach meets the objectives, there is great merit in keeping things simple. However, you should realize that the principles we have discussed about history matching are applicable no matter what degree of model sophistication is used.

PROBLEMS

10.1 The following data are taken from a volumetric, undersaturated reservoir. Calculate the relative permeability ratio k_g/k_o at each pressure, and plot versus total liquid saturation.

Connate water, $S_w = 25\%$

Initial oil in place = 150 MMSTB

$B_{oi} = 1.552$ bbl/STB

p psia	R SCF/STB	N_p MMSTB	B_o bbl/STB	B_g bbl/SCF	R_{so} SCF/STB	μ_o/μ_g
4000	903	3.75	1.500	0.000796	820	31.1
3500	1410	13.50	1.430	0.000857	660	37.1
3000	2230	20.70	1.385	0.000930	580	42.5
2500	3162	27.00	1.348	0.00115	520	50.8
2000	3620	32.30	1.310	0.00145	450	61.2
1500	3990	37.50	1.272	0.00216	380	77.3

10.2 Discuss the effect of the following on the relative permeability ratios calculated from production data:

(a) Error in the calculated value of initial oil in place.

(b) Error in the value of the connate water.

(c) Effect of a small but unaccounted for water drive.

(d) Effect of gravitational segregation both where the high gas-oil ratio wells are shut in and where they are not.

(e) Unequal reservoir depletion.

(f) Presence of a gas cap.

10.3 For the data that follow and that are given in Figs. 10.15 to 10.17 and the fluid property data presented in the chapter, perform a history match on the production data in Figures 10.18 to 10.21, using the computer program in Table 10.1. Use the new permeability ratio data plotted in Figures 10.8 and 10.13 to fine-tune the match.

Laboratory core permeability measurements

Well	Average Absolute Permeability to Air(md)	
	Zone 1	Zone 2
5–6	5.1	4.0
8–16	8.3	6.8
9–13	11.1	6.0
14–12	8.1	7.6

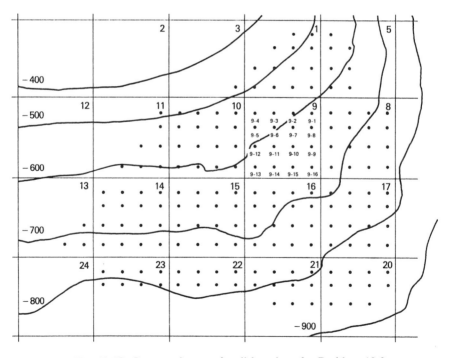

Fig. 10.15. Structural map of well locations for Problem 10.3.

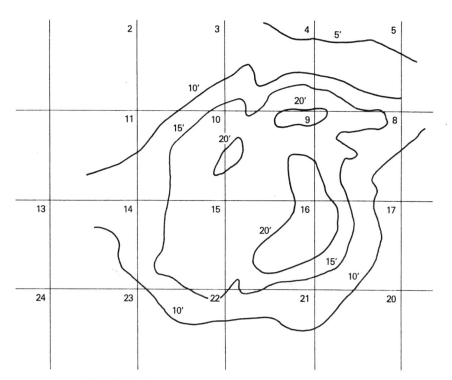

Fig. 10.16. Isopach map of Zone 1 for Problem 10.3.

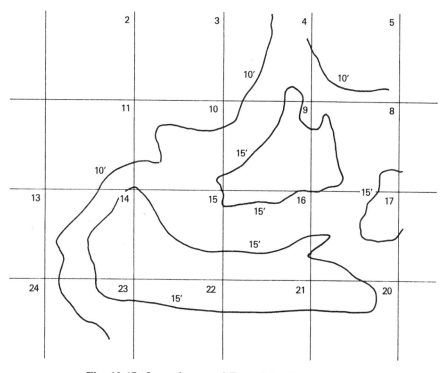

Fig. 10.17. Isopach map of Zone 2 for Problem 10.3.

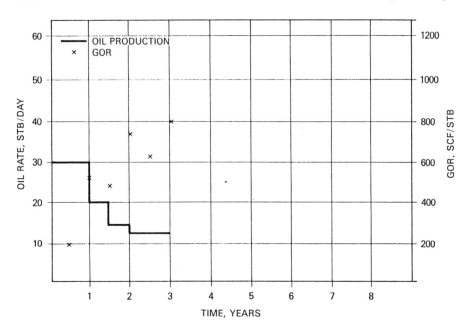

Fig. 10.18. Actual oil production and instantaneous GOR for Wells 5–6 for Problem 10.3.

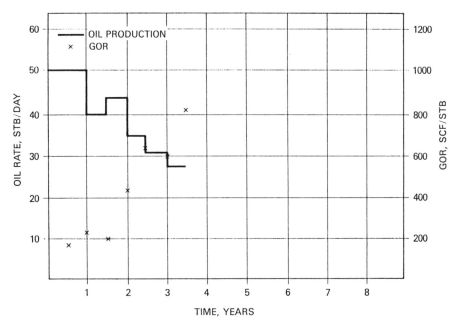

Fig. 10.19. Actual oil production and instantaneous GOR for Wells 8–16 for Problem 10.3.

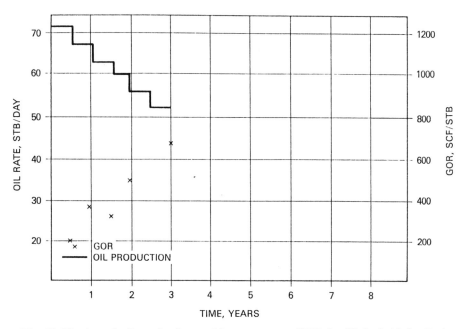

Fig. 10.20. Actual oil production and instantaneous GOR for Wells 9–13 for Problem 10.3.

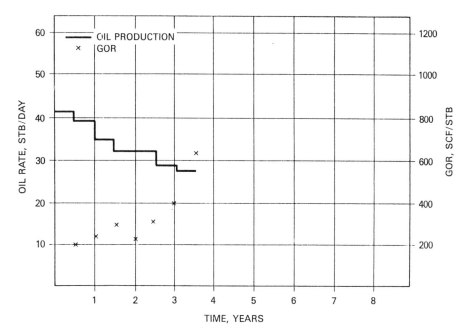

Fig. 10.21. Actual oil production and instantaneous GOR for Wells 14–12 for Problem 10.3.

10.4 Write a computer program that uses the Muskat method discussed in Chapter 9 in place of the Schilthuis method used in Chapter 10 to perform the history match on the data in Chapter 10.

10.5 Write a computer program that uses the Tarner method discussed in Chapter 9 in place of the Schilthuis method used in Chapter 10 to perform the history match on the data in Chapter 10.

REFERENCES

[1] A. W. McCray, *Petroleum Evaluations and Economic Decisions.* Englewood Cliffs, NJ: Prentice Hall, 1975.

[2] H. B. Crichlow, *Modern Reservoir Engineering—A Simulation Approach.* Englewood Cliffs, NJ: Prentice Hall, 1977.

[3] P. H. Yang and A. T. Watson, "Automatic History Matching With Variable-Metric Methods," *Society of Petroleum Engineering Reservoir Engineering Jour.,* (August 1988), 995.

[4] Personal contact with D. W. Green.

[5] T. D. Mueller, J. E. Warren, and W. J. West, "Analysis of Reservoir Performance K_g/K_o Curves and a Laboratory K_g/K_o Curve Measured on a Core Sample," *Trans.* AIME (1955), **204,** 128.

Index